TIDAL ENERGY

TIDAL ENERGY

Roger Henri Charlier, Ph.D., Litt.D., Sc.D.
Professor, Northeastern Illinois University
Professor in the extraordinary
Vrÿe Universiteit Brussel

VNR VAN NOSTRAND REINHOLD COMPANY
NEW YORK   CINCINNATI   TORONTO   LONDON   MELBOURNE

Copyright © 1982 by Van Nostrand Reinhold Company Inc.

Library of Congress Catalog Card Number: 81-13111
ISBN: 0-442-24425-8

All rights reserved. No part of this work covered by the copyright hereon may be reproduced or used in any form or by any means—graphic, electronic, or mechanical, including photocopying, recording, taping, or information storage and retrieval systems—without permission of the publisher.

Manufactured in the United States of America

Published by Van Nostrand Reinhold Company Inc.
135 West 50th Street, New York, N.Y. 10020

Van Nostrand Reinhold Publishing
1410 Birchmount Road
Scarborough, Ontario M1P 2E7, Canada

Van Nostrand Reinhold Australia Pty. Ltd.
17 Queen Street
Mitcham, Victoria 3132, Australia

Van Nostrand Reinhold Company Limited
Molly Millars Lane
Wokingham, Berkshire, England

15 14 13 12 11 10 9 8 7 6 5 4 3 2 1

Library of Congress Cataloging in Publication Data

Charlier, Roger Henri.
   Tidal energy.

   Bibliography: p.
   Includes index.
   1. Tidal power. I. Title.
TC147.C58   621.31′2134   81-13111
ISBN 0-442-24425-8   AACR2

*To Patricia, Connie and Jac,
who found me too often behind
my typewriter, yet always
encouraged me.*

Preface

For the last twenty years or so, I have followed with steadily increasing interest developments in the utilization of the nonliving resources of the ocean. Among these, the harnessing of ocean energies appears as a promising additional source of energy. Remarkably, this topic has fascinated men for a very long time; hundreds of patterns have been taken out, for instance, to extract and put to work the energy dissipated by waves. But tidal-derived power is certainly the oldest area of endeavor. Tide mills provided mechanical power for centuries; scores of scientists, engineers, and inventors have nurtured dreams of large electricity plants "fueled" by the tides since the beginning of this century.

The literature dealing with tidal power is voluminous, yet so far only two attempts were made to actually build tide-powered electricity plants, one in Brittany (France) and one in Maine (USA); both efforts were aborted because of financial problems. Electricité de France finally constructed a sizable facility fifteen years ago on the Rance River estuary, whose technology was used in 1969 to build a small experimental plant on Kislaya Guba (USSR). Both stations have performed well; yet, except for quite small tidal power plants in China, these achievements did not lead the way to a multiplication of such facilities.

I have found a keen interest for tidal energy among the public at large, evidenced, for instance, by the frequent articles dealing with the subject that have appeared in a wide array of periodicals. Yet I did not find a book that covered the entire range of issues pertaining to tidal power. Gibrat's *"L'énergie des marées,"* now fifteen years off the presses, came closest to filling the need. *"Tidal Power"* (1972), edited by Gray and Gashus, provided a valuable compendium of individual papers on the topic. It was the absence of a comprehensive volume that spurred me to write *"Tidal Energy."* I was furthermore motivated by my sincere belief that the tides should not be overlooked as a valuable supplementory source of energy for industrial and developing nations alike.

Roger H. Charlier
Chicago, Illinois

Acknowledgments

It would not have been possible to successfully bring together the information, contained in this volume without the help of numerous colleagues and researchers. To name all of them would require several pages, and I hope that they will accept as my expression of deep gratitude the global assurance that I am greatly indebted to them. Typists and secretaries have taken on the burden of shaping up the voluminous notes gathered over a span of more than 15 years into a finished manuscript which Ms. Alberta Gordon edited with utmost care. They, too, have earned my sincere appreciation.

I have mentioned directly in the text researchers who took time out to personally communicate with me, providing me with unpublished data and personal views. David Mappin and associates, and Professor T. S. Shaw made it possible to include an in-depth analysis of the Severn Estuary project. D. H. Saunders provided me with the latest information on the Kimberleys Tidal Power project and the Maunsell report. C. Lebarbier generously allowed me to use several passages from his manuscript on the Rance River power station. I am grateful to Rex Wailes who provided excellent photographs of tide mills; his classic article on the tide mills of England and Wales was excerpted for details on the old British mills. The Bay of Fundy Tidal Power Project Report has proven invaluable for the discussion of the Canadian plans and large segments of it have been incorporated into the text; the writings, communications, and calls from R. H. Clark of the Canadian Inland Waters Directorate provided the ultimate updating for our chapter on Canada's projects.

Electricité de France deserves a special mention for supplying the writer with a rich selection of photographs of the Rance River and the Tidal Power Station.

Finally a word of thanks to authors and publishers who allowed me to reproduce or interpret illustrations.

Contents

Preface / vii
1. Introduction: Energy from the Ocean / 1
2. From Tide Mills to Tidal Power Plants / 52
3. The Tides / 75
4. The Electricity Generation / 111
5. Economics of Tidal Power / 137
6. Tidal Power Around the World / 149
7. The Tidal Power Plant of the Rance / 184
8. The Kislaya Guba Plant / 206
9. United States Projects / 219
10. Tidal Power and Canada / 251
11. The Severn Project / 275
12. Environmental Impact / 288
13. The Future of Tidal Power / 303

Glossary / 321
General Bibliography / 329
Index / 347

TIDAL ENERGY

1
Introduction: Energy from the Ocean

Though palpable evidence abounds demonstrating that the energy crisis is still very much with us, without those long lines at gas stations we can easily be lulled into a false sense of sufficiency. It is important to be aware that we import about 50% of the oil needed for our domestic consumption, and that current sources of production span the globe, from Alaska to the North Sea and from South America to the Western Pacific. It must be understood that domestic (American) and international aspects of the energy problem are intertwined. Many industrialized democracies are in an even more precarious situation than the United States, with the obvious exceptions of Great Britain and Norway—and yet Great Britain, blessed with an important oil discovery near its shores, is at the forefront of the search to harness alternative energy sources.

To cope with the energy crisis and to forestall a halt to if not a recess of, the economic development, it will be necessary to cut the growth of energy consumption and reduce demand (in non-oil-producing countries especially) for imported oil. Failure to do so may well lead to a sudden overwhelming of the entire international economic system.

Curtailing oil consumption, however, will probably not suffice by itself; the search for and the development of alternative sources of energy are of paramount importance. Emphasis has been placed on a steady increase of nuclear energy production. United States policy is directed toward use of the light water reactor and deferral of the early commercialization of technologies not yet fully developed or lacking safeguards against potential misuse, in particular the breeder reactor. Yet nuclear power proliferation raises grave environmental concerns and has led, worldwide, to vehement protests from groups concerned with the potential danger involved with its use, its wastes, and its malfunctions.

More attention is consequently in order for development of renewable energy resources. Solar water heating is already competitive in prices with electricity and gas in some geographical areas of the United States. Costs of photovoltaic cells decline, making this a not-so-exotic power source. By the turn of the century, wind and geothermal energy could provide a non-negligible amount of power; steam from the earth is already tapped in Japan, Indonesia, New Zealand, the United States, Iceland, Italy, and elsewhere (Fig. 1.1).

But the ocean, too, offers considerable energy resources that can, at least partially, be tapped. The sources are numerous: electromagnetic and kinetic energy, salinity power, biochemical energy, thermal exchanges, wave power, current energy, and, of course, tidal power. Besides all this, were the world to switch, as is frequently suggested, to a hydrogen economy, the ocean would be an inexhaustible source (Fig. 1-2).

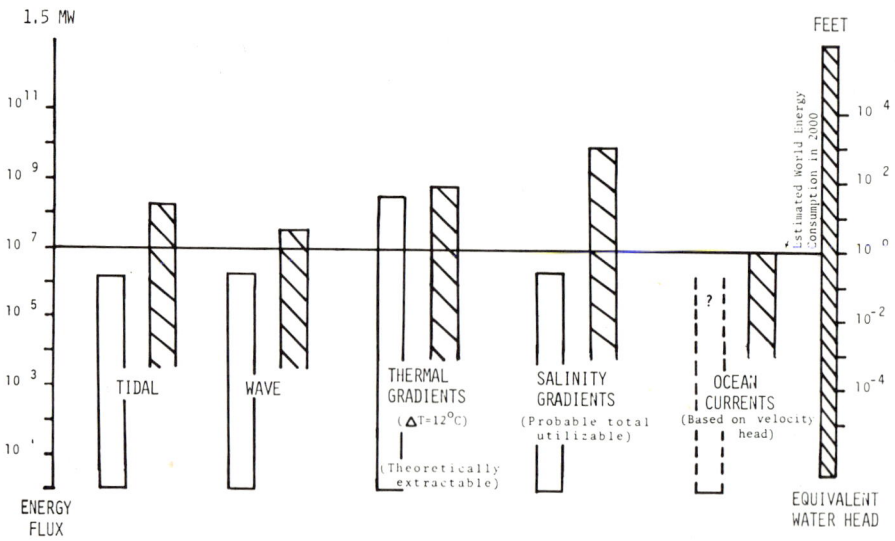

Fig. 1.1. Ocean power. (Adapted from Wick and Isaacs, "Utilization of the energy from salinity gradients," Wave and Salinity Conversion Workshop Proceedings, Univ. of Delaware, 1976)

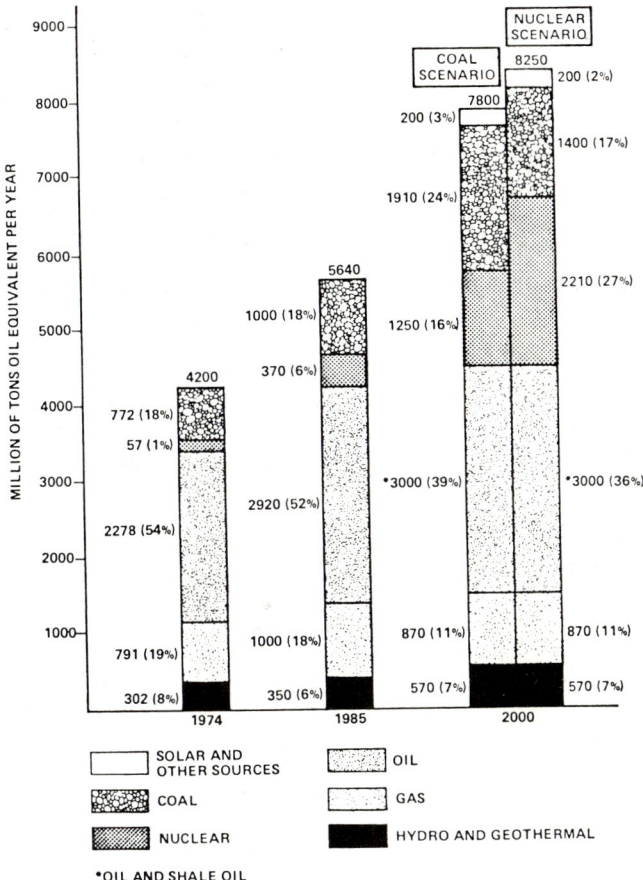

Fig. 1.2. World energy consumption. (Adapted from "Energy global prospects 1985–2000," Report of the Workshop on Alternate Energy Strategies, WAES, C. L. Wilson, Project Director (M.I.T.). New York: McGraw-Hill, 1977)

OCEAN ENERGY POTENTIAL

Some years ago, an oceanographer assessing the flurry of bills in the United States Congress, the spiraling number of companies showing interest in actual exploitation of the seas, the first attempts at pinpointing national goals and problems in oceanography, and the risk of losing valuable undersea resources to countries ahead in sea exploitation technology, stated: "The time has come to shift emphasis from pure oceanographic research to oceanographic projects that have useful goals."

It seems that, if a total shift has not yet taken place, considerable progress is being made daily in the field of what could be called "applied oceanography." This view remains accurate, one may believe, even though at the December 1968 conference on "Oceanography as an Investment," reluctance to intensive investment was voiced, principally due to the domination of the field by industrial giants and the lack of confidence in the small and unknown companies vying for a morsel of the market.

Oceanography, a venerable science, is at the threshold of a new era. Man has still an innate fear of a medium that is strange to him, yet the promise that the sea holds is as large as its area is wide. The Greeks gave the ocean a considerable importance and were derided for it by other nations. Today this importance is no longer considered disproportionate. We are turning to the sea for an ever-increasing number of purposes, including, perhaps, for our ultimate source of food for survival of the earth's exploding population.

The future holds momentous changes in store. The ocean is due to be radically transformed, with fish herded and raised in offshore pens, and with fields of kelp, seaweed, and the like tended by farmers living in sea-bottom dwellings. And perhaps equally dramatic will be climate control, as yet in an embryonic stage. The Technical Management Planning Organization of Santa Barbara, California, estimates that United States energy needs in electricity could be supplied by 12 nuclear generators, and that one such station located at the top of Mount Wilson in Los Angeles would produce enough heat, as a by-product, to lift the inversion layer hanging over the city some 5790 m (19,000 ft) and rid the city from smog; meanwhile, a sea breeze aspirated in the space thus produced could bring in rain in quantities sufficient to transform the desert stretching from Los Angeles to Las Vegas into productive land.

The future is certain to hold many extraordinary and dramatic changes, with the application of extraordinary ideas. And, in this vein, why not call on the ocean for energy?

Mechanical energy dissipated by and in the seas has fascinated physicists the world over. Wave energy lost by waves striking shorelines was calculated by H. Lacombe to be about 6×10^{-7} watts/cm^2; to this should be added the energy dissipated by wave viscosity, offshore, about 3×10^{-8} watts/cm^2. Friction in currents was evaluated by B. Saint-Guily to be 3×10^{-7} watts/cm^2 and tide friction, according to Jeffreys,

Table 1-1. Total Estimated Ocean Power (in MW) Based on One Year's Utilization, Exclusive of Conventional Oil, Gas, and Coal Offshore Resources.

TYPE OF POWER	TOTAL POTENTIAL	MW FOR 30 YEARS	PRESENT USE
Ocean thermal (OTEC)	10,000,000		0
Ocean waves	500,000		negligible
Ocean currents	50,000		0
Ocean tidal	200,000		248
Ocean winds	170,000		negligible
Ocean salinity gradient	3,540,000		0
Ocean bioness conversion	770,000		negligible
Offshore geothermal (U.S. only)	3,000,000	100,000	0
Offshore shale resources	194,000,000	6,466,000	0
Offshore tar sand resources	38,800,000	1,293,000	0
Offshore uranium resources	77,200,000	2,573,000	0
Total	328,230,000	10,432,000	248

Estimated Capabilities in MW (All Types)	
Total U.S. power	2,000,000
Total U.S. electrical power	440,000
Total world power	8,200,000
Projected needed world power by 2000	15,000,000
Current utilization of ocean power (all types)	983,148

uses up 3×10^{-7} watts/cm². Thermal energy passed on from ocean to atmosphere amounts to approximately 0.8×10^{-2} watts/cm²; the thermal flow from depth layers of the earth has been calculated at 6×10^{-6} watts/cm².

ELECTROMAGNETIC OCEAN ENERGY

Electromagnetic energy has received the least attention of all ocean energies. A paper read by Le Grand at the *Quatrièmes Journées de l'Hydraulique* (1957) concluded disappointingly that the ocean's electromagnetic power is negligible as compared to the total energy present. Yet this energy manifests itself in different forms. The ocean can be considered a concentration battery due to temperature and ion concentration variations. When a salinity difference of 10^{-3} and a temperature change of between 4°C and 5°C occur, an electromotive force of 1 millivolt develops; nevertheless, salinity and temperature gradients could create such electrical currents. However, salinity and thalassothermal power can be tapped under different circumstances, as we will see below.

Electrical currents can be inducted by seawater movements in the earth's magnetic field along the vertical component. Telluric currents can be created in the ocean, and coastlines and sea floors can act as galvanic elements when in contact with seawater.

Magnetohydrodynamics applied to the oceanic milieu reveal that electromagnetic forces are only about 10^{-7} of the Coriolis force; electromagnetic induction creates electromotive forces when waves develop; and, finally, magnetic storms often generate substantial electrical currents in seawater masses.

None of these energy sources, however, has been given serious consideration for tapping.

SALINITY POWER

Gerald Wick reported at the 6th Convocation of *Pacem in Maribus* (Okinawa, Tokyo, 1975) on the September 1974 San Francisco meeting which reexamined the possibility of tapping the salinity gradient between fresh water and seawater, and also between any two bodies of water of inequal salinity. Few, if any, large-scale attempts have actually been made to transform the energy dissipated at the interface of fresh and salt water; osmotic pressure differences and salinity gradient power, though recognized, have not retained the attention of scientists as have tidal, wave, and thermal power.

British studies showed that the energy released by the contact of fresh and salt water is equal to that of a river going over a 250-m-high waterfall (Charlier, 1969); the power of rivers such as the Amazon, the Brahmaputra, and the Congo, as they enter the oceans, is staggering. (Fig. 1.3) Batteries which let fresh water and salt water run through alternate cells have been developed for some time, and cells with maximal ratings of 1 watt have been manufactured. To achieve more power, several cells can be paralleled, but the scheme is bulky because each unit weighs 24½ kg. Using free-flowing seawater as an electrolyte, Lockheed researchers built an experimental 180-milliwatt unit, which functioned for several years. The cell is inexpensive, has a long life, needs neither waterproofing nor a pressure case, and has an energy output rating exceeding 80 watt-hr/lb (Table 1-2).

In fact, dry-charge primary batteries using seawater are in rather

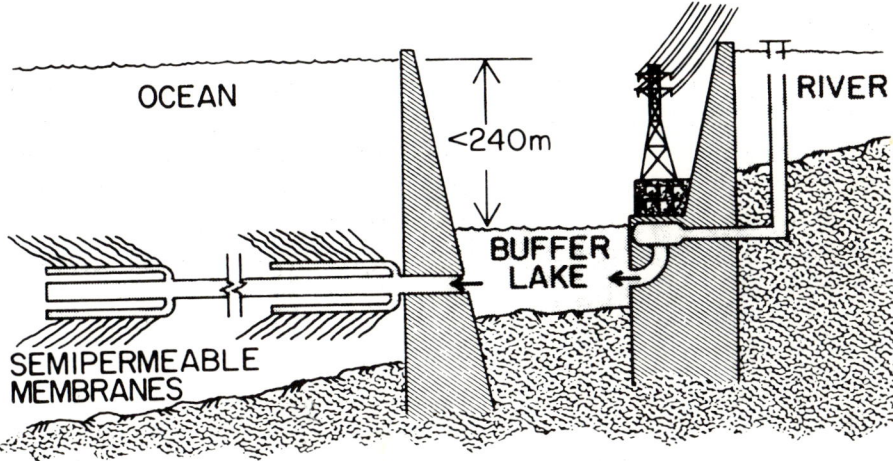

Fig. 1.3. Estuarine salinity gradient energy converter. (Wick and Isaacs, "Utilization of the energy from salinity gradients," Wave and Salinity Energy Conversion Workshop Proceedings, Univ. of Delaware, 1976)

Table 1-2. Ocean Power: Magnitude; Implementation Dates.

TYPE	THEORETICAL POWERS COMMONLY FOUND IN THE LITERATURE	POSSIBLE TECHNICAL DATES	POSSIBLE SOCIAL DATES
Thermal gradients (OTEC)	$40,000 \times 10^6$ MW	1990	2010
Salinity gradients	$1,400 \times 10^6$ MW	2000	2050
Marine bioconversion	10×10^6 MW	1985–1990	2000
Marine currents	5×10^6 MW	1990	2020
Tides	3×10^6 MW	1977	1990
Ocean waves	2.5×10^6 MW	1985	1995
Offshore winds	$> 20 \times 10^6$ MW	1985	1995

common use. Immediately activated upon immersion in water, they are useful in emergency situations and for marine safety devices. They have, incidentally, been used aboard the bathyscapth *Trieste* at depths of 610 m and can probably function at even greater depths.

This large source of energy is, furthermore, as are other ocean energy sources, renewable, because the sun causes water evaporation, leading to subsequent precipitation.

Salinity Energy

Though financially prohibitive with current technology, osmotic effects could be used directly. River water could be channeled through hydroelectric turbines into a reservoir, in pretty much the same way as the ebb current is led through the turbines in a tidal power plant and into the reservoir or retaining bay. The reservoir, however, in this case, should be some hundred meters below sea level; the difference in height should be approximately 240m (osmotic equilibrium level). At any rate, this must provide the driving force for the discharge of fresh water directly into the seawater through semipermeable membranes (Wick and Isaacs, 1975).

The membrane (for instance, hollow fibers) appears to be the costliest element in plant operation, though problems related to bio-fouling, silting, concentration polarization, and impact on aquatic life need also to be resolved (Sourirajan, 1970).

Pattle showed that reverse electrodialysis cells put in series can generate a substantial amount of voltage (1954). Such cells utilize the potential created by the contact of fresh and seawater separated by a charged membrane (approximately 80 millivolts at the interface). According to Wick (1975), common reversible electrodes are too expensive, but cells in series would only require them at the end of the stack; membranes are also very expensive, but their life could be extended by reverse electrodialysis where salt added to the dilute solution leads to increased conductivity instead of a membrane attack. Experience gathered in electrode protection in desalination and tidal power plants may prove useful in developing a satisfactory scheme (Le Grand and Lambert, 1962).

Renewed interest may be shown for a scheme experimented with by Katchalsky, using collagen, which expands and contracts when in contact with salt water (1970). Slowness of response to contact is the main objection to such a system.

The amount of heat generated by the irreversible mixing of fresh and seawater is estimated at only $0.5°C$; to make its tapping worthwhile, concentration could be attempted through an intermediate reaction. This can be done by separating some salts by means of a semipermeable membrane. Water can then become heated to a far greater extent.

Anticipating the later discussion of thalassothermal energy, we must consider Claude's proposal to use water vapor pressure differences of deep cold water and surface warm water in tropical seas (1930). If seawater were used as one reservoir and brine as another, the efficiency would be about five times greater at 70°C than that of a system using all seawater at a 12°C temperature difference. "In order to extract energy from salinity gradients, some sort of membrane is necessary. Water surfaces are by far the cheapest." (Wick, 1976). The exchange surfaces can be enlarged by using spray nozzles in both reservoirs (Fig. 1.4).

The salinity gradient energy converter (thalassohalitic energy) based on osmosis could be used on salt lakes as well as in the ocean, opening sites in the Caspian and Dead Seas, and in Grand Salt Lake (Utah), for instance. The systems could be coupled with solar energy plants or electrochemical concentration cells, and used near ponds and marshes subject to evaporation.

Proponents of such schemes are confident in implementation, although technical problems (e.g., cost-reducing of membranes) remain to be solved, further environmental impact studies should be conducted, and, at least in the United States, funding of research is very modest.

Implementation

The obstacles to implementation of any of the envisioned schemes tend to be discouraging in view of difficulties by no means swept out of the

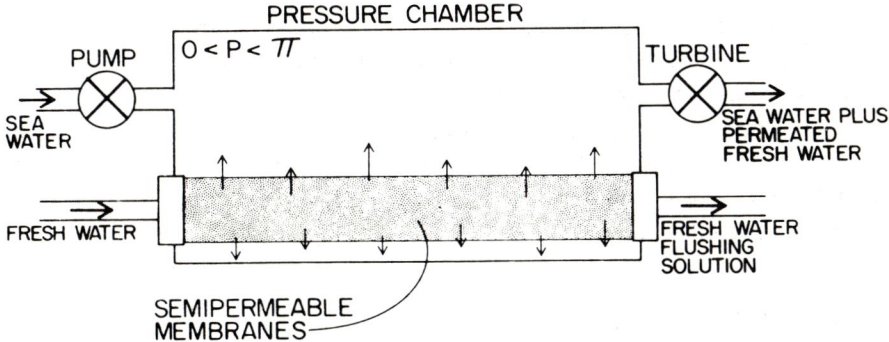

Fig. 1.4. Salinity gradient energy conversion. (Wick and Isaacs, "Utilization of the energy from salinity gradients," Wave and Salinity Energy Conversion Workshop Proceedings, Univ. of Delaware, 1976)

way by contemporary technology. Perhaps oddly, the discarded Claude scheme for ocean thermal power, or its modified version, appears to be holding (at least for the immediate future) considerable promise.

Besides a prohibitively high cost for many of the five schemes outlined above, technical difficulties include reducing the number of electrodes or using irreversible ones. It would also be possible to use seawater hydrogen freed as a by-product of the latter alternative, which would somewhat compensate for a loss in power. Finally, a toroidal stack-up of cells would enable the plant to dispose of the need for electrodes.

Neutral membranes, even though less efficient, are less expensive than anion-permeable ones and may be the answer to cost reduction in that sector. Polarization—namely, production of gases or even valuable chemicals—can be reduced, as shown in Israel, by periodical current reversal (a fraction of a second) to "flush off" such products from the membrane.

Power loss results from the energy used in pumping the water and in the cells. Ways to counteract this include the use of narrow cells, widening the access channel of seawater, or increasing seawater's salinity, at the risk, however, in this latter instance, of speeding up membrane deterioration (Weinstein and Leitz, 1975). Ion exchange capsules inside dilute cells increase surface area and reduce distance but decrease flow through the channel (Wick, 1976).

Corrosion, fouling, and silting are important factors requiring attention; they had been important factors considered for tidal power plants. Thermal "pollution" is negligible, estimated at merely $0.5°C$, but protective devices must be developed to prevent marine animals from being sucked into seawater inlet pipes—as in tidal power plants—and estuarine environmental protection must be provided for.

It has been pointed out that small tidal power plants might prove most useful and affordable for developing nations. Low efficiency and small-sized salinity power schemes may likewise be of considerable value to remote—and, particularly, developing—areas.

Geographical Sites

A reasonable search for suitable sites on which to build tidal power plants has been conducted, although much work remains to be done in order to refine and complete the listing (Charlier, 1970). Far less has been done for salinity power station locations. Yet this may be as important a task as the further improvement of the technology involved. Wick and Loeb (1975) have provided two theoretical examples: the discharge of the Columbia River in the Western United States amounts to 6,600 m^3/sec and, at 30% efficiency, half of such a flow could produce 2,300,000 kW, about equal to the ocean thermal power from a similar water volume. However, electricity production would be greater if river water and salt water came into contact. The Dead Sea brine and the Jordan River fresh water coming into contact provide an osmotic pressure that corresponds to a 3,100-m-high dam (Table 1-2).

Most promising geographical locations are thus those at the mouths of powerful rivers, hypersaline basins, and salt pans which dot desert coasts.

There is, as for tidal and thalassothermal plants, ample justification for a geographical study of sites and markets.

OCEAN BIOENERGY

Ocean energy sources under most active consideration are all physically generated and only recently has some attention been given to chemically-biologically generated power. The American Gas Association has funded a research project to examine the potential of marine plants conversion to substitute natural gas. Additional funds for the project have been supplied by ERDA's Office of Solar Energy. General Electric is currently working on a marine biomass methane production scheme. This is only one of several programs dealing with alternate production methods of natural gas; besides kelp conversion, the undertaking would also yield residual by-products that could be processed into feeds and fertilizers. North has discussed the "Marine Farm Project" (Mitsui *et al.*, 1978) and, in Asiatic countries, mass culture of *Chlorella* is under study (Tsukada in Mitsui *et al.*, 1978). Mitsui proposed, as a potential energy source, the hydrogen produced by marine photosynthetic organisms, and Kurita proposed its photoproduction from water (in Mitsui *et al.*, 1978) (Fig. 1-5).

Recently, the popular French magazine *Match* published an article asking whether the world might be "devoured by giant algae" and went on to say that, while the French would not conduct further research on *Macrocystis pyrifera* (the fast-growing, rapidly regenerating, giant

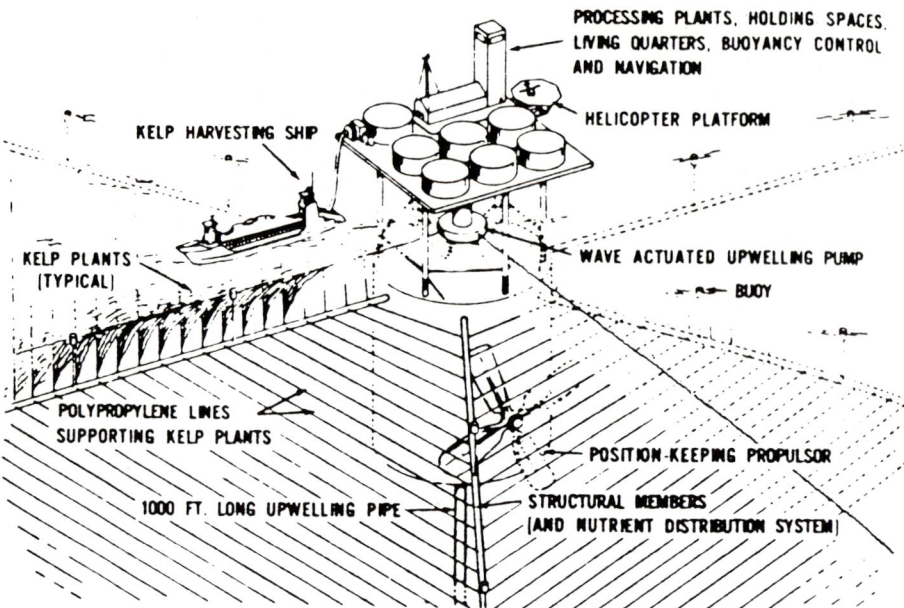

Fig. 1.5. Conceptual design of ocean food and energy farm unit. (Wilcox, "Ocean farming," Proc. Conf. on Capturing the Sun through Bioconversion, Washington, D.C., 1976)

kelp), Americans would. Americans and Mexicans harvest kelp along the coasts of lower California.

This brown alga normally inhabits coastal nutrient-rich regions, but a marine test farm constructed some 8 km off Corona del Mar (California) will attempt to grow the alga on an artificial substrate of depths of 18–24 m by generating an artificial upwelling of nutrient-laden water located below the farm anchored at 150–300 m. The task of developing the process of converting kelp into methane is entrusted to the Chicago-based Institute of Gas Technology; this is the same Institute that proposed in 1973 to extract hydrogen from seawater and switch from a traditional economy to an "hydrogen economy," an idea often brought to the fore in connection with ocean thermal and tidal power tapping.

The basic concept is to retrieve and store solar energy captured by photosynthesis. However, because average value of capture of solar energy by plants is estimated at only 2 or 3% of 500 cal/cm^2/day, huge farm areas are necessary, which only the ocean can provide economically; furthermore, the productivity of the ocean is higher than that of the land. Of the brown algae, *Macrocystis* is the largest, the heaviest, and the most widely distributed. A perennial, it probably absorbs as much as 99% of the incident sunlight through its layers of blades floating at and near the water surface. The alga thrives in mesothermal climates of the southern hemisphere, particularly along South American coasts, in the northeastern Pacific and northern Atlantic. Furthermore, the giant kelp resists rough waters, attaches itself easily, regenerates rapidly if only the top part of the plant is harvested, and can be mechanically harvested.

Various substrates were considered—even scrap tires—and a tension grid was ultimately retained; it supports loads by axial tension, has flexible cable members, and is a netlike structure. The single module—acronym QAM (*Q*uarter *A*cre *M*odule Farm)—apparently will be nonpolluting and reasonably easy to deploy and maintain. Waters at depths of 150–300 m contain the necessary nutrients absent in surface conditions at the offshore site and will be brought up to the artificial substrate level of 18–24 m.

Kelp harvesting has been done for a long time and specialized vessels are equipped with a mechanism that shears the plant at a depth of 1 m below the surface; there are 4–6 harvests a year. The economic feasibility study of the project will have to take into account whether the traditional depth of cut and frequency of harvesting should be maintained or modified, possibly increasing the size of the harvest. United Aircraft Research Laboratories is considering seeding kelp along a 130-km tract in the Pacific and another one in the Sargasso Sea. Theoretically, the two tracts together could meet 12% of United States energy needs (Table 1-2).

Costing

As with other alternate sources of energy, marine biomass energy production faces economic problems, principally lack of data on capital and operating costs for farm structure and support subsystems and on the commercial value of residues and by-products. The cost of the methane produced has been estimated to be comparable to that of methane produced by coal gasification, but the results are based on speculations (Flowers and Bryce, 1977).

Implementation

Once harvested, the kelp must be used for methane production; alternatives under study are anaerobic digestion, a process already familiar; the possibility of hydrogasification, which is barely in the primary stages of investigation; and, finally, thermal conversion techniques (Fig. 1.6).

The General Electric scheme emphasizes bioconversion and an inoculum development program to isolate and characterize the marine microorganisms utilizing kelp as a growth medium. After pretreatment, ground kelp would be put in septic tanks, where bacterial action would produce methane.

Provided the scheme is economically feasible, a pilot scale marine farm and gas production factory will be built; assuming that the pilot plant shows acceptable costs of construction, operation, and maintenance, and a satisfactory yield, a commercial facility could become a reality. Solar energy bioconversion may also hold promise for developing countries, as discussed by Horstmann (in Mitsui *et al.*, 1978).

The potential of this type of energy is large, but substitute natural gas production requires large areas of the ocean. In view of this, one may prefer to think in providing more for regional needs than in terms of large energy production. At this time, the United States is engaged in developing within 5–10 years a 40,500-hectare ocean farm. Such a farm would yield 1.6 quadrillion BTU (403 quadrillion calories) of food and 16 quadrillion BTU (56 trillion cubic meters) of methane.

Research pertaining to biomass conversions is, or has been, conducted by the University of California at Davis, the Institute of Gas Technology, the California Institute of Technology, and the General Electric Corporation, with funding from the U.S. Department of Energy and the American Gas Association. Woods Hole Oceanographic Institution studies whether various macroscopic algae species could be profitably cultivated, in some cases possibly using wastewaters as nutrients; these cultures could then be used for energy conversion.

As a first step toward development of the more ambitious farms, an experimental ocean farm was positioned on a polypropylene raft of about 2.8 hectares anchored at 1 km off San Clemente Island in water 100 m deep. Some 150 *Macrocystis pyrifera* thrived well for about six months (1974–1975), after which the raft became a total loss due to accidents. A new farm exists off Corona Del Mar, but it covers only 0.1

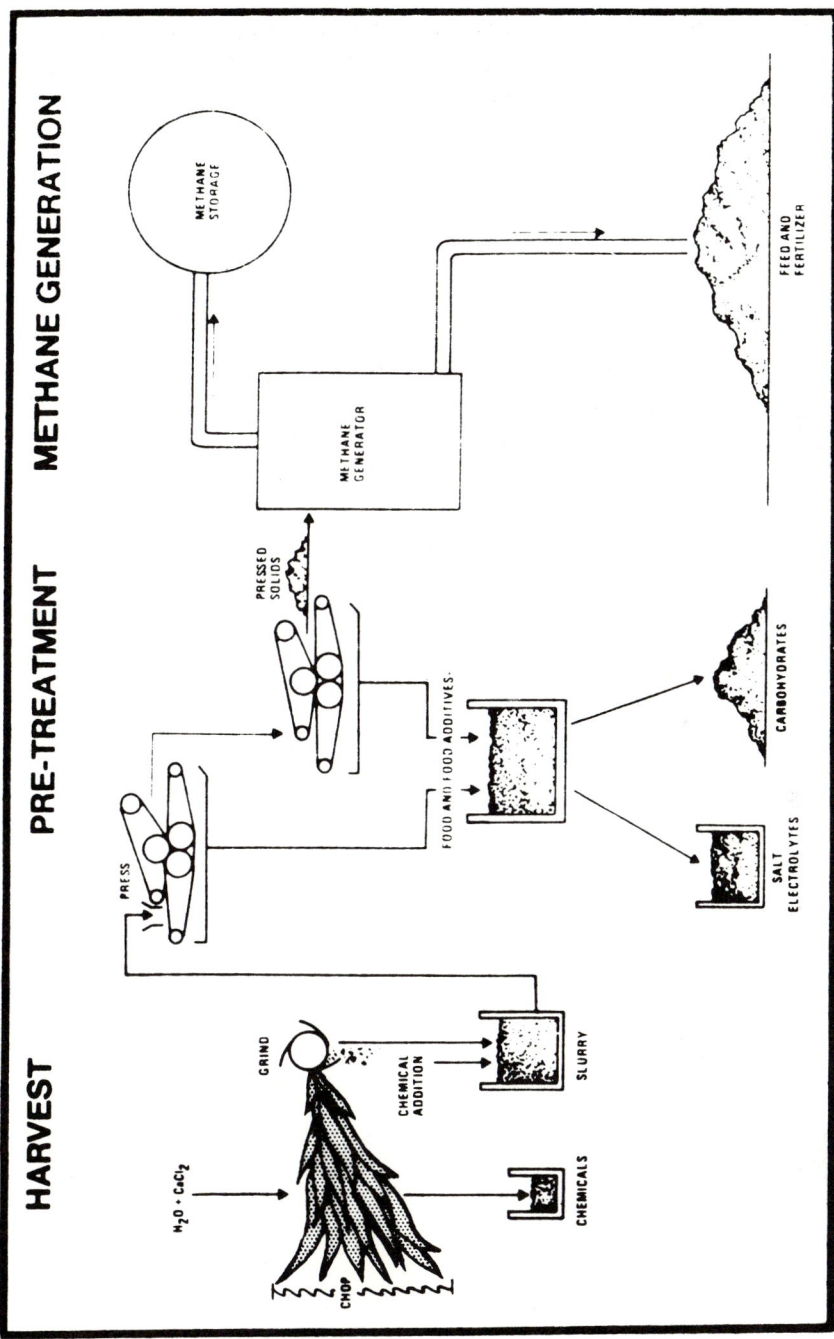

Fig. 1.6. Kelp bioconversion process. (Wilcox, "Ocean farming," Proc. Conf. on Capturing the Sun through Bioconversion, Washington, D.C., 1976)

hectare. A subsequent phase foresees the positioning of a farm in the Atlantic Ocean and another in the Pacific Ocean, each covering 45 hectares. Besides *macrocystis*, such kelp as *ecklonia* and *laminaria* could be used.

THERMAL ENERGY

The oceans are responsible for converting solar energy into thermal gradients. The solar radiant energy is most efficiently absorbed by the ocean, and it is commonly estimated that oceans and the atmosphere above them absorb some 80-billion MWt, or 50,000 times the amount used daily by man in 1977. (Table 1-2) The concentration of solar energy due to the difference of temperature between surface waters and deep waters in the tropics is considerable. It is this difference that Arsène d'Arsonval and Georges Claude tried to utilize.

George Claude made several attempts, the first one using thermal pollution caused by manufacturing plants' wastewaters poured into the River Meuse near Ougrée (Belgium). He made further attempts off the coast of Brazil, as well as in Cuba. But credit should also go to Charles Beau and N. Nizeri, both Public Works engineers, who founded the Société de l'Energie des Mers, which actually built the first full-scale operating plant utilizing the temperature differences between deep and surface ocean waters. Actually, this was merely the culminating effort of a long series of unsuccessful attempts stretching over the 20-year span separating the two world wars.

The idea of using an intermediate fuel to resolve several technical problems is not as recent as currently assumed. The idea of an intermediate fuel was picked up by the American Campbell, who suggested in *Engineering News* (1913) to use liquefied gas, and a bit later (1923) by the Milanese Boggia and Marius Dorning, who studied tapping thermal energy in the lakes Bracciano and Bolsena.

Differences of temperature between deep and surface ocean waters often reach 40°F (4.4°C), which allows propane to boil and condenses thermal fluids. Such fluids, in the state of pressurized vapor, will drive a vapor turbine electric generator. What is needed is a partially submerged vessel, with a boiler, turbine, electric generator, condensers, and auxiliary equipment. There are, of course, no fuel costs.

There are basically two types of heliothalasso systems: one, already susceptible to implementation, extracts solar energy stored in ocean surface layers; the other would use concentrating mirrors placed on a floating platform to focus incident solar energy on a boiler. Engines placed on such a float could tap the thermal difference potential of Gulf Stream surface waters and the very cold waters 300 m below.

The first full-scale attempt at building an ocean thermal plant was initiated in 1942 by the French Ministry of Colonies and the Centre National de la Recherche Scientifique; after six years of study, construction of a plant was finally decided. The technical aspects are abundantly described in a report published in the *Proceedings of the Fourth World Power Conference* (1952). Basically, in this project, the thermodynamic cycle consisted of evaporating, under vacuum, part of the warm surface waters at 82°F (28°C), encountered in tropical areas. The steam is taken in by the condenser, itself cooled by the colder—46°F (8°C)—deep waters, and on its way the steam proceeds through a turbine that drives an electric generator. At that time, sea thermal energy compared favorably, from an economic viewpoint, with hydroelectric energy, the more so since the energy could be directly used for an evaporating plant for chemical industries.

The site chosen was Abidjan in today's Republic of Ivory Coast. Here, fresh water was in greater need than electrical energy; a project combining both aims was ideal. The major problems arose with the immersion of the cold water adduction pipe; repeated experiments with a large-diameter duct all failed. When Nizeri and Adam (1952) used articulated joints for the duct and anti-wave floaters to hold it up suspended on cables held by winches, the operation met with success. If the plant nevertheless went out of business, this was due to the facts that conventional power plants did produce cheaper electricity at that time, that the ducts suffered repeated ruptures, and that turbines required large dimensions. It is noted, in this regard, that Hydronautics' system calls for no less than six 23-m (78-ft) turbines using temperature differences of 27°F (15°C), while the French engineers felt no difference of less than 30°F (20°C) should be considered; hence, Hydronautics' system would be suitable for a wider geographic range (Fig. 1.7).

Since Nizeri and Adam's scheme, the matter has been reexamined in France by Gougenheim and Romanovsky (1957) and by Daric (1957). Howe conducted studies dealing with the size and cost of turbines at the University of California at Berkeley; combination solar-thermal energy schemes were proposed by Barjot (in Sternmann, 1971),

18 TIDAL ENERGY

Fig. 1.7. OTEC model. *(Source: Lockheed Corp.)*

Masson (1955), and Gomella (1966). Considerable thought has been given to tapping ocean thermal differences in the United States. Heronemus (1972), following the work of the Andersons, suggested (1974) a scheme which envisions a series of such plants spaced in an area 15 miles east-to-west by 550 miles south-to-north along the western portion of the Gulf Stream. The electricity produced, according to Heronemus, could be transmitted to virtually any location at competitive cost. The plants would use a closed Rankin cycle with ammonia or propane, for instance, as a cooling intermediate fluid, which could function with a 30°6F (17°C) temperature difference. The power plant itself would be semi-submerged and contain multiple units. If J. H. Anderson, and his son J. H. Anderson, Jr., founders of Sea Solar Power, Inc., have pioneered many of the current ideas since 1964 and kept on refining them since, credit for proposals in thalassothermal development goes also to Lavi and Zener (1975). Zener, then Chief Scientist at Westinghouse, already studied in 1965 the practical aspects of tapping ocean thermal power; Lavi and Zener are reexamining the open cycle option while the Division of Solar Energy in the government's Energy Research and Development Administration was forging ahead with the closed cycle system. Finally, it should be pointed out that Lavi and Zener's views are parallel with the Lockheed's study conclusions, though there are divergences where water pipe and heat exchanger designs are concerned.

Gérard and Roels published in 1970 in the *Journal of the Marine Technology Society* a paper concerning ocean water as a resource. In it, they explored using upwellings in the Virgin Islands as a cold water source to provide a temperature difference with the surface waters, in order to produce electricity. The study also involved fertilization of the sea. Roels picked up the subject again, in collaboration with Othmer (*Science*, October 12, 1973); they feared that the various designs involving vertical suction pipes suspended from vessels or platforms, submerged power cables, and fresh water lines carrying products to shore would augment the difficulties for controlled mariculture, which they proposed to link with energy production. They also discussed in detail the engineering design made by Alemco, Inc., for their particular site.

OTEC

We refer to a thalassothermal plant as an OTEC, an *o*cean *t*hermal *e*nergy *c*onverter. Federal government guidelines require that the plant last at least 40 years, that it be accessible and easily maintained, that it present a minimum technical risk, and that it generate at least 100 MW at the lowest cost per kW-hr using current technology. Numerous proposals have been introduced and several sites suggested. On the average, the government allocated less than 6% of its budget to non-nuclear and non-fossil fuel research. In fact, only 1.8% of the energy program's budget was occasionally earmarked for all aspects of solar energy research and development. Total funding of OTEC for the period 1977–1986 is $524,563,000, quite an increase compared to $85,000 in 1972, and $8,500,000 in 1976. In 1977, $14,500,000 was allocated. This amount ought to be compared to $900 million for solar thermal energy, $500 million for wind energy, and $100 million for photovoltaic energy. OTEC's share is thus about one-fifth of the total. (And yet, one week's import of oil by the United States required about $600 million in 1977!) (Fig. 1.7)

Hydronautics' design is an open-cycle plant using steam produced by seawater driving large-diameter turbines. The structure would have a diameter of about 25 m and be 12 m high, and Hydronautics foresees a 600-m-long cold water adduction pipe 11 m in diameter. Production would reach 100 MW.

TRW would build a concrete cylindrical vessel displacing 212,000 tons with a cold water adduction pipe 1,300 m long and 15 m in diameter. Based on technology used for the North Sea Brent platform, it uses ammonia as an intermediate fuel.

The Lockheed model, which was displayed at Okinawa's Ocean Expo '75, also uses ammonia as an intermediate fuel. Semi-submersible, it is a 260,000-ton concrete platform, submerged for 300 of its 320-m height. Production would reach 160 MW (Fig. 1.8).

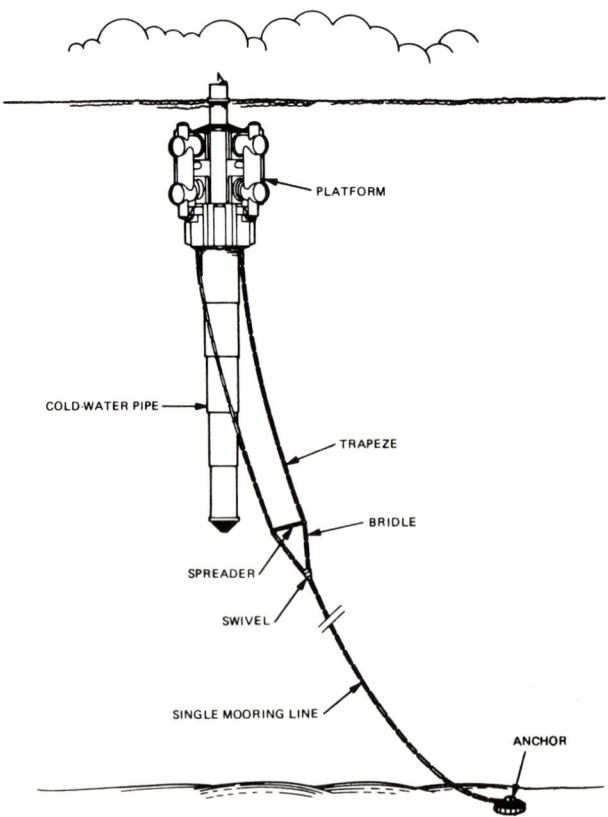

Fig. 1.8. Full view of Lockheed OTEC plant model. *(Source: Lockheed Corp.)*

Sea Solar Power has constructed a working model using a patented (R-12/31) intermediate fluid (freon). The floating structure is mostly below water level so as to shelter it from storms and waves; the hull measures about 110 m. A pipe, about 11½ m in diameter, taps cold waters at depths ranging from 600–1,300 m.

Further studies and proposals have been made: the Carnegie-Mellon University design uses ammonia as an intermediate fluid, while the University of Massachusetts researchers favor propane. The latter is a plant 120 m high, 205 m long, and 150 m wide, with an adduction pipe 325 m long. The structure would be moored offshore. An innovation of the Carnegie-Mellon University proposal is that it would use the thalassothermal power to generate hydrogen fuel by electrolysis, an eminently transportable fuel, and solving the thorny problem of energy transmission networks.

Still another scheme deserves mention. It has been developed by the Johns Hopkins University Applied Physics Laboratory; however, instead of requiring only current technology (an ERDA guideline), it calls for technological innovations in the heat exchanger, the hull design, and the construction. This proposal also suggests producing the ammonia at sea, aboard the OTEC plant-ships located in tropical waters. Production could start with a 100-MW_e plant and be gradually brought up to 500 MW_e.

Costing Comparisons and Implementation Plans

The two schemes to be assessed are the open cycle and the closed cycle. In the open cycle, arrangements are made above sea level, and direct contact heat transfers take place. Seawater is used. In contrast, the closed cycle does not directly use seawater, but instead generates power by using the vapors of an intermediary working fluid. In the closed cycle, heat exchangers, usually of expensive metal, are required. This second scheme has a larger heat transfer surface, good conductivity, anti-fouling provisions, and a high corrosion-resistance.

Power costs of the Rance River tidal power plant are comparable to those for nuclear power generation, and thus oceanic energy can generate electrical power at costs comparable to those projected for the early 1980s. As Griffin (1977) has pointed out, the capital costs of open-cycle OTEC power generation, in the 10–20-MW_e range ($1,360–1,700/$kW_e$), convert to hourly fixed charges of 27–30 mills/kW-hr when computed in the same manner as the closed cycle. Costs per kW adjusted to $\Delta T = 18.9\,°C$ are 1,900 for Lockheed's scheme, 1,650 for TRW's, 1,630 for Carnegie-Mellon's, 900 for Johns Hopkins, and 524 for the University of Massachusetts design (Lavi). If we compare costs for current electricity methods, fixed charges in mills/kW-hr are 16 and 14 for pressurized water reactors and desulfurized coal-powered plants, respectively, versus 40 baseline and 21 production for TRW's scheme, and 32 for France's tidal plant; however, fuel costs amount to 11 and 14 for the water reactor and the coal plant, versus nil for the others.

Currently, a barge formerly used with the Hughes Corporation's Glomar Explorer, and rechristened OTEC-1, is being prepared and adapted to serve in 1981 as a 5-MW_e pilot plant off Hawaii, Florida, or Puerto Rico, or in the Gulf of Mexico.

A recent evaluation of the environmental impact of the open ocean, on heat exchanger tubes and on a 1-MW_e component test platform, will lead to a decision on eventual construction of a 25-MW_e plant by 1984.

Geographical Sites

The Gulf Stream off Key West, Florida, near Miami, or off Cape Canaveral, which flows above a much colder countercurrent, would offer excellent possibilities. The temperature gradient over a vertical column of 1,000 m varies between $16\,°C$ and $22\,°C$. Full utilization would provide 180 million GW-hr of electricity annually (Metz, 1973).

This appears very promising, although transmission costs present a serious handicap beyond 1,000 km of the generating site. Thus, the number of offshore sites is, for now, rather limited. Plans for utilizing solar resources (not specifying whether on land or at sea) were reported in 1976, by only six coastal nations: Nigeria, India, Israel, France, Australia, and the United States. Besides the United States, France continues modest efforts in the area of sea solar power.

The Gulf Coast of the United States has also been considered, but further population and industry concentration there might have an unfavorable environmental impact. This is not true near Keohole Point, close to Kailua (Hawaii); here, a study by Bathen (1975) indicates that a cost-competitive OTEC plant with a 35-MW capacity could be constructed and apparently without unacceptable environmental consequences; however, transmission problems for either an offshore or onshore plant must be resolved.

Other locations which offer suitable conditions include the south coast of Puerto Rico, the north coast of St. Croix (one of the U.S. Virgin Islands), and several Pacific U.S. islands in Micronesia. There has also been some consideration given to the Atlantic Underwater Testing and Evaluation Center in the Bahamas. Besides underdeveloped areas in the Caribbean, sites exist in Central America and along the northern and central coasts of South America. In Asia, Taiwan, the Philippines, Indonesia (particularly Borneo and New Guinea), Ceylon (Sri Lanka), and southern Arabia qualify. In Africa, locations stretch from Mauretania to Zaire and from Djibouti to Madagascar. Australia also has suitable sites. We know from Claude's experiments (1930) that excellent sites exist off the coasts of Cuba and Brazil. Barjot, in the midsixties, even suggested that the waters of the Arctic Ocean could serve as the warm fluid and that the polar air above could serve as the cold fluid, providing a 50°C temperature differential. As Sternmann (1971) has pointed out, Barjot-type plants, which require only short cold water adduction pipes, could be built along the coasts of Alaska, Canada, Greenland, Siberia, and Scandinavia, and in the Arctic and Antarctic areas; provided, naturally, that consumers justify the construction.

The Japanese Science and Technology Agency and the Ministry of International Trade and Industry are conducting thalassothermal research, while in the private sector Japan's sea thermal power resources are spurring on work by electrical appliance manufacturers such as Toshiba.

Clearly, besides putting the final touches on the technology and solving the transmission problems—unless hydrogen production is decided upon—there is ample justification for a geographical study of sites and markets.

The idea of harnessing ocean thermal energy has been examined for nearly a century, but less than ten years ago the interest was still mostly academic. Current and future energy needs have sparked a keener interest. The success of thalassothermal plants no longer depends on its feasibility, but on circumventing some remaining construction, operation, and possibly environmental problems. We already have come a long way since the Abidjan plant, because engineering firms are now confident that, in addition to saving millions of barrels of fossil fuel, ocean thermal plants can compete cost-wise with conventional electricity-producing plants.

At this time, the Lockheed proposal could lead to an operating thermal plant by 1986, with a pilot-plant functioning already in 1981. Similarly, TRW set a target date of 1981 for an onshore pilot facility of 2.5 MW; in their view, a goal to be attained, if ocean thermal conversion is to be significant is 20,000 MW_e installed by 1990. The year 1981 is also proposed by Johns Hopkins University to put a plant ship to work (100 MW); by the end of 1982, it would put a first commercial-size plant into operation (300-500 MW).

The deadline of 1981 will not be met by Carnegie-Mellon because of a delayed start, but it still could soon complete a 100-MW pilot plant and, about four years thereafter, a 100-200-MW commercial plant. The University of Massachusetts also plans a 10-MW pilot plant, to be followed by a commercial facility, and its program would culminate in the acquisition of 40-160,000 MW within 15 years. The U.S. Department of Energy's own plan includes a 5-MW pilot plant by 1980-1981 (complete power cycle) and a 100-MW demonstration plant in 1983-1984; commercial plants would eventually be put into service by 1990.

Technology exists and operating principles are well understood, but all systems under consideration are still plagued by some engineering and non-negligible cost problems. Ocean thermal energy conversion is a continuous source of energy because the solar energy conversion is continuous and it is stored in surface ocean waters undergoing very small temperature variations.

THE WAVES

Waves, if sufficiently regular and powerful, could provide an acceptable source of energy. Normally of low height and only occurring a few seconds apart, waves pose some problems; the difficulties are created by storms. The characteristics of ocean waves place constraints upon converting devices. Wick and Castel (1978) have made field tests with several models of wave power generators; their experimental results support their view that recently developed wave pumps are suitable for power generation in a variety of circumstances. Experimental tryouts have been attempted for a long time: in 1902, Alva Reynolds built a wharf and suspended panels underneath it; the force of the waves was transmitted to a wheel attached to an electric generator. His collaborators and successors made the headlines when they tested a wave motor off Pacifica (California) in 1971. Von Arx (1979) would amplify the energy effect of the surf by focusing it on a "horn" designed to accept the surge; a head of tens of meters could thus be created and the water stored to supply hydroelectric turbogenerators. Since a minimum of sediment transport is necessary to absorb incoming wave energy, a large number of such horns could upset the balance (Figs. 1.9, 1.10).

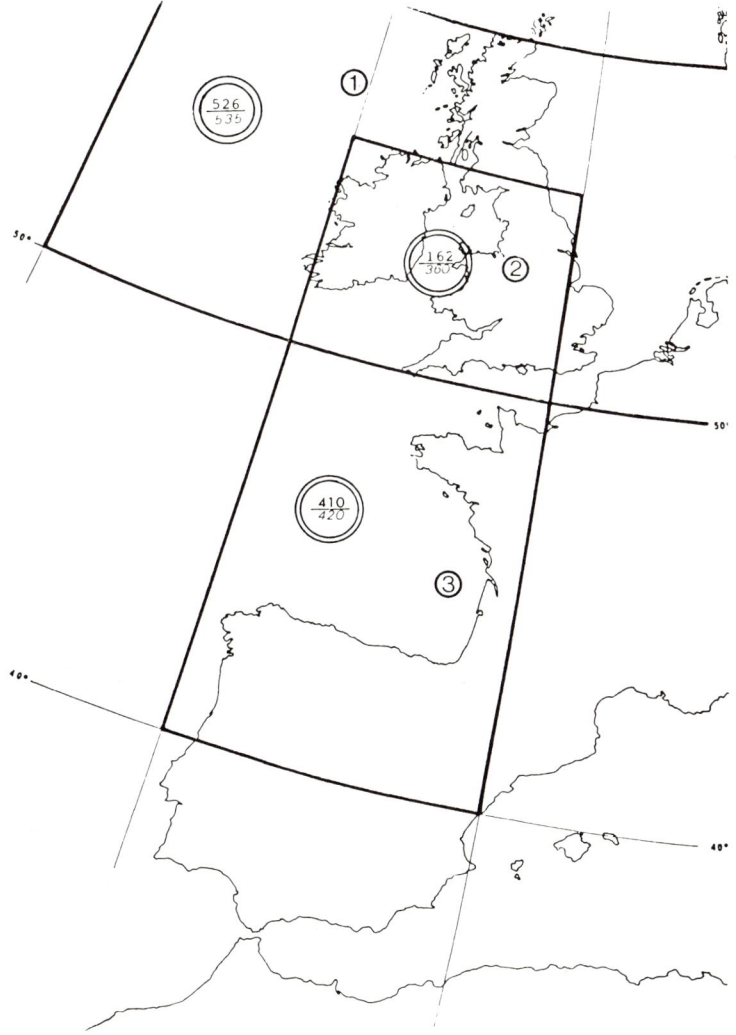

Fig. 1.9. Annual wave energy. *(Source: Eurocean, 1978)*

26   TIDAL ENERGY

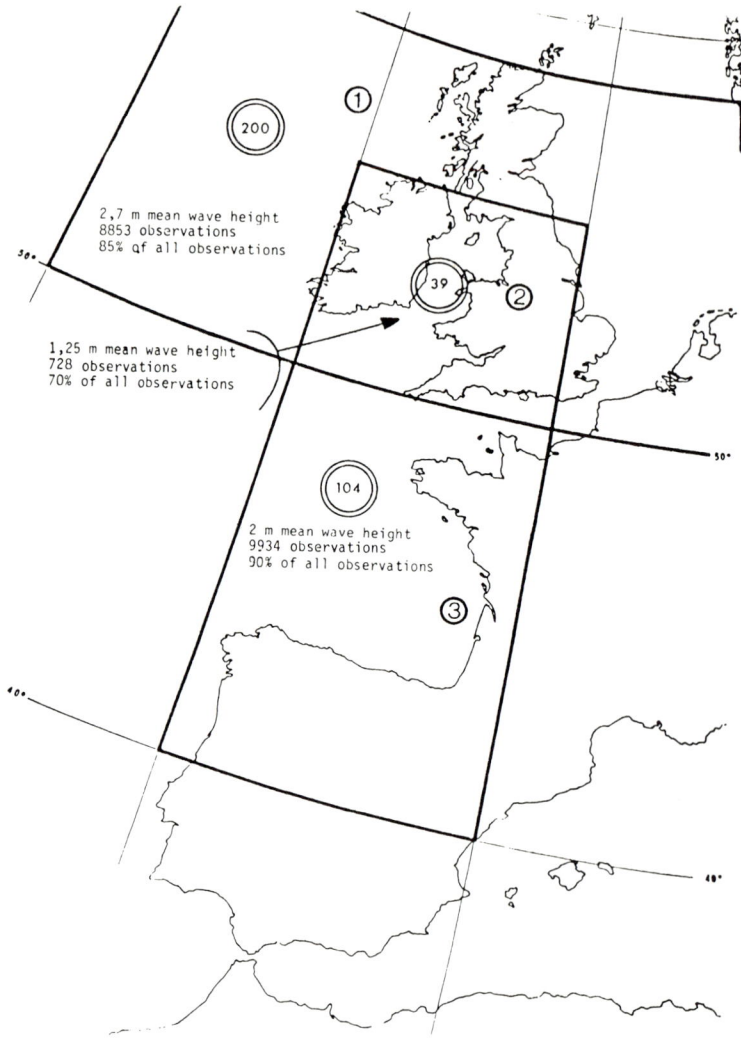

Fig. 1.10. Trimestrial (September–November) wave energy. *(Source: Eurocean, 1978)*

Slow-moving internal waves at the interface between waters of different density or temperature possess considerable power as evidenced by the "diving ballast and trim" records of submarines; the problem is to concentrate their energy. Von Arx suggested the use of submarine canyons and appropriate bottom topography features as horn or receiver.

Capturing ocean (and why not lake?) energy dissipated by waves is clearly not a recent idea. Besides the string of lights using wave energy on the Pacifica pier, some at-sea devices use wave-energy. Such energy could be a valuable supplementary source of power. Hundreds of patents have been taken out; the Japanese pressed into service, in 1978, a huge barge which, in fact, is an electricity plant. Clean, safe, and of negligible environmental impact, implementation of any of the schemes is stalled by the usual problem of cost (Table 1-2).

From a technological viewpoint, wave energy capturing devices can utilize rise and fall motion of the wave to activate a turbine; utilize the rolling motion of the wave by vanes or cames that drive turbines; and concentrate waves in a converging channel, thus building up a hydraulic head which then operates a turbine. Motors could also be operated by the varying slope of the wave surface and by the impetus of the wave rolling up the beach.

Leishman and Scobie, reporting in 1976 for the U.K. National Engineering Laboratory of the Department of Industry, concluded that large-scale production of energy from sea waves is technically feasible and could be achieved by the development of existing technology. It has been claimed that harnessed wave energy along Great Britain's coastlines could furnish most of the power currently obtained from petroleum-fired plants.

Devices and Environmental Effects

Devices for trapping breaking wave energy include converging ramps leading to a natural or man-made reservoir wherefrom the water flows back to the sea after passing through low-pressure turbines, and devices that are set in motion by the wave impact itself. Converging channels supplying a basin constituting the forebay for a conventional low-head power station seem to provide the highest output of any scheme proposed, yet presently uneconomical. Power systems conducted successful small-scale tests with a concrete trough parallel to the shorelines in which a pliable strip filled with hydraulic fluid is secured. Submerged at 7-m depth where wave shape plays no role, the strip breaks the waves, undergoes the hydrostatic pressure of the water mass above, and transmits it to a hydraulic accumulator through the hydraulic fluid. Pressure is stored by an accumulator until a specific magnitude is attained, at which point it delivers it to a fluid dynamo-connected motor.

28 TIDAL ENERGY

Wave power is pollution-free; like wind and sun, it is a renewable resource widely available and requiring no fuel. Installations could provide simultaneously coast and harbor protection, while power units can be coupled to desalination plants.

Lybrand Smith (in *Ocean Industry*) calculated that 33 CV/m of sea wave front are furnished by a wave less than 2 m high, which represents more than 31,500 CV/km; in 9-m-deep water, this becomes 204,120 CV/km and roughly 150,000 kW/km (242,000 kW/mile).

British Studies

Great Britain has one of the world's most favored coastlines for tapping wave power. Energy contained in Great Britain's waves is variously estimated at from 40 kW/m to 70 kW/m, depending on geographical site. Fortunately, this energy is available throughout the year and seems to coincide with the seasonal pattern of electricity demand. It has been estimated that 50% of the total need for electrical power could be produced by harnessing waves along a 1,000-km ocean stretch (Figs. 1.11, 1.12).

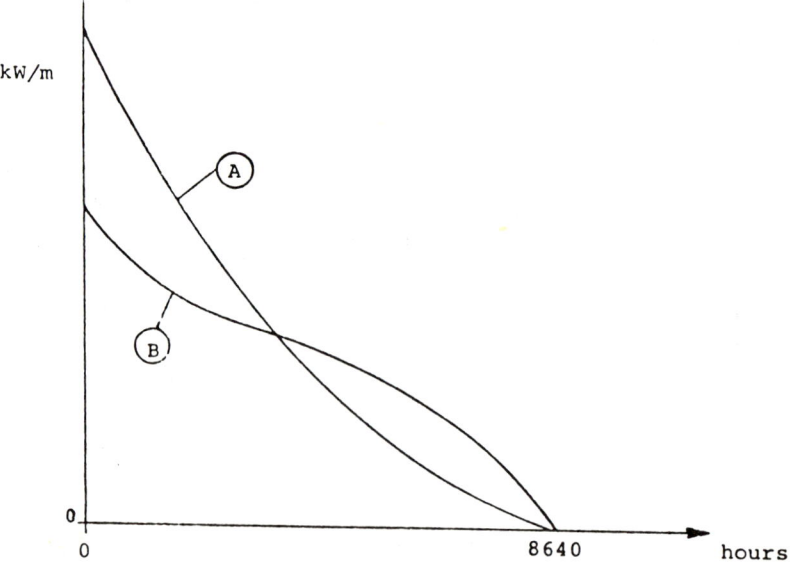

Fig. 1.11. Number of hours in a year during which wave power per unit length exceeds the individual value. *(Source: Eurocean, 1978)*

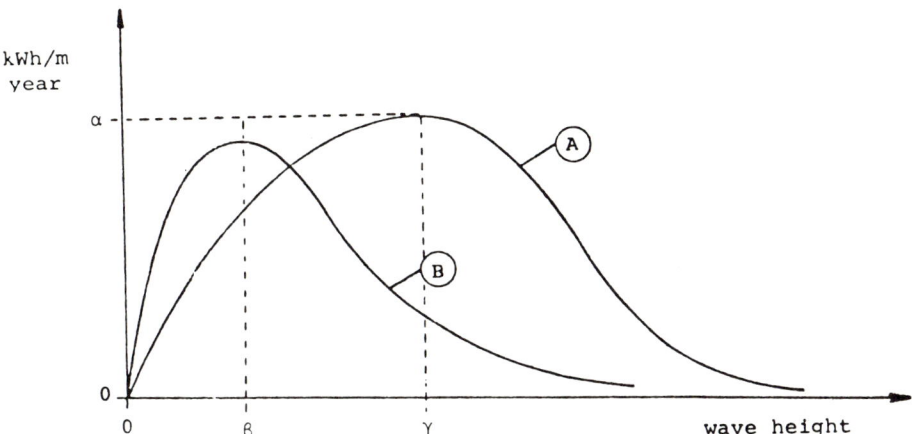

Fig. 1.12. Total energy carried by waves during one year for each wavelength. *(Source: Eurocean, 1978)*

Fig. 1.13. Wave energy machine used on a California beach. *(Source: Los Angeles Times, 1978)*

30 TIDAL ENERGY

Among the numerous devices that have been proposed for wave energy transformation, the British government is selecting four for funded research. The Russell Rectifier is a structure with high- and low-level reservoirs that is exposed to waves. Waves drive seawater in the high-level reservoirs and extract it from the low-level reservoirs, separated from the sea by vertical non-return flaps. The Salter duck, developed by Salter (1974) after many years at Edinburgh University's

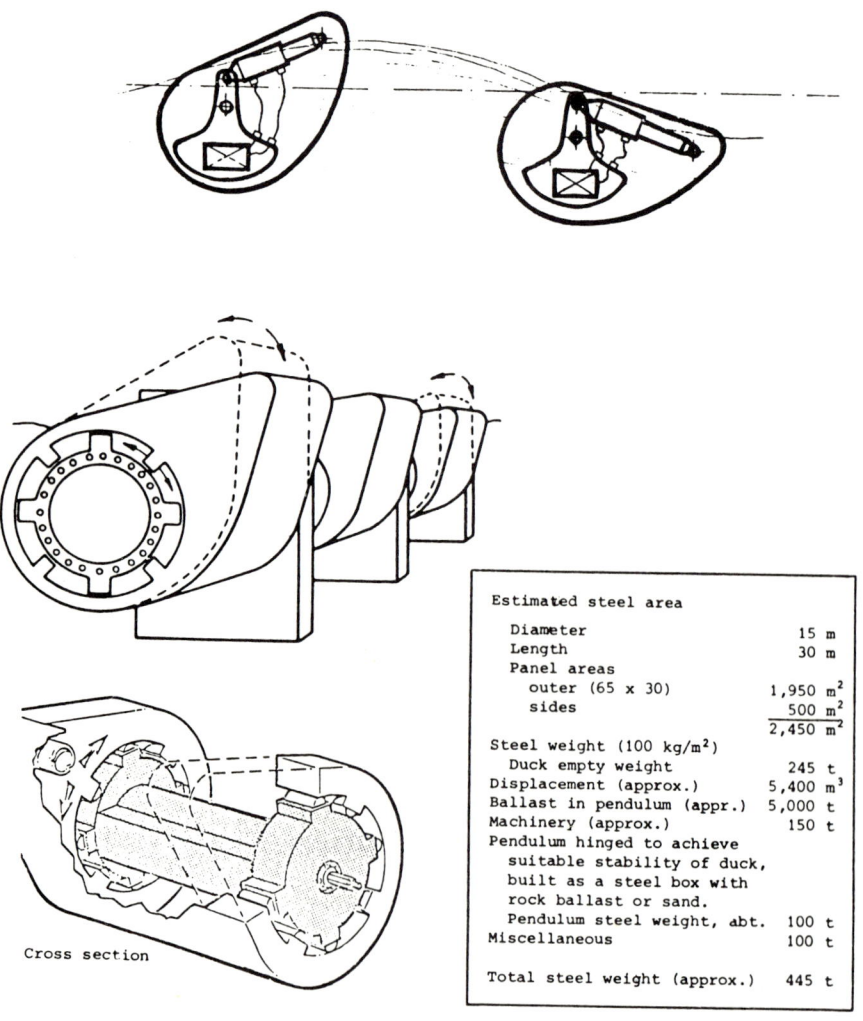

Estimated steel area	
Diameter	15 m
Length	30 m
Panel areas	
outer (65 x 30)	1,950 m²
sides	500 m²
	2,450 m²
Steel weight (100 kg/m²)	
Duck empty weight	245 t
Displacement (approx.)	5,400 m³
Ballast in pendulum (appr.)	5,000 t
Machinery (approx.)	150 t
Pendulum hinged to achieve suitable stability of duck, built as a steel box with rock ballast or sand.	
Pendulum steel weight, abt.	100 t
Miscellaneous	100 t
Total steel weight (approx.)	445 t

Cross section

Fig. 1.14. Schematic view of Salter nodding duck system. *(Source: Eurocean, 1978)*

Bionic Research Laboratory, is an oscillating vane, so shaped that it permits extraction of a high percentage of incident wave energy (Figs. 1.13, 1.14, 1.15).

Wavepower, Ltd., pursues work with a series of rafts, separated by hydraulic motor-pump combinations that convert the raft motion energy into high pressure in a fluid. The system is designated as "contouring rafts." Pistons containing cylinders hinge three rafts together; these pistons are forced to pump water by the flexing of the hinges in the waves. A small operational size model of such rafts has been put to work in the English Channel and produces 1 kW, thus proving feasibility. According to hovercraft designer Sir Christopher Cockerell, a cluster of 300 rafts would produce the same amount of electricity as a large (300-MW) conventional power station. The British National Engineering Laboratory is involved with the Air Pressure Ring Buoy; a study of floating breakwaters in Japan revealed that wave height could be significantly reduced by shaping breakwaters like inverted boxes, while causing the wave motion in the box to work on the air, forcing the air in and out of orifices in the top of the box (Fig. 1.16).

Salter (1974) also considered energy storage. Electrolytic production of hydrogen from seawater appears to him promising in that respect. He suggests self-propelled installation which could move in line ahead, a low drag condition, into the Atlantic, turn abreast to the waves, and

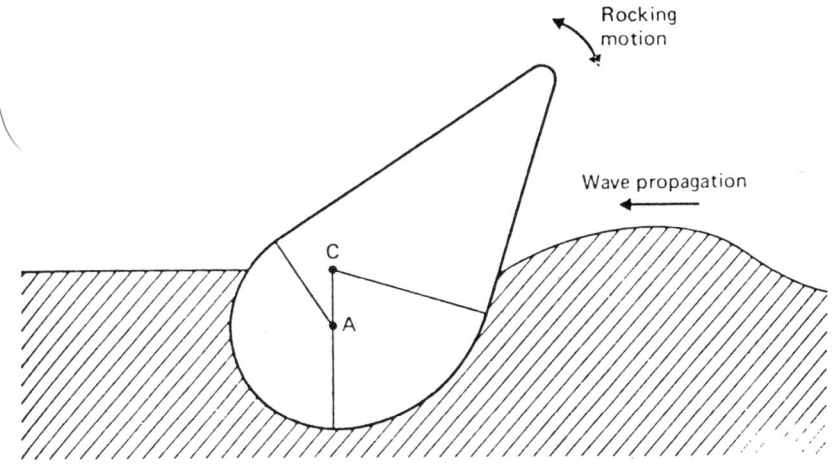

Fig. 1.15. Artist's concept of operating nodding duck. (Salter et al., "Wave power," Proc. Energy of the Ocean: Fact or Fancy? Raleigh, N.C., 1976)

32 TIDAL ENERGY

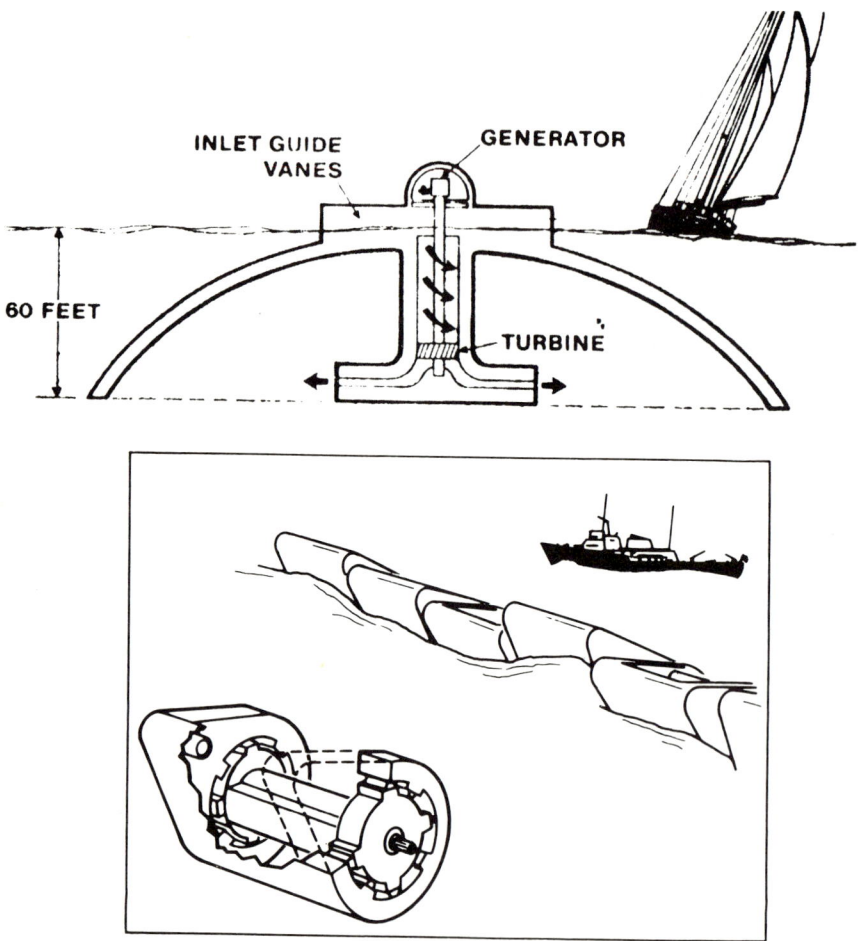

Fig. 1.16. Contouring rafts concept of wave-energy conversion. *(Source: Ocean Industry, August 1976)*

be slowly driven back by wind and wave thrust, storing hydrogen on the way. Once near shore, most hydrogen could be unloaded at a terminal, keeping enough in the installation to get out to sea again; among the advantages of this mobility, is elimination of mooring problems. The best location for such a wave power station would be the approaches to the Hebrides.

Implementation and Models

Wave motion has been tapped for air column excitation, bell clapping for navigational aids that whistle or ring, and a wave-activated pendulum that drives a spring generator have been patented. Besides the already mentioned Reynolds attempt, not many experiments have been conducted: Coyne made some tryouts near Brest (France) in 1926, and more sophisticated experiments were conducted at Point Pescade and Sidi Ferruch (Algeria). Dhaille (1957) computed that, utilizing waves of 1–4 m of amplitude only, 17,000 kW-hr/m of shoreline could be produced annually at Ain Diab (Casablanca, Morocco).

The Ryokuseisha TG-2 wave-activated generator uses wave motion to compress an enclosed air supply that rotates an air turbine directly connected to an electric generator. It is used for ocean survey instruments, lighthouses, buoys' beams, etc. Isaacs, Castel, and Wick (1976) tested several models of wave power generators and concluded that devices can be successfully developed for power generation. A form of solar energy resulting from the interaction of the ocean surface and winds has been estimated, for the global ocean, to reach between 2.5 and 2.7 million kW; Lacombe (1957) calculated that the amount of mechanical energy dissipated by waves breaking upon shorelines amounts to 6.10^{-7} W/cm^2, and the amount lost by viscosity in the open ocean, per second, per cm^2 of surface, and at 3% steepness, is about 3.10^{-3} W/cm^2. It is quite evident that only a fraction of this power can be recaptured.

Problems attached to the harnessing of wave energy include low head, storms, and aperiodicity, making floats and bellows, for instance, inefficient "direct" converters. In recent tests, however, the pressure head of 2-m waves was increased 9 and 20 times using pumps respectively 61 and 91 m long (Isaacs et al., 1976). Translating this in terms of actual electricity production for a tradewinds geographical area—thus with relatively steady winds—a 152-m long pipe with a diameter of 91.4 cm would produce 50 kW.

A Scripps Institution of Oceanography model, by no means the only one nor the first (Baird, 1968), is a 61-m-long free-flowing pipe with a 20.32-cm diameter (8-in.) supported at the top by a surface buoy and having, at 6.1 m below the surface, a unidirectional flow valve; it was tested in a sea with waves 1.83 m high on periods of 6–8 seconds.

Japan has marketed a "wave power" engine and exported it to France, Canada, and the United States. The small power output air turbine is being utilized for buoys and lighthouses. However, under development since 1976 is a large ocean wave electricity-generating device of 2,000 kW. Construction has been started on a buoy or barge measuring 80 m in length and 12 m in width and which is 5 m high. The Mitsui Engineering and Shipbuilding group has developed technology to generate electricity by the use of revolved propellers: attached to the tip of a buoy, they utilize both upward and downward movements of ocean waves.

In August 1978, they placed at 4 km (2.15 nautical miles) off Honshu Island a 500-ton experimental station; it has regularly produced 125 kW of power (Figs. 1.14, 1.15, 1.16).

Economic Aspects

From an economic viewpoint, the greatest advantage of wave-generated power is the same as that of thermal or tidal power: fuel is free! Plant size is reasonable and, as in other ocean-powered devices, the larger the number of units, the lower the per-unit cost. Isaacs and his co-workers estimated, in 1975, the cost of a 20-kW experimental station at $2,000/kW. However, a 91-m plant with a 4.5-m diameter will have "a capital cost of roughly 300 kW-hr," but with pump maintenance costs still unknown, thus leaving actual economic aspects to be assessed (Morello et al., 1975).

One often overlooked aspect of wave power is its significance for isolated sites and, in particular, for developing nations. Assuming use of the power close to the generating site, and thus avoiding long transmission costs, no new technology is needed. As with other alternative energy sources, hydrogen production could provide the solution to the problem of long distance transmission of large amounts of power.

OCEAN CURRENTS

Ocean currents result from the complex interaction of several factors: earth rotation, wind, salinity and temperature distributions, geometric shapes of basins, and, regionally, tidal effects. No major facility exists to harness the currents, but fluid motion energy recovery has been long practiced; the fluid is not forced to pass through a narrow channel as is done with a hydroelectric plant, so only a part of the energy is recovered. The technology exists, and hence current energy from the Florida Current or the Kuroshio, for instance, could be recovered (Table 1-2).

Installation costs of such a scheme are very high, and no implementation is foreseeable until less expensive structures than the ones on the drawing board can be devised. But competitiveness will increase when underwater free-stream turbines are improved.

In 1950, both Bouteloup and Romanovsky held that sea current power schemes could, in some sites, be successfully put into operation. And, in 1957, Remenieras and Smagghe suggested using aerogenerators. They listed several geographical locations suitable for plant construction, but because the technology is still somewhat lagging behind as far as generator installation is concerned, sites should be selected more for their sheltered or estuarine location than for current speed. Because of low velocity and current position migration, there has been little interest in current schemes until recently, when von Arx, Stewart, and Apel (1974) suggested that a cluster of rather large turbines placed in the Florida Straits, where the Gulf Stream flows, could furnish roughly a million kilowatts on a year-round basis; this amount is the equivalent of about two very large nuclear plants.

An attempt, now terminated, to tap an ocean current, was carried out in Iceland on the Breidefjord with a small pump driven by tidal and current flow.

There is a large resource available in the kinetic energy in the Florida current, approximately equivalent to that produced by 25 1,000-MW power plants.

At a workshop held in February 1974, various propellers and turbines were discussed, and it was concluded that apparently the Florida Current energy could be used to produce electricity at competitive prices by the 1980s. Current power, it was suggested, could be a supplementary source of electricity in all areas where rather fast near-shore currents flow.

The "water low velocity energy converter" proposed by Gerard Steelman (1974) would place an electric generator on an offshore anchored ship where the current would drive a wheel placed below the ship. A cable equipped with umbrella-like devices would catch the current, as the umbrellas would open when facing the current and close up when not facing it.

A vertical axis turbine appears particularly suited in areas where energy gradient is large and flow directions vary; the Kaplan and Propeller type turbines are perhaps better suited elsewhere; although of about equal efficiency, the propeller type has the advantage of requiring substantially less structure.

A possible problem evolving from current power plants is that of navigational hazards in heavily traveled areas and possible deflections of climatic influences of the currents.

TIDAL CURRENTS

Tidal current power has been far less often mentioned than tidal power as a potential source of electrical energy. Some researchers point out that it could be tapped far less expensively. Tidal currents could be tapped both in the sea environment and in tidal rivers and streams. Taken as an example, if only 10% of tidal current power was put to work in Great Britain, 1.5 GW would be available (or about 6% of that country's current needs). At Reading University, P. L. Froenkel and P. J. Musgrove (Musgrove, 1979) have carried on research in this area, particularly with developing countries in mind. The annual average power (in GW) for the currents through the North Channel in the Irish Sea is 3.6; for the Pentland Firth (between Scotland and the Orkneys), 6.1; between the Isle of Wight and Cherbourg (France), 3.3; between the Orkneys and the Shetlands, 1.7; and through the Alderney Race, 1.6. This totals to 15.3; 10% of this is more than the Severn scheme would produce.

Current power has been examined in the United States and it is estimated that harnessing the Florida current would yield 25 GW. More recently, the Department of Energy awarded a $46,000 grant to a California company to build a test-scale model of its "aqua power barge," which uses a high-impulse, low-head turbine. The barge can be used to tap current power in tidal rivers and along coasts. With a current of 6 knots (3.15 m/sec), 50 kW of installed power could be produced. The Golden Gate Bridge in San Francisco could extract all its needed electricity from the 8-knot (4.2-m/sec) tidal currents that pass underneath.

An "aqua power barge" (Fig. 1.17) measuring about 15 by 6 m could save, as a 5-MW station, between 30,000 and 70,000 barrels of oil

OCEAN CURRENT ENERGY

"AQUA-POWER BARGE" UTILIZES CURRENTS IN OCEANS, RIVERS AND STREAMS TO PRODUCE TURBINE-POWERED ENERGY

CONCEPTUAL BASELINE - POWER BARGE

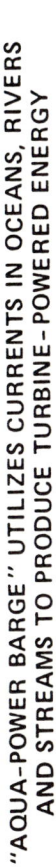

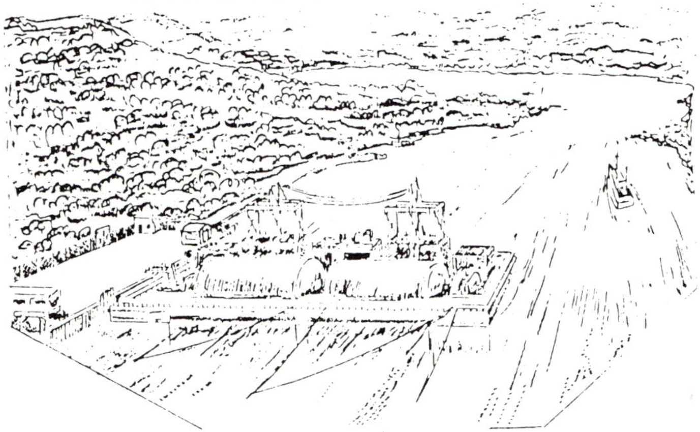

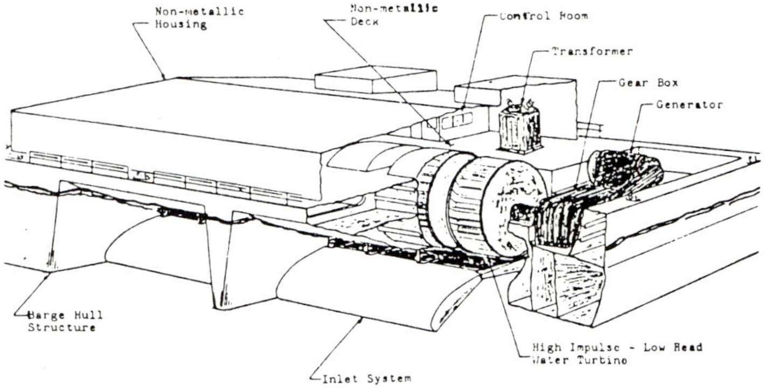

▲ Fluid Energy Systems' "Aqua-Power Barge" as it would appear situated in a river with sufficient current to produce power (three to seven knots).

Fig. 1.17. Aqua power barge utilizing current energy. (*Source: Ocean Energy Liaison*, June 1979)

yearly. This is a modest undertaking when compared to the giant Coriolis project.

Froenkel and Musgrove (Musgrove, 1979) believe that the most straightforward system would be the underwater equivalent of a windmill, using vertical axis rotors. Diameters of the rotors would possibly exceed 100 m; with blades 50 m long and a current speed of 4 knots (2.1 m/sec), a 10-MW output could be expected.

Estimated direct cost of a 10-MW rotor is about $10 million, but plant longevity actually should be taken into consideration, which reduces the true cost. The scheme best suited for rentability would be to anchor the rotors, but keep them suspended in midwater to avoid wave effects, and let them drive hydraulic pumps. Electricity conversion would occur at a central facility servicing several rotors. By spacing rotors over some distance, the dephasing due to tidal variation in time could be somewhat compensated. Because of relative low cost, additional turbines could be put in the system as needs warrant. In arid areas without rural electrification, such systems could be used to lift water for irrigation and human consumption.

Tidal currents are purely alternating; their maximum velocity occurs at high and low water. The motion is uniform from surface to bottom, except for wave interference at surface, and increases with distance. The tidal currents are difficult to observe because of superimposition of other currents, so that an exclusive picture of the tidal motion requires complex extensive data.

Tidal currents of major importance are encountered on the North Siberian Shelf in the Arctic Ocean, the Channel, the Irish Sea, the Skagerrak-Kattegat, the Hebrides, the Mosken, the Gulf of Mexico, the Gulf of St. Lawrence, the Bay of Fundy, the Amazon River, the LaPlata River, and in the Strait of Magellan and near South Africa in the Atlantic Ocean. In the Mediterranean Sea, there are the currents of the Gibraltar, Messina, Sicily, and Bosporus straits, and the Aegean Sea.

In the Pacific Ocean, main currents include the Taiwan, Tsushima, and Tsugaru straits, the Kuril Islands, and the Gulf of California, and those along the coasts of Northwest Australia and Southern Chile. As for the Indian Ocean, the Bab-el-Mandeb, Hermuz, and East Indian Archipelago straits and the currents of the West Australian coast should be mentioned.

SUBMARINE GEOTHERMAL ENERGY

Geothermal energy, the natural heat of the earth, increases with depth due to the rocks' natural radioactivity. Extractable where highly concentrated and shallow located, it is transferred to the surface by water or steam. One estimation holds, for instance, that major geothermal systems exist every 20 km on the mid-Atlantic Ridge and every 3 km on the East Pacific rise (Table 1-2).

The only station in production in the United States is the Geysers in Northern California; some 520 MW are thus created. Another 16 MW in Oregon and Idaho are not converted into electricity. Total world electrical generating capacity is almost 740 MW, and another 560 MW is under construction. But these are all land-based plants, even though the sea floor potential is considerable. Favorable zones exist in the Gulf of Mexico, the Indian Ocean, and the North Sea, among others.

The North Sea basin contains some of the highest temperature gradients recorded near the British shores, and a preliminary governmental study tends to indicate that extraction costs run below those of importing petroleum. However, tapping this resource would entail boring holes as deep as 8 km. The U.S. Geological Survey has indicated that under the waters of the Gulf of Mexico, a geothermal-recoverable potential may exist that exceeds 100,000 MW power for three decades.

Wallace and his co-workers (1978) made an assessment of geopressured-geothermal energy in pore waters of sedimentary rocks, both onshore and offshore, in the Northern Gulf of Mexico Basin. They conducted their investigations to a depth of 6.86 km. They found, in waters in sandstone, from which the initial production would be drawn, a thermal energy of $11,000 \times 10^{18}$ joules. Assuming methane saturation of the water, the total dissolved methane would amount to $1,614 \times 10^{12}$ hectoliters, equivalent to $6,000 \times 10^{18}$ joules of thermal energy. These figures are based on an energy equivalent of methane of 3.73×10^7 joules/m^3. Under controlled development, with limited pressure and subsidence, some 104×10^{18} joules of thermal energy, and 56×10^{18} joules of methane energy equivalent to this would be available on the Federal outer continental shelf area. If depletion of reservoir pressure were permitted instead, thermal energy would amount to $1,080 \times 10^{18}$ joules, and methane to 580×10^{18} joules.

The electricity produceable from the thermal energy, at a conversion efficiency of 8%, ranges from 8,860 MW$_e$, for 30 years, with the controlled development plan, to 92,570 MW$_e$ permitting depletion of reservoir pressure.

Geothermal energy is nonrenewable. Many favorable locations have been found during drilling for oil and gas; additional ones could be discovered, but specific technology must be adapted. Platform technology used for oil search could probably be used, but costs are likely to run higher because of heat and corrosion wear on equipment. Blueprints exist for floating geothermal plants and for plants built on the sea floor.

Environmental considerations may well hamper the harnessing of offshore geothermal power because of air, thermal, water, and noise pollution, waste heat, and possibly seismic and subsidence activity.

OTHER SOURCES

Not mentioned thus far are tidal and wind energy, fossil fuels, and hydrogen extracted from the ocean. Tidal energy has been omitted because it will be discussed in detail in the following chapters; hydrogen and deuterium technologies are still very much at early research stages. Deuterium can be used for nuclear energy, and its availability in seawater could act as an incentive to locate generating stations offshore. Hydrogen can be extracted from seawater; it can be easily stored and transported, and a hydrogen economy has been proposed by some. Fossil fuels are abundant underneath the ocean bed, and modern technology has developed means to drill at ever increasing depths for oil. Twelve years ago, 20% of the world's petroleum production was of submarine origin and forecasts are that by the 1980s the amount will be 30%, with some two billion tons extracted from the ocean bottom by 1990. Natural gas world reserves from offshore sources are estimated at over 1,000 trillion cubic meters. The problem, however, is that these are depletable sources of energy and, while all authors do not agree as to how much oil and gas there is left, and how many years the supply can last at current, and future, consumption rates, eventually we will run out of these fossil fuels.

Even before the Christian era, mines were worked by the Greeks beneath the sea at Laurium. Coal has been extracted from submarine mines in many countries, including Great Britain, France, Chile, Japan, Canada, and the United States. Some two million tons were extracted in 1965. Finally, methane deposits are being tapped in the North Sea. But, again, these energy sources are exhaustible.

The offshore plant has certain advantages over the land-based one: expensive real estate does not have to be acquired, there is an ample supply of cooling water, and capital costs can be lowered by the use of standard-design facilities; on-land construction and transportation to site is another cost-saving device. A mariculture undertaking may provide valuable side-benefits as, for instance, on the coast of Kyushu (Japan) in Kagoshima Bay.

Basic technology exists, but adaptation to offshore locations still requires research; some sophisticated proposals have been made, such as injecting water into molten rock, with ensuing production of hydrogen or methane gas, both transportable.

Offshore winds are inexhaustible. Matters pertaining to tapping this source of energy have received renewed attention. Although resources are considerable and the United States (for instance) could find an ample supply of energy in winds off the Gulf of Maine coast, no effort at harnessing the energy has been made. Research is, however, in progress. Problems faced by wind energy include, principally, the design of a technology which would permit its conversion at a cost-competitive level (Table 1-3).

Extraction of wind energy could be achieved offshore in the zones of most favorable regimes. Heronemus proposed an offshore Wind Power System (OWPS) which would use this energy with the wind generator platform doubling as a self-propelled tankship equipped to generate and liquefy electronically produced hydrogen (Fig. 1.18).

Among the favorable areas close to shore and markets are the Great Bahama Bank and other West Indies locations, the United States-Canada North Atlantic shelf (e.g., the Gulf of Maine, the Georges Bank, Prince Edward Island, and Nova Scotia), the Irish Sea, the Irish coast, the Baltic (Swedish and Finnish coasts), the shelf near the Kuriles and Aleutian Islands, the Falklands, in South America, the shelves of South Africa (including Namibia), and the shelf South of Australia.

The plant complex would include a 2-MW wind turbine, over 100 wind stations, a distillation-electrolyzer (to generate hydrogen gas using a wind unit), over 80 wind units, and an offshore collection system (a compressor, a receiving station, storage tanks, an onshore electricity-generating terminal, and distribution stations).

Some consideration will have to be given to the effect of hurricanes

42　TIDAL ENERGY

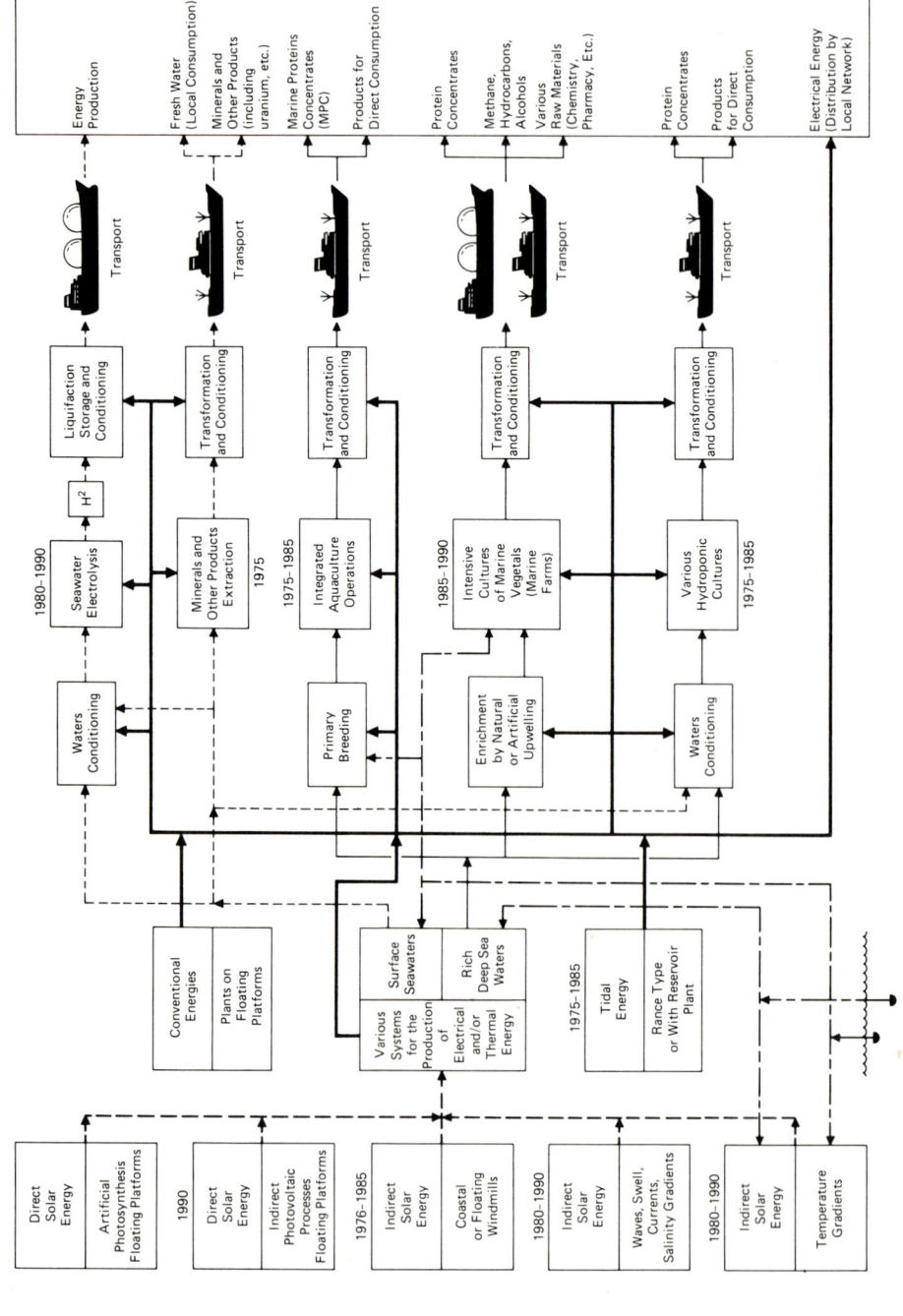

Table 1-3. Potentialities of the various energetic resources of the sea. *Source: Constans, Eurocean, 1978.*

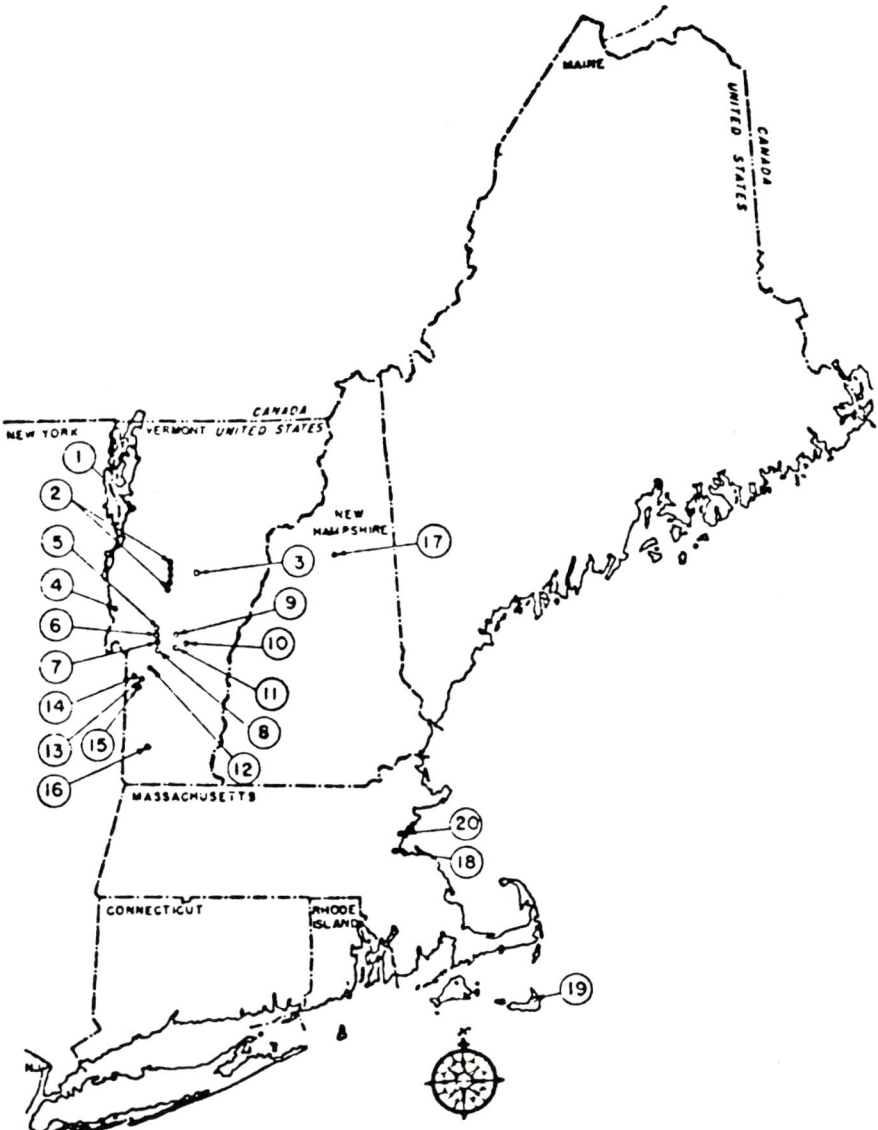

Fig. 1.18. Wind survey results used in wind turbine site selection. *(Source: Eurocean, 1978)*

Table 1-4. Power from the Wind.

SITE	ANEMOMETER ELEVATIONS			PERIOD OF OPERATION		DAYS OF OPERATION
	ABOVE GROUND	ABOVE TREES	ABOVE SEA LEVEL	FROM	TO	
Pico Peak	40	10	4007	Jan. 31, 1941	March 24, 1941	52
	80	50	4047	July 2, 1940	April 30, 1941	301
	110	80	4077	Jan. 31, 1941	March 24, 1941	52
Glastenbury	80	40	3840	March 2, 1941	Sept. 10, 1941	212
Seragg	53	45	2633	May 31, 1940	Oct. 10, 1940	133
	66	38	2646	May 31, 1940	Oct. 10, 1940	133
	78	70	2658	May 31, 1940	Dec. 1, 1940	185
	108	100	2688	May 31, 1940	Oct. 10, 1940	133
Herrick	80	80	2640	June 1, 1940	March 20, 1941	292
Cluttenden	80	30	2460	April 28, 1940	Nov. 30, 1940	219
East Mountain	80	40	2200	May 10, 1940	Aug. 10, 1940	85
Seward	59	13	2139	July 11, 1940	Feb. 2, 1941	206
	70	24	2130	July 11, 1940	Feb. 2, 1941	206
	80	34	2160	July 11, 1940	Feb. 2, 1941	206
	110	64	2190	July 11, 1940	Dec. 30, 1940	172
Biddie Knob Proper	75	61	2085	July 14, 1940	Dec. 1, 1940	140
Grandpa's Knob	50	50	2040	June 9, 1940	June 20, 1941	376
	64	64	2154	June 9, 1940	May 20, 1941	345
	80	80	2070	June 9, 1940	June 20, 1941	376
	110	110	2100	June 9, 1940	Dec. 18, 1940	192
	40	40	2030	Apr. 11, 1941	Dec. 31, 1945	1724
	80	80	2070	Apr. 10, 1941	Dec. 1, 1941	235
	120	120	2110	Apr. 1, 1941	Dec. 31, 1945	1731
	150	150	2140	Apr. 3, 1941	Dec. 1, 1941	242
	185	185	2175	Apr. 2, 1941	Dec. 31, 1945	1733
Biddie Knob 1	50	4	1975	July 13, 1940	Dec. 31, 1940	171
	64	18	1989	July 13, 1940	Dec. 31, 1940	171
	80	34	2005	May 3, 1940	Dec. 31, 1940	242
	110	64	2035	July 13, 1940	Dec. 31, 1940	171
Moose Horn	76	46	1921	March 30, 1940	June 19, 1940	81
Middle	76	32	1836	Apr. 18, 1940	June 15, 1940	58
Pond	47	3	1407	Apr. 26, 1940	May 18, 1940	22
	64	20	1424	Apr. 26, 1940	May 18, 1940	22
	80	36	1440	Apr. 9, 1940	May 25, 1940	46
	110	66	1510	Apr. 26, 1940	May 18, 1940	22
Crown Point	151	—	240	Jan. 1, 1941	March 1, 1941	60

on the structures, the hazard to navigation constituted by stations, units, and turbines positioned at 67 and 103 m above sea level, and costs. With a peak load of 24 GW, a plant would cost over $22 billion (159 TW-hr/yr) and electricity 2.8¢/kW-hr.

Finally, another ocean energy, although not exactly "extracted" from the sea, is provided by radioactive wastes. The Siemens Corporation (U.S.) made some preliminary studies on radioactive decay heat from strontium-90; this heat can be converted in an undersea plant using thermoelectric conductors. Current models produce 1–150 W, at 5¢ (U.S.)/W, and need no attention for five years. Land-based plants produce 1 W for 25 U.S. mills. Emission remains below 200 rem/hr.

Capturing the energy dissipated by the sea has fascinated and tantalized man since remote times, though interest has soared in recent years because of a hunger for new resources of power. Use of the sea for energy production can be traced back to the Greeks, who sought to use tides in Euripus channel separating the islands of Euboea and Beotia, where, according to legend, Aristotle committed suicide when he could not solve the problems of tide harnessing and tide delay around the island. But tidal currents were used in ancient times to operate water mills near Chalcis, on Euboa, and in the Euripos Strait, and other tidal mills harnessed energy near Agostoli, on the island of Cephalonia, in the Ionian Sea, and (nearer to the present) in Dover and in St. Malo in Brittany.

The sea, a source of energy, may help man to satisfy his thirst for power; he has domesticated marine energy here and there in mostly modest schemes since antiquity. As to whether man will extend his quest and ultimately increase the number of ocean thermal, tidal, wave, and other power plants will depend less on the sea itself and the abilities of engineers than on the capital necessary for building these plants, the competition from other sources of energy, and the continued rise of oil prices. As to demand, it might be safe to assume that during the coming decades demand will exceed supply and will trigger an anxious search for alternative power sources and provide a renewed impetus to projects for plants harnessing the energy of the sea.

ERDA AND OCEAN ENERGY

The Energy Research and Development Administration was absorbed, through administrative reorganization, by the subsequent Department

ment of Energy. It has shown interest, to a greatly varying extent, in most of the technologies we have reviewed, and also in tidal power and offshore winds.

Where *salinity gradients* are concerned, the Department's program aims at identifying geographical locations and the magnitude of this resource. Research is conducted in the areas of conversion and delivery, economic aspects, and the design and test of conversion systems, particularly the development of durable membranes. No estimate of the potential power has been furnished, though guesses run in the range of hundreds of megawatts per year.

Open ocean and shoreline salt water/brackish water systems receive Department of Energy support in the *biomass energy* program. Currently, various seaweed species are being screened for yield, potential growth, nutrient assimilation, fermentability, and economic rentability. Furthermore, potential sites, energy balances, operating concepts, and selection of systems for design and costing are under study. The Department is working on the development of a prototype test platform for growth studies of oceanic biomass, in cooperation with the American Gas Association. Experimental work is being conducted at Woods Hole, in Fort Pierce (Florida) and at the California Institute of Technology. Theoretically, gases could provide the equivalent of as much as ten quads per year.

Apparently, major support is provided to the *Ocean Thermal Energy Conversion* process; it is classified as a solar technology inasmuch as the oceans can be considered as the largest solar energy collectors, providing a continuous renewable source of substantial quantities of non-polluting energy. Particular interest in OTEC plants is to be ascribed to their ability to operate year-round, day and night, requiring no storage facilities, and being suited to baseload electric utility operation. Studies are pursued in regard to legal and institutional questions associated with OTEC plant operations in United States waters, in domestic waters of other countries, and in international waters. In fact, these studies will be relevant for other ocean power sources as well. The Department of Energy estimates that ocean thermal resources could satisfy the needs of the United States' regions where they would be built (Hawaii, Puerto Rico, the Gulf Coast, and the Southeast).

The Maritime Administration of the United States provided, in 1974, technical support in the evaluation and selection of proposals for the initial studies conducted by the Lockheed and TRW Corporations. It participated in the final report and analysis. In addition, the Administration sponsored a technical and economic feasibility study of the Johns Hopkins University plant ship concept, and provides low-level technical support to the Department of Energy.

Finally, the National Oceanographic and Atmospheric Administration provides support to the Department of Energy by conducting studies in biofouling, corrosion, environmental problems, engineering aspects, modular design, mooring, and station-keeping.

In terms of dollars, the National Science Foundation dedicated $84,100 to sea thermal power in 1972 and increased the amount gradually to $3 million in 1975; when ERDA took over, $8.2 million was granted. Of this amount, $328,188 went to Lockheed Missiles and Space Company, $391,427 to TRW Systems, $190,000 to Carnegie-Mellon University, and $169,800 to the University of Massachusetts to develop its plant design. Smaller research organizations, such as Sea Solar Power, received only $31,000 for model testing and $39,100 to study a cold water adduction conduit.

As far as *wave power* is concerned, the agency seems only to "maintain close contact and coordination with the British program"; the United Kingdom has allocated $5 million for a two-year study of wave energy; they added another $5.5 million for 1978–1979. The Japanese committed $5 million for 1980–1981. Though sites in the Atlantic and Pacific oceans are favorable, only the shores of the northern half of the Pacific are considered as favorable for wave power harnessing in the United States, and potential production is estimated at 5,000 MW. The conditions are quite different elsewhere, as, for instance, in Great Britain, which has outstanding wave power resources.

Interest is similarly limited for *currents*. Only the Florida current holds promise. Furthermore, technology is by no means perfected for other types of ocean energy, and environmental protection requires special consideration. Preliminary studies on "tailor-made" solutions for the areas considered are presently underway. Net electrical power that would be available for the United States from this source would run about 1,000 MW, or the power of an average plant.

Wind power is studied on land. Of course, many of the conclusions will be valid for offshore winds. At this time, comparative economic potential of offshore design concepts are the topic of studies. The coastal areas of the northeastern United States offer the most favorable sites.

The Department of Energy commissioned the Boston company of

Stone and Webster to make a survey of favorable tidal power sites and pursue a follow-up study. The two major schemes that could be constructed in the United States (one in Maine, the other in Alaska) could provide 4,000 MW. The latest assessment holds that the Passamaquoddy Bay project is now cost-effective and that an entirely U.S.-based undertaking not only is feasible but may be preferable as it avoids the necessity of international treaties. In fact, Canada could quite possibly not be interested any longer in a joint scheme because its own alternatives are even more favorable than those of the United States. *Tidal power* is the only ocean energy resource that the United States ever attempted to harness: work was actually started when Franklin D. Roosevelt was president. It was never completed; attempts to revive it under President Kennedy did not succeed. It is also the only ocean energy actually put to work since classical times and which produces electricity at this time.

For the fiscal year 1979, requests were made for $32.5 million to support OTEC, while other ocean energy sources were funded to a mere fraction of that amount: together, wave, current, and salinity ocean power only were allocated $1.2 million; biomass alone received $1.5 million, and offshore wind $100,000. It is recognized that tidal power, the only actually producing type of ocean energy, is beyond the research and development stage, yet no development funds have been requested!

The Congress' Office of Technology Assessment made a lengthy study over two years, but by the time the report was finally released, several problems listed for OTEC implementation, for instance, had been solved. Conclusions reached by the Department of Energy differ on both OTEC and tidal power: the Department feels that these two sources of ocean energy can be competitive with traditional energy sources by the year 2000.

REFERENCES

Generalities and Electromagnetic Power

Gibrat, R., 1953. L'énergie des marées, *Bull. Soc. Franç. d'Electricité* **VII**, *No. 3*:283–332; and 1962. Source de l'énergie des marées: Énergie cinétique ou énergie thermique du soleil? *La Houille Blanche* **XV**, *No. 2*:255–266.

Jeffreys, J., 1920. Tidal friction in shallow seas, *Philos. Trans.* **239**.

Legendre, R., 1949. Les ressources énergétiques de la mer, *Bull. Inst. Oceanogr. Monaco* **947**:1–16.

Le Grand, Y., 1957. Energie électromagnétique des océans, *Quatrièmes Journées de l'Hydraulique (1956).* Comptes Rendus: *La Houille Blanche* **I**:225–228.

Salinity Differences

Charlier, R. H., 1970. Harnessing the energies of the ocean, *Mar. Techn. Soc. J.* **3**, *No. 3*:13–32; **3**, *No. 4*:59–81.

Claude, G., 1930. Power from the tropical seas, *Mechanical Engineering* **52**, *No. 12*: 1039–1044.

Le Grand, R. and Lambert, M., 1962. Mesures électrochimiques appliquées à l'étude de la protection cathodique des ouvrages de la Rance, *La Houille Blanche* **XV**, *No. 2*: 177–186.

Loeb, S., 1975. Osmotic power plants, *Science* **189**:654–656.

Norman, R., 1974. Water salination: A source of energy, *Science* **186**:352–355.

Pattle, R. E., 1954. Production of electric power by mixing fresh and salt water in the hydro-electric pile, *Nature* **174**:660–666.

Sourirajan, S., 1970. *Reverse Osmosis*. New York: Academic Press.

Sussman, M. V. and Katchalsky, A., 1970. Mechano-chemical turbine: A new power cycle, *Science* **167**:45–49.

Weinstein, J. and Leits, F., 1975. The dialytic battery: Electric power from differences in salinity, *Science* **191**:557–559.

Wick, G. and Isaacs, J. D., 1976. Salinity power, *Proc. Symposia of Expo '75*:153–165.

Biomass Energy

Flowers, A. and Bryce, A. J., 1977. Energy from marine biomass. *Sea Technology* **18**, *No. 10*:18–21.

Mitsui, A., Miyachi, S., San Pietro, A., and Tamura, S. (Eds.), 1978. *Biological Solar Energy Conversion*. New York: Academic Press.

1977. Kelp farm planned for methane project, *Ocean Industry* **12**, *No. 10*:67–68.

Thermal Differences

Bathen, K. H., 1975. Discussion of oceanographic and socio-economic impact of a nearshore ocean thermal energy conversion pilot plant in Hawaii, *Look Lab/Hawaii* **5**, *No. 2*:15–31.

Beau, Ch., 1955. L'état actuel des études et travaux en vue de la construction d'une centrale électrique thermique des mers en Côte d'Ivoire, *Industrie & Travaux d'Outre-Mer* **III**, *No. 17*:222–223.

Boucherot, P., 1928. Utilisation rationelle de l'eau glacée du fond des océans, *Rev. Universitaire des Mines* **5**:205–214.

Claude, G., 1930. Power from the tropical seas, *Mechanical Engineering* **52**, *No. 12*: 1039–1044.

Gérard, R. D. and Roels, O. A., 1970. Deep ocean water as a resource for combined mariculture, power and fresh water production, *Mar. Techn. Soc. J.* **4**, *No. 5*:69–79.

Gomella, C. 1966. *La Soif du Monde et le Dessalement des Eaux*. Paris: A. Colin.

Gougenheim, A. and Romanovsky, V., 1957. Les remontées d'eau profondes, *La Houille Blanche* **II** (4ᵉ., J. de l'Hydraulique): 712–719.

Griffin, O. M., 1977. Power from the oceans' thermal gradients, *Sea Technology* **18**, *No. 8*:11–15, 38–40.

Lavi, A. and Zener, C., 1975. Solar sea power plants—Electric power from the ocean thermal difference, *Naval Eng. J.* **87**, *No. 2*:33–46.

Metz, W. D., 1973. Ocean temperature gradients: Solar power from the sea, *Science* **180**:1266–1267.

Mulcahy, M., 1977. Ocean thermal energy conversion is one of ERDA's exciting new programs, *Sea Technology* **18**, No. 8:16-18.

Othmer, D. F. and Roels, O. A., 1973. Power, fresh water and food from cold, deep sea water, *Science* **182**:121-124.

Sevette, P., 1958. *L'énergie dans les Pays en Voie de Développement*. Paris: Commiss. Econ. pour l'Europe des Nations Unies.

Weeden, S. L., 1975. Thermal energy from the oceans, *Ocean Industry* **10**, No. 9: 219-228.

Committee on Science and Aeronautics, U.S. House of Representatives, 1974. *Solar Sea Thermal Energy*. Washington, D.C.: U.S. Government Printing Office.

Wave Energy

Baird, W. F., 1968. *On Means of Utilizing the Energy of Wind Waves*. Kingston, Ontario: Queen's University, Dept. of Civil Engineering.

Charlier, R. H., 1970. Harnessing the energies of the ocean, *Marine Technology Society Journal* **3**, No. 3:12-32, **3**, No. 4:59-81.

Dhaille, R., 1957. Technique et rentabilité des dièdres à houle, IVes *Journées de l'Hydraulique (N°Sp. La Houille Blanche)* **II**:421-429.

Isaacs, J. D., Castel, D., and Wick, G. L., 1976. Utilization of the energy in ocean waves, *Ocean Engineering* **3**, No. 4:175-187.

Isaacs, J. D. and Seymour, R. J., 1973. The ocean as a power resource, *Intl. J. Environmental Studies* **4**, No. 2:201-205.

Isaacs, J. D. and Wiegel, R. L., 1949. The measurement of wave height by means of a float in an open-ended pipe, *Transact. Amer. Geophys. Union Proc.* **30**, No. 4: 501-506.

Lacombe, P., 1957. *Les Énergies de la Mer*. Paris: Presses Universitaires de France.

Legendre, R., 1949. Les ressources énergétiques de la mer, *Bulletin Inst. Océanograph. de Monaco* **947**:1-16.

Masuda, Y., 1971. Wave-activated generator, *Symposium International sur l'Exploitation des Océans (Bordeaux), Comptes-Rendus*.

McCormick, M. E., 1974. Analysis of a wave-conversion buoy, *J. Hydronautics* **8**, No. 3:77-82.

Morello, A., Taralli, C., and McConnell, J., 1975. Power transmission by way of submarine cable, *Proc. Annual Offshore Technology Conf. (Houston)* **VII**, Paper OTC 2257.

1970, New concept for harnessing ocean waves, *Ocean Industry* (August):13-14.

Parenty, H. and Vandamme, G., 1920. Utilisation de la force des marées et du choc des vagues, *Comptes-Rendus Acad. Sci. (Fr.)* **171**:896-898.

Romanovsky, V., 1950. *La Mer, Source d'Énergie*. Paris: Presses Universitaires de France.

Salter, S. H., 1974. Wave power, *Nature* **249**:720-724.

Swift-Hook, D. T., Count, B. M., Glendenning, I., and Salter, S., 1975. Characteristics of a rocking wave power device, *Nature* **254**:504-506.

Valembois, J., 1957. Possibilités de captage de l'énergie de la houle au moyen de résonateurs, IVe *Journées de l'Hydraulique; N° Spéc. La Houille Blanche* **II**:418-420.

Vincent, M., 1924. *Réflexions sur l'Utilisation Future des Énergies Naturelles: Vagues, Chutes Hydrauliques et Barométriques, Énergie Solaire*. Paris: Fischbacher.

Wick, G. S. and Castel, D., 1978, The Isaacs wave-energy pump: field tests off the coast of Kaneche Bay, Hawaii, *Ocean Engineering* **5**:235-242.

Woolley, M. and Platts, J., 1975. Energy on the crest of a wave, *New Scientist* **66**: 241–243.
1976. Britain launches wave power research program, *Ocean Industry* **11**, No. 8:64–65.

Current Power

Bouteloup, J., 1950. *Vagues, Marées, Courants Marins.* Paris: Presses Universitaires de France.
Duing, W. W., 1974. Synoptic studies of transients in the Florida Current, *Proc. MacArthur Workshop on Feasibility of Extracting Usable Energy from the Florida Current* (Palm Beach Shores, Florida, 2/27–3/1, 1974).
Remenieras, G. and Smagghe, P., 1957. Sur la possibilité d'utiliser l'énergie des courants marins au moyen de machines à aérogénérateurs, *IVe Journées de l'Hydraulique, N° Spéc. La Houille Blanche* **II**:532–539.
Romanovsky, V., 1950. *La Mer, Source d'Énergie.* Paris: Presses Universitaires.
Sheets, H. E., 1975. Power generation from ocean currents, *Naval Eng. J.* **87**, No. 2: 47–56.
Steelman, G. E., 1974, An invention designed to convert ocean currents into useable power, in: Stewart, H. B., Jr. (Ed.), *Proc. MacArthur Workshop on the Feasibility of Extracting Useable Energy from the Florida Current. NOAA Atlantic Oceanogr. and Meteorol. Lab., Miami,* pp. 258–277.
Stewart, H. B., Jr., 1974. Current from the current, *Maritimes* **XVII** (Summer):38–41.
von Arx, W. S., Stewart, H. B., Jr. and Apel, J. R., 1974, The Florida current as a potential source of useable energy in: Stewart, H. B., Jr. (Ed.), *Proc. MacArthur Workshop on the Feasibility of Extracting Useable Energy from the Florida Current. NOAA Atlantic Oceanogr. and Meteorol. Lab., Miami,* pp. 91–103.

Geothermal Energy

Denton, J. C. and Dunlop, D. D. 1973. *Geothermal Resources Research: Geothermal Energy.* Stanford, California: Stanford University Press.
Harper, M. L., 1971. Approximate geothermal gradients in the North Sea Basin, *Nature* **230** (March 26): 236.
Jacoby, C. H. and Paul, D. K., 1974. Salt domes as a source of geothermal energy, *Mining Engineering* (May): 34–39.
Palmer, H. D., Forns, J. N., and Green, J., 1975. Exploitation of sea-floor geothermal resources: Multiple-use concept, *Proceedings, U.N. Symp. Development and Use of Geothermal Resources (San Francisco, California)* **II**, No. 3 (May 20–29): 2241–2247.
Wallace, R. H., Jr., Kraemer, T. F., Taylor, R. E., and Wesselman, J. B., 1978. Assessment of geopressured-geothermal resources in the northern Gulf of Mexico Basin, in: Muffler, L. J. P. (Ed.), *Assessment of Geothermal Resources of the United States, 1978.* Arlington, Virginia: U.S. Geological Survey (Circular #790), pp. 132–163.
Weeden, S. L., 1976. Geopressured geothermal energy—Will it work? *Ocean Industry* **XI**, No. 5:119–122.
Williams, D. L., 1976. Submarine geothermal resources, *J. Volcan. Geotherm. Research* **1**. (June):85–100.

Wind Energy

Heronemus, H. E., 1972. Power from the offshore winds, *Meet. Mar. Techn. Soc.* **VIII**: 435–466; and 1974. Using two renewables, *Oceans* **XVII** (Summer): 20–27.

2
From Tide Mills to Tidal Power Plants

Man has always been attracted to water, be it river, lake, or sea. Whether merely contemplating the sea or living near it, one cannot help but be awed by the inexorable regularity and the mighty power of the tides. And man has undoubtedly yearned to become the master of that power and to put it to work for himself. Apparently, tidal energy has been used by coast dwellers for at least ten centuries, and probably even longer; history records the existence of a tide mill before the year 1000. It had been the first ocean energy to be tapped as an individual undertaking.

Tide mills are conventional water mills using the tidal current as a source of power. Though some depend on tide alone, others use a proportion of fresh water from an impounded stream; the principle is the same in all cases. The idea is similar to that at the basis of the gigantic schemes proposed for locations in various parts of the world and implemented in France, the Soviet Union, and China. It involves a retaining basin, a dam, and sluice gates; the basin fills as the tide flows in.

First a stream is dammed where it enters an estuary. Alternatively, a pond can be created by enclosing an area in a tidal creek. The dam has automatic sluice gates set in the wall. The water rises with the tides and the pressure thus exerted forces the gates open, and the pond—or basin—fills. As the tidal current is reversed, water pressure inside the pond exceeds that outside and the gates are pushed closed. The same effect is attained with flap valves.

Water flowing from one level to a lower level can produce "work" which can be calculated using the formula work = force × distance; the force is gravity, the work is potential energy. The speed with which the flow or displacement occurs creates kinetic energy, some of which is lost by friction. This friction includes mutual water molecule friction, and friction of the water molecules against the conduits or other bodies.

Several centuries ago, Richard Carew wrote in his *Survey of Cornwall*, "Amongst other commodities afforded by the sea, the inhabitants make use of divers his creeks, for griste-milles, by thwarting a bancke from side to side, in which a floud gate is placed with two leaves: these the flowing tyde openeth, and after full sea, the weight of the ebbe closeth fast, which no other force can doe; and so the imprisoned water payeth the ransom of dryving an undershoote wheele for his enlargement." While in the English language the earliest written reference to a machine exploiting tidal energy is a tidal mill in Dover harbor, such mills had already been in use during classical times near Chalcis and Cephalonia in Greece.

In bygone times, the most common idea for putting tidal power to work was the so-called float method: the incoming tide was used to raise a floating mass, which, when falling back to its original position, could do useful work. More often, however, tide mills included rotating paddle wheels mounted on a shaft and activated by ebb and flood; power was transmitted by a shaft.

A third system would let air contained in a conduit of metal or concrete be compressed by the incoming tide, thus providing compressed air power, and indeed some thought is being given today to this basic idea. Finally, a fourth system includes damming in part of the sea, thus providing a basin which fills at incoming tide. The water is released at low tide through turbines either back to the sea or to another basin. This is the concept followed for the Rance plant.

Popular interest, and even concern, for tide mills has grown in recent years. When an old tide mill was "rediscovered" near Plougastel in Brittany, and an enterprising businessman divulged his plans to turn it into a restaurant "with atmosphere," indignation was such that plans to build the inn were dropped, and instead the mill was restored. New England now also boasts of a working tidal mill that attracts hundreds of tourists each year.

Tide mills were in use in Spain in medieval times, near Vigo, and in Russia in the eighteenth century. In the nineteenth century, tide mills were still at work in Italy, and near Hamburg, tidal power was used to pump sewage as late as 1880.

Tide mills using the tidal current are basically similar to river water mills. The water is diverted, in part, to a channel and forced to fall upon the parts of a "waterwheel." The mill could be equipped with an overshot-, undershot-, or midshot-wheel, with each type having its particular advantages. The overshot-wheel has the highest yield, but its diameter is limited by the height of the fall; the undershot-wheel, with a smaller yield, has no diameter limitation and was already in use during Roman times. The floating mills on the Danube, as well as ships' waterwheels, were undershot-wheels. The midshot-wheel combined the characteristics of both other types.

These tide mills had a rotation speed of only 10–15 rpm, but turbine type mills, which made their appearance in the nineteenth century, made speeds up to 150 rpm and had a much higher yield. In the ethnographic museum-village of Eterat in Bulgaria, water mills provide all the energy needed, including electricity. The disappearance of water mills during the current century, and even before, was due principally to increased use of the wind energy, introduction of steam engines, abolition of the millers' privileges in the wake of the French Revolution, and, along rivers, the demands of an expanding agriculture that wanted to divert water for irrigation.

Actually, the first waterwheel seems to have been put to work by the Persians in 200 B.C., when they needed a system to irrigate their fields. They let a flowing river drive a wheel to whose circumference they attached buckets. These buckets scooped up the water, and that way water was emptied onto fields as buckets reached the top. Refinements of their invention are to be credited to the Romans, who were reported to have used waterwheels as early as 85 B.C. to provide mechanical energy to grind grain. A toothed gear for their wheels is mentioned in Vitrivius' *De Architectura*. But fear of angering the particular gods of a waterway added to the potential consequences of putting slaves out of work and creating unemployment, delayed use of waterwheels improved or not, at least until Emperor Constantine embraced Christianity and freed the slaves, though, probably, they were used by the same Romans far earlier in Britain.

CLASSICAL AND MEDIEVAL PERIOD

Harnessing the energy of the sea is a problem that has continuously fascinated and tempted men for centuries. If the subject is broached today with more insistence than some decades ago, the reason is, perhaps, as Claude Arnaud put it, that the world suffers from kilowatt hunger (or should that be thirst?). An added incentive undoubtedly, is the completion by the French of a tidal power plant and the construction, by the Soviets, of a pilot plant on the White Sea. The Rance River barrage brought to reality dreams nurtured for more than half a century, though construction of tidal power plants was begun in 1928 at the Aber-Wrac'h, on the estuary of a small river, some 25 km north of Brest (France) and in 1935 at Passamaquoddy (U.S./Canada). Both these early projects came to a standstill due to lack of financing.

The power of the sea has, however, been put to use since remote classical times. The Greeks made attempts to take advantage of the tides in the Euripus, a rather narrow channel between Boeotia and the Isle of Euboea, near Thebae, Marathon, and Plataea, where Aristotle is reputed to have committed suicide. Near Chalcis (on Euboea), on the Evripos Strait, water mills used, in bygone times, the tidal currents, and other tidal mills provided energy in Cephalonia, near Agostoli and the Ionian Sea.

Tide mills were in use on the Channel, at the entrance to Dover harbor, and constituted, in fact, a danger for navigation. Others were operating in Brittany—for instance, near Saint Malo—and although most of them are derelict today, some were not yet abandoned when the Rance plant was inaugurated.

Floating tide mills were in use on the Danube in early times. In the Rance River alone, 14 tidal mills furnished the Breton millers with energy, once per tide, thus realizing a simple effect cycle. Of these, all but a few are today merely ruins. In England and Wales, no less than 23 mills were still in existence at the end of 1940. Ten remained in use and two had been converted for use by another type of power; the others were either derelict or no longer at work as mills.

Such mills have existed for centuries in England and Wales, and one is mentioned in Dover harbor in the *Domesday Book* (1066). Among the oldest ones, in the British Isles, the Bromley-by-Bow mill, in the London region, built around 1135, was functioning during World War II. The mill of Woodbridge, in Suffolk, on the Deben estuary, dating from 1170, was still in use, according to the latest report, in 1961. It stands on the quay beside the power mill and was then served by a pond of three hectares and a 1.8-m head. The wooden wheel drove four pairs of stones on the first floor, all controlled by a single pair of governors driven from the 56-cm-diameter oak upright shaft.

ENGLAND AND WALES

Few of us are probably aware that the highly sung London Bridge, now a tourist attraction in the American Southwest, had a tidal mill used for pumping water in 1580. The 6-m waterwheels installed under the arches provided part of London's water supply from 1682 to 1849. In the eighteenth century, four wheels, each occupying one arch, made up the system. There were, in total, 20 double and 32 single acting pumps. They delivered, at normal tide, 4.660 hectoliters of water per hour, though about 100 hectoliters were lost because of leakage. Maitland, in his *History of London*, published in 1782, described the mill as follows: "Each wheel was 20 feet [6 m] diameter by 14 feet [4.27 m] wide, with twenty-six floats and mounted on a shaft 19 feet [5.8 m] long by 3 feet [0.91 m] diameter. Each shaft ran on brasses fitted to levers 16 feet [4.88 m] long, which could be raised up and down by a geared windlass according to the state of the tide." In the eighteenth century, there were four mills driving a total of 52 pumps. The wheels also ran in either direction depending on the tidal current.

Fig. 2.1. Tide mill, St. Osyth, Essex, U.K., built in 1491, derelict. (Rex Wailes, *Tide Mills in England and Wales*, 1940)

FROM TIDE MILLS TO TIDAL POWER PLANTS 57

Fig. 2.2. East Medina mill, Whippingham, Isle of Wight. (Rex Wailes, *Tide Mills in England and Wales,* 1940)

Working mills, 30 years ago, were found in Suffolk, Essex, Sussex, Hampshire, Pembrokeshire, London, and the Isle of Wight (Figs. 2.1–2.6).

The East Greenwich mill, located approximately halfway between Blackwall Turmel and Woolwich, in the London region, is perhaps a precursor of the contemporary double effect Rance River plant. Its wheel revolved in a wooden frame which rose and fell with the tide. Planking hinged to the floor of the frame and to the base of the beams projected towards the mill pond and would not let the water brought in by the tide through. This forced the planking up and kept the wheel at working level. At each end of the wheel, a ring of cogs engaged bevels which could be disengaged by means of a lever so that one would func-

Fig. 2.3. Carew, Pembrokeshire, tide mill on Carew River. (Rex Wailes, *Tide Mills in England and Wales,* 1940)

tion, and the other not, according to the direction of tidal flow. Each shaft had two bevels, and, as emphasized by Gregory in his 1826 edition of *Mechanics* (pp. 35–37), a lever made it possible to disengage one bevel and switch to the other with each change of tide.

The mill, parallel to the river, had a 3.35-m-diameter wheel which was nearly 8 m wide.

At the head of St. Osyth Creek, in Essex, the local abbey had built a mill before 1491 and, while seemingly restored in 1730, it kept functioning until 1930. Built on a bridge, it involved a retaining pond of 12 hectares. Barrow Hill Mill was pulled down and Osyth Mill disappeared just before World War II, but the foundations of two others are still in evidence at Battle Bridge on the Couch. Built in 1772, they probably succeeded an earlier mill on the same site and were dismantled at the turn of the century. On the Roman River, stands Fingringhoe Mill, built around 1530, which could be put back to work if the sluice were repaired. Still functioning in 1940, Stambridge Mill, on the Roach River, near Rochford, is 250 years old; its wheel is about 9.9 m wide and has a diameter of approximately 5.5 m. Mounted in a wooden shaft, it drives three pairs of stones. (See Figs. 2.1 and 2.6)

Near Bromley-by-Bow, three mills were built in the twelfth century; of these, the Clock Mill was rebuilt in 1817 and has a clock dated from 1753. Situated on the Lea River, it has three wheels, each about 6 m in diameter and each providing between 20 and 25 hp. The mill runs by water for 7–8 hours at a time. The oak arms of the wheels are over 150 years old and the teeth last 40–50 years.

Birdham Mill, at the mouth of the Chichester Canal, in Sussex, built in 1768, worked until 1935; it is used as a motor-boat repair shop, and its 12-hectare pond as a dock. Corn came by ship from Southampton, and 150-ton vessels could reach the mill. Of late, only the wheat stones and one grist mill were worked. Two wheels, each with 24 floats mounted on 46-cm oak shafts, were served by a 13-hectare pond, giving a normal 3.65-m head, and worked five and a half hours. Also on the

Fig. 2.4. Pembroke, Pembrokeshire, tide mill on Pembroke River. (Rex Wailes, *Tide Mills in England and Wales*, 1940)

60 TIDAL ENERGY

Chichester, but at the head of its channel, remain the foundations of Salt Mill. Here, once a worker was pinned in bed by a ship's bowsprit, which drove through the wall of the millhouse in a storm (Fig. 2.5).

Slipper Mill, near Emsworth, had been repaired in 1735, and probably also 200 years later; it was fully operational on the eve of World War II. Nearby, but in Hampshire, Quay Mill, built around 1740, stopped working in 1920, but remains a picturesque historical landmark. On the river Test, Eling tide mill, constructed on a toll bridge, has automatic gates incorporated in four hand-operated sluices. The tide enters through flap valves instead of by lock gate. A mill on the Beaulieu River uses fresh water and the tide. A sack hoist, however, is operated with an electric motor.

Rex Wailes mentions that Wootton Bridge was a working mill on the Isle of Wight before World War II. This mill uses tidal waters as well as those from Blackbridge Brook. Each tide provides about four hours

Fig. 2.5. Sluice gate of Birdham tide mill. (Sussex, U.K.). (Rex Wailes, *Tide Mills in England and Wales,* 1940)

Fig. 2.6. St. Osyth tide mill, Essex. Stones, wheat cleaner, and sack hoist.

of power. Budshead Mill, in Devonshire, has its two original wheels of 4.27-m diameter, each 2.5 m wide, replaced by two water turbines developing a maximum of 50 and 28 hp, respectively. They were supplied from a pond of 2.4 hectares with a head varying from 2.74 to 4.27 m giving from 1½ to 4½ hours of work per tide.

Very little remains of the numerous Cornwall tide mills; of Hayle Mill, a four-floor stone building, the foundations are still visible. In the middle of the eighteenth century, copper smelting works were established here. The mill pond was created by closing off a branch of the creek, and advantage was also taken of an island, much of it copper slag, possibly even artificially built, so that it would form one side of the retaining pool. The mill eventually closed because no labor force could be found that would agree to work when the tides provided power at different times of night and day.

None of the Anglesey mills seem to have still been working in this century. Felin Wen was a corn mill. At Bidston, in Cheshire, an iron slitting mill and a corn mill once used the tides to provide power. The slitting mill, worked by two 5.5-m-diameter wheels, had three double and three single rolling mills and two furnaces. In the eighteenth century, the mills were also used for rasping and chipping of dyeing woods and the manufacture of tobacco stems.

The Pembroke area became a stronghold and a base of operations against Ireland for the Normans. Great prosperity came at the end of the thirteenth century. It remained an important place and during the Civil War was the major foothold of the Parliament forces. Tide mills were built here and near Carew, some five miles northeast. Carew castle, originally built in 1102, was rebuilt in successive centuries but now stands in ruin. Pembrokeshire was settled in part by Flemings, who were already familiar with tide mills and windmills.

Both the Pembroke Mill and the Carew Mill, situated (respectively) on the Pembroke and Carew rivers, are semi-tidal. They were apparently built by Frenchmen. These Pembrokeshire mills are quite large: the first one has five stories, and uses fresh and tidal water (Figs. 2.3, 2.4).

Waterwheels were wooden in tide mills, though some were spoked wheels with iron hubs. Wooden buckets, straight wood floats, and Poncelet floats made of iron were used. The wheel shafts, of wood and iron, varied considerably in size.

All these mills were of local importance. Often, the entire industry of a village centered on the mill. Mills were used to grind grain, pump water, crush oats, and provide mechanical power for other operations. However, the mechanical energy produced was always modest and generally for local use, varying between 30 and 100 kW. This was certainly a factor in the demise, toward the beginning of the twentieth century, of the tide mill. Its problems were compounded by the construction of electric motors, the possibility of long distance power transmission, and the advent of the era of power economics.

Huge waterwheels were put to work in river mills. On the Isle of Man, a 21-m-diameter wheel, weighing 100 tons and producing 172 kW, was used to pump water from a lead mine, and another one was put to work near Halifax and generated 190 kW. Apparently, the

United States held the record of power produced from water power, as one wheel is reported to have furnished 7.5 MW.

At the November 30, 1940 meeting of the Junior Institution of Engineers in London, N.L. Ablett underscored the quality of tidal mills not to consume irreplaceable stores of energy, but a continuously available source of energy. Their gradual disappearance is a loss, but more important, their restoration could perhaps refloat decaying village industries.

Whether any of the last functioning ten tidal mills operating in 1940 is still in use now is not known. Tidal mills tried through use of waterwheels, water pressurization, air compression, lifted platforms to put to work the tides' potential energy, the tidal currents' kinetic energy, or both. They died of technological obsolescence, unable to compete with the lesser priced energy from thermal and river plants.

NINETEENTH AND TWENTIETH CENTURIES

Though no tidal power station was included in the Dutch "Delta-plan," tide mills did function in Zuid-Holland and Zeeland as early as 1200. The Dutch colonists built such mills in the early seventeenth century near New York, and one of these, on Spring Creek, was still at work in 1899. The first tidal mill in operation in the United States had been built in Salem, Massachusetts in 1635, though some claim such a mill was already at work in 1617.

Slade's Spice Mill, still in existence, was built in 1734, at Chelsea, Massachusetts, to grind spices. The trapped water provided up to 50 hp equivalent to 375 kW. Prior to 1800, at least two small tide mills operated in Passamaquoddy Bay; they were grist mills.

A mill, powered in part by tidal energy, had been built in Canada in 1607 at Port Royal under instructions of the Sieur de Pontrincourt.

In Rhode Island, a particularly large tide mill was constructed in the eighteenth century: it used 20-ton wheels, 8 m wide and 3.35 m in diameter. A Maine mill operated a sawmill.

A tide-powered motor was built by the city of Santa Cruz (California), reported the *Revue Scientifique* (1902), around 1898. Several other mills were still in operation in New York State at the beginning of the twentieth century (*Revue Scientifique*, 1899). These tidal mills, all on Long Island, were located at Garrettson's Creek, near Canarsie and on Spring Creek.

As one zips along the Van Wyck Expressway, one never gives it any thought that a Van Wyck built between 1793 and 1797 a tide mill near Huntington, on Long Island (New York). There were several such mills that dotted the shores and coves of Long Island Sound in the eighteenth and early nineteenth centuries. Today there are not even ten mills in more or less good condition along the Atlantic seaboard. Abraham Van Wyck built the tide mill on Mill Cove Pond, where the Lefferts family operated it until about 1893. Known today as the Van Wyck-Lefferts Tide Mill, it had been constructed on the southern end of a dam across the cove. Tide gates had been inserted in the middle of the dam; the gates were hinged at the top so that the in-flowing tide would be let through and fill the pond. The gates were closed as the ebb current started by the shift in flow. Next to the mill, a sluice gate could be opened to let the impounded water out; this outlet was made available as soon as the head between pond and harbor was sufficiently great. As the water sought its way out of the pond, it drove the undershot waterwheel, providing roughly six hours of operation. The mill's machinery for milling is housed in a separate structure within the mill's building, which also provides room for storage.

Bernard Forest de Bélidor is usually credited for having focused, in "modern times," the attention of the scientific world upon the energy potential dissipated by tides. He discussed harnessing of tides in the course he taught at the French Military Academy, the Ecole d'Artillerie et du Génie, and published his views in 1737 in his *Traité d'Architecture Hydraulique*. This work is perhaps at the origin of a surge of interest in the tapping of tidal energy and of the numerous patents taken out in the nineteenth and early twentieth centuries. Bélidor developed a scheme for the utilization of tidal cycles to produce continuous power by the use of two tidal basins. Decœur, in 1890, took out a patent for a hydraulic machine with a new turbine model, providing for the continuous use of the tidal force, aiming to build it in the Seine estuary. Since then, several patents were taken out in France, such as by Caquot and Defour for the "perfected utilization of the tidal energy" in 1937, and also in Great Britain.

According to Wickert (1956), the oldest treatise on utilization of tides predates by several centuries Bernard Forest de Bélidor, the work of the Italian Mariano (1438). The Brooklyn mill, still in existence, was built after the plans of Veranzio, another compatriot of Dante. Though most early patents were taken out in France, even researchers from countries with modest tidal amplitudes viewed with interest this energy potential; Pein, a German, published some plans for a tidal power plant in the North Sea, in 1912.

The Germans indeed considered twice building such power plants: once in 1912 at Busum (Schleswig-Holstein) and once again in 1954 at Wilhelmshaven, located on a bay west of the Weser River estuary. The Wilhelmshaven project had unusual economic appeal because of its location in the Wilhelmshaven, Cuxhaven, Bremerhaven, Bremen, and Oldenburg pentagon, a complex of more than two million people, close to one of the major European harbors.

Even more recent is the attempt to build the tidal electric power plant on the Aber Wrac'h, close to where the Amoco Cadiz was wrecked in 1978. The effort, started in 1928, came to naught plagued by insufficient funds and the economic crash. Close-by sites, like the Minquiers and the Brest, currently have several proponents.

A large number of books recommending the industrial use of tidal power were published shortly after World War I; for instance, by Maynard (1919), Cattaneo (no date), Bigourdan (1920), Boisnier (1921), and Fichot (1923). Furthermore, since 1918, numerous articles on the topic had appeared in such periodicals as *Nature*, *Génie Civil*, and *Science et Industrie*, while statistical studies had been released in reports of the French ministries of Public Works and Commerce. At the forefront of tidal power research stood also Claude, as well a proponent of ocean thermal (or thalassothermal) power.

Going beyond the problem of harnessing tidal energy, Parenty and Vandamme (1920) presented at the Academy of Sciences of France a report which dealt not only with the use of tidal power but also with that of waves shock, and Rigaud (1926) made a complete survey of all energy sources. The French seem to have been fascinated by the various ways to produce energy and compare all sources to coal using polychromatic descriptions to differentiate them: black coal (bituminous), white coal (waterfalls), green coal (rivers), golden coal (solar radiation), colorless coal (atomic power), and blue coal (ocean energy).

In England, Davey published his now classical views on tidal power (1923) but, as is so often the case with modern scientific "visionaries," his "ideas" were ridiculed. Twice, predictions made by Norman Davey were mocked: he prophesized that gas turbines would supersede all other heat engines and that tidal power could be harnessed. Facts have vindicated him.

The rather large number of studies published in France and dealing with the energy from the sea certainly bears witness to the intense interest in that country in harnessing ocean power. However, they are not symptomatic of a lack of projects and plans in other countries—to name but a few, Great Britain, Argentina, Canada, the United States, Australia, and the Soviet Union have shown interest. Yet perhaps the most comprehensive book on tidal power of the preceding decade is Robert Gibrat's (1966), considered by most as the father of tidal power utilization schemes.

In the Soviet Union, Lyakhnitskii published a book in 1923, reviewing the U.S.S.R. tidal power potential and tidal power concepts. The Mezen estuary was again studied in 1935 (by Poteryakhin) and, four years later, Bernshtein made a concrete proposal for the Kislaya Bay. With several collaborators, he examined possibilities for the Mezen estuary, the Koloi River, the White Sea, Lumbovskaya (Sorokin, 1959), and the Sea of Okhotsk. The ultimately constructed Kislaya facility is the work of Bernshtein and a team of engineers.

The Comision Honoraria para el Estudio de Captacion de las Mareas Patagonicas began in 1923 to examine such Argentinian sites as the gulfs of San Jose, San Jorge, San Julian, and Santa Cruz. More than a dozen sites are favorable for tidal power harnessing. In Chile, at least four locations (Banco Direccion, Bahia Posesion, Punta Catalina, and Dungeness) have either been studied or offer good conditions.

In Germany, a working plant at Busum was dismantled at the onset of World War I, and World War II stymied research undertaken by Siemens-Schuckert at several locations in the North Sea.

PROJECTS IN CANADA AND THE UNITED STATES

The harnessing of tidal power is, as we have seen, not new to America. Tidal power plants differ, however, from the tide mills in that their operation must be a sustained and reliable one. They differ from most fluvial hydroelectric plants, among other ways, in that their potential average hydraulic head is rather small, but, on the other hand, very large quantities of water, accurately predictable, will remain available for very long periods of time.

The Bay of Fundy, with its exceptional tides, and the surrounding terrain configuration, offers excellent tidal power plant development possibilities. The world's strongest tidal amplitudes are encountered in the Bay of Fundy, in which the Memramcook and Petitcodiac rivers debouch; here, tides reach a range of 16 m. Though tides are only 7.2 m in Cobscook Bay, most proposals considered the coupling of Passamaquoddy and Cobscook bays, where the joint basins would provide a pool of 366 km^2.

Various schemes and designs have been under study over the last 50 years in the Passamaquoddy region. The Passamaquoddy Bay's tides have ranges reaching 15 m. If no power plant has actually been built, the cause lies with economic rather than engineering factors. As in Australia, the distance between plant and eventual consumer is considerable; however, since transmission is no longer a problem and demand for power is on the upswing, interest in the project revived. This new situation, it was said ten years ago, would remove the reported opposition of the New England power industries which feared competition from a tidal power plant in an area where electricity rates were probably the highest in the nation.

Prior to 1800, and until 1825, two or three small tidal mills operated in Passamaquoddy Bay. They were small operations and had disappeared long before the turn of the century. One hundred years later, Dexter P. Cooper made applications to build tidal power plants: in 1924, 1926, 1928, and 1933. Each plan amended or completed the preceding one. Some plans of Cooper were for international plants at Passamaquoddy and Cobscook bays, others for plants wholly within the State of Maine. His efforts to secure a federal loan were, however, turned down. The 1933 application dealt with a two-pool project completely located within United States territory; it was denied in September 1935, because of the Federal Government's initiation of its own project.

The General Electric Company surveyed the possibilities in 1927. Production of 1,594 million kW-hr was foreseen for a construction cost of $124,690,000, not a tenth of the contemporary cost. Hence, as with the British schemes, had the plant been constructed, it would have paid for itself a long time ago.

In 1935, President Franklin D. Roosevelt allocated $7,000,000 for construction of the so-called Passamaquoddy tidal power plant. The U.S. Army Corps of Engineers was entrusted with the task, and work was started the same year under the Corps' direction, on an initial Cobscook Bay project, in Maine (Fig. 2.7).

Though U.S. Department of Energy interest in tidal power appears

Fig. 2.7. President Franklin Delano Roosevelt visits the display model of the Passamaquoddy tidal power plant project on July 30, 1936.

today still quite limited, such is not the case for its various counterparts, and one may hope that a new impetus may come from Canadian, British, French, and Soviet research.

WHAT IS TIDAL POWER?

Tidal power can be of local major importance, and even for regions at considerable distance from the coast, because now long distance transmission lines have been devised and energy storage is being perfected. This importance increases considerably when the tidal power scheme is used in conjunction with other power sources in a regional grid, or even a national grid as it is in France.

The selection of a suitable site remains the primordial step in developing tidal energy utilization. Until recently, only locations with high tidal ranges could be considered, but technological advances may well reduce that factor's impact.

Gibrat (1906), Bernshtein (1939), and Mosonyi (1963) have calculated the potential energy of tidal power sites. Only the value assigned to the constant K varies. Gibrat assigned it a value of 1.97, Bernshtein of 1.97, and Mosonyi of 1.92. Expressing the average tidal range in meters and representing it by R (Gibrat uses the average range of the equinoctial spring tide), and the basin area in square kilometers (A), the potential energy (E_p) in kilowatt-hours/year becomes

$$E_p = K \cdot 10^6 AR^2. \qquad [2\text{-}1]$$

A utilization factor must be introduced to get an approximation of the actual usable energy; this factor varies according to the type of scheme used (see below). The factors proposed by Bernshtein are as follows.

Single basin, single tide	0.224
Single basin, double tide	0.34
Double basin, single tide	0.224
Double basin, single tide, with reverse pumping	0.277
Double basin, single tide, with pumps	0.234
Double basin, double tide	0.210

Mosonyi, for instance, assumes a utilization factor of 0.30, while Bernshtein goes as low as 0.13 for Decoeur's scheme of a double basin with a power house in a separate dam. Introducing Bernshtein's most favorable factor of 0.34 in Eq. 2-1, we find that the actual developable energy (E_{act}) is given by Eqs. 2-2 and 2-3.

$$E_{act} = 1.97 \times 0.34 \times 10^6 AR^2 \qquad [2\text{-}2]$$

$$E_{act} = 0.67 \times 10^6 \times AR^2 \text{ kW-hr/yr} \qquad [2\text{-}3]$$

Accuracy of this calculation depends on the basin's sufficient depth and a tide range sufficiently high, and provided the dimension of the basin in the direction of the travel of the tide's wave is equal or larger than the tide wave's length. Resonance can play a modifying role (see below). If the basin's bottom is partially uncovered during a tidal cycle, the water area (the so-called "basin area") is changed and a coefficient of output reduction must be introduced to reflect variation of the value of A in Eq. 2-1. On the other hand, the new techniques of electrotechnical equipment may reduce the importance of the range (R).

Gibrat established also a "site value": the ratio of length of dam needed to natural energy. This ratio is represented by a coefficient k. Natural energy is the annual theoretical production (kW-hr) of unit turbines operating in both directions. The smaller this ratio, the more desirable the site; it is 0.36 for both the Rance and Mont St. Michel on the French side of the Channel, while the Seven River location in Great Britain rates 0.87, and the ratio is 0.92 for the United States 1950 Passamaquoddy project. Coefficient k is computed for the theoretical situation where one could for each tide, keep getting energy as the basin empties to its very lowest level. Advance knowledge of tide level variations makes possible an accurate computation of power output of a planned tidal station. Though indicative, this site ratio should not be the

Fig. 2.8. Low tide in a Maine bay.

Fig. 2.9. High tide in a Maine bay.

only basis on which to reach a decision on a project's feasibility. Finally, the rating of schemes in function of power output per length unit of dam must be "tempered" by taking into consideration that too narrow an entry may cause constriction unless sufficient depth exists (Figs. 2.8, 2.9).

There is some similarity between a tidal power plant and a hydroelectric plant. However, contrary to the river scheme, the tidal power plant has to cope with a two-directional flow and with salt water; tides are predictable and consistent, but rivers sometimes go through flood and drought periods. In the existing French and Soviet plants, an upstream basin or retaining bay was created by cutting an estuary or bay with a dam. Water rushes through opened sluices at incoming tide, and the flow is stopped when high tide is reached. The difference of elevation between water in the basin and at sea level creates the "head" available for generating power. By using reversible blades, turbine generation of electricity is made possible at ebb and flow tides.

By judiciously choosing times of opening and closing of tide gates to create this hydraulic head, generation of power can be synchronized with peak-demand periods of the day, even when these do not coincide with tide peaks that are in phase with the moon. On those days when tide and demand peak coincide, a still higher head is reached and still more power is generated. The head can be even further increased by reversing power units, thus temporarily turning the turbine generator unit into a pump-motor. The problem of the low head of a power plant was solved when the bulb turbine were developed; an ogive-shaped steel shell containing an alternator and a Kaplan turbine, it is placed in a horizontal hydraulic duct and entirely surrounded by water, a shaft providing communication with the engine room of a power plant. The bulb group regulates flow in both directions and acts as turbine and pump. The alternator is directly coupled to the turbine.

The tidal energy itself is derived from the earth's inherent force: the earth's rotation within the disturbing field of sun and moon compares to the movement of an alternator's rotor in the field of a stator. In coastal seas, where depths hardly reach 100 m, friction dissipates almost all the energy radiated as progressive waves from the main stationary waves in the open ocean, where the tide-generating force is strongest and friction least. Though tides are precisely predictable, they are not exactly in phase with sun and moon movement because tidal waves are distorted by land masses, narrow water passages, and shallow depths. The tidal function, in semi-diurnal tides, follows a naturally occurring behavior with a period of approximately 12 hours and 25 minutes between successive high tides.

The plant can operate in different modes, depending on timing, quantity, and type of electricity desired; among these are maximum power production, peak-hour production, and long time duration. The plant's major components include the dam, the retaining basin, and the link to the grid; the dam houses a powerhouse and consists further of dykes connecting with the natural embayment abutment, and a sluiceway. The scheme can involve one or more basins. Existing plants all use a single basin, and Canadian plans opted in favor of the same approach; the Half Moon Cove project in Maine (U.S.) leans toward the more complex multiple-basin system, selected when dependable power versus intermittent power is desired. In a two-pool scheme, basins can be linked or paired. In the first type, turbines are located between the two pools and, in the second, production is coordinated using one basin as the high pool and the other as the low pool. In the latter system, production is additive and no mechanical interfacing is needed. Some schemes involve pumping to "overfill" and to "over-empty" the basin.

How important is tidal power? This can be answered in a variety of ways. Were the U.S.S.R. to harness its tidal energy, it could dispense with any other source of power and would have power to spare. The rise and fall of tides results in a dissipation of energy that can be roughly estimated at 3.10^9 kW, of which 10^9 kW are spent in shallow seas. However, to be useful for electricity production, the tidal amplitude, or usable head, must be rather great, and the geographical site suitable. This eliminates by contemporary estimates about 80% of the theoretically available tidal energy, leaving 2.10^8 kW harnessable in principle. But this amount is further reduced by current technological limitations, and probably 350 TW-hr per annum remain available. If the tidal amplitude is less than 5 m, the location is presently considered unsuitable.

In 1969, the U.S. National Academy of Sciences report, "Resources and Man," held that sites throughout the world represent a power potential of 13,000 MW. On a regional basis, due to concentration, this potential becomes more significant. The conservative estimate of the U.S. National Petroleum Council allocates to tidal power a significant future role. If such power would only generate 50 billion kW-hr, this would save 2.6 billion barrels of oil over 30 years, saving at least $10 billion. A British tidal power plant on the Severn River could produce 8–14 TW-hr, or 4–7% of the 1973 British annual consumption (Wilson 1973).

These considerations, increasing power demand and the rising cost of petroleum, should motivate a new look at harnessing tidal energy. Currently, American energy amounts to 95% from fossil fuels, 4% from hydropower, 1% from nuclear fission, and less than 0.1% from geothermal sources. American consumption quadrupled in 25 years (1950–1975), but Japanese consumption rose 270% in just five years (1970–1975).

REFERENCES

Bigourdan, G., 1920. Un moyen économique d'utiliser la force des marées, *Comptes-Rendus de l'Académie des Sciences* **171**:211–212.
Boisnier, G., 1921. *Utilisation de l'Énergie des Marées*. Paris: Annales des Ponts et Chaussées.

Caquot, A. and Defour, 1937. *Utilisation Perfectionnée de l'Énergie des Marées*. Paris: Presses Universitaires de France.
Cattaneo, F. *La Transformazione della Forze del Mare in Energia Elettrica*. Genoa.
Claude, G., 1921. Sur l'utilisation de l'énergie des marées, *Rev. Génér. de l'Electricité* **10**:627–631.
Creek, H. Tidal mill near Boston, *Civil Engineering* **22**:840–841.
Davey, N., 1923. *Studies in Tidal Power*. London: Constable.
de Bélidor, B. F., 1787. *Architecture Hydraulique*. Paris: Conservatoire National.
Fichot, E., 1923. *Les Marées et leur Utilisation Industrielle*. Paris: Gauthier-Villars.
Fox, W., Brooks, B., and Tyrwhitt, J., 1976. *The Mill*. Toronto: McClelland and Stewart.
Gibrat, R., 1906. *L'Énergie des Marées*. Paris: Presses Universitaires de France.
Gregory, 1826. *Mechanics* (4th Ed.). London.
Maitland, W., 1782. *History of London* I. London.
Mariano, 1438. *Utilization of Tidal Power*. Siena (in Latin).
Maynard, F., 1918. Étude sur l'utilisation des marées pour la production de la force motrice, *Rev. Génér. de l'Electricité* **XLI** (November 9).
Meerwarth, K., 1963. *Wasserkraftmachinen, ein Einführung in Wesen, Bau und Berechnung von Wasserkraftmachinen and Wasserkraftanlagen*. Berlin: Springer-Verlag.
Moreau, M., 1931. *Étude sur l'Utilisation des Marées en France*. Paris: Delagrave.
Mosonyi, E., 1963. Utilizable power in seas and oceans, in: *Water Power Development* **I**, Chapter 4. Moscow: State Publishing House.
Parenty, H. and Vandamme, G., 1920. Utilisation de la force des marées et du choc des vagues de la mer, *Comptes-Rendus de l'Académie des Sciences* **171**:896–898.
Poteryakhin, I. A., 1935. *GES na Prilivno-Otlivnykh Kolebaniyak v Ust'e r. Mezeni* (a HEP utilizing the tidal oscillations in the Mesen estuary). Moscow: I.S.I., thesis.
Rigaud, M., 1926. *Les Réserves d'Énergie*. Paris: Gauthier-Villars.
Tuttel, J., 1978. Watermolens, eeuwenoud en eeuwig boeiend, *Aard en Kosmos* **8**, **9**, and **10**.
Wailes, R., 1941. Tide mills in England and Wales, *Junior Institution of Engineers, Journal and Record of Transactions* **51**:91–114.
Wickert, G., 1956. Tidal power, *Water Power* **8**, *No. 6*: 221–225; **8**, *No. 7*:259–263.
1902. Le moteur à marée en Californie, *Revue Scientifique IV* **17**, *No. 8*:253.
1899. Les moulins à marée de New York, *Revue Scientifique IV* **II**, *No. 1*:30.
1934. A tidal power project at Avonmouth, *Energelicheskoc Obozvenie* **7**.
1929. *Utilizacion de las Mareas de las Costas Patagonicas*. Buenos Aires.

3
The Tides

The rolling in of a tide may recall the sudden surge of a river, particularly when the tidal bore rushes into the channel with the thundering noise of a hundred horses' hooves stampeding. But that is where the similarity stops. The tide is far more complex than the regime of a river; it results from gravitational pull and the earth's rotation: the moon and the sun exert gravitational forces upon the earth, itself subject to its own rotational force.

THE TIDE

The tidal phenomenon is the periodic motion of the waters of the sea, caused by celestial bodies, principally the moon and the sun, upon different parts of the rotating earth.

Tide and tidal currents must be differentiated, for the relation between them is not simple, nor is it everywhere the same. To avoid misunderstanding, the oceanographer uses the word *tide* for the vertical rise and fall of the water, and *current* for the horizontal flow. The tide rises and falls; the tidal current floods and ebbs. In its rise and fall, the tide is accompanied by a periodic horizontal movement of the water called *tidal current*. The two movements, tide and tidal current, are intimately related, and form parts of the same phenomenon brought about principally by the tide-producing forces of the sun and moon.

The gravitational force decreases as the square of the distance increases, and the moon is much closer to the earth than is the sun. Therefore, regardless of the sun's larger mass, its effect upon the earth is less than half that of the moon. Indeed, according to Newton's law of gravitation, the force producing the tide is proportional to the quotient of the masses and inversely proportional to the square of the distance separating these masses. Earth, moon, and sun are not constantly in the same relative position to one another.

76 TIDAL ENERGY

Because of the greater proximity of the moon, tides on earth follow the lunar cycle, and their amplitude varies according to the relative position of the earth, the moon, and the sun. The "pull" of the sun and the moon, each on its own, minutely counteracts earth's gravity and bulges are created in the oceans, which rise in a direction opposite to, but facing that of, the resultant force of the pulls of the sun and the moon. The bulges are created by the centrifugal force acting in a direction away from the earth-moon axis of rotation. Amplitude or tidal range is the difference between the level of high and low tide; the term *resultant* is the theoretical force, the effect and direction of which are equivalent to two or more forces acting independently upon the same mass. As for the centrifugal force, it is the tendency of a mass subject to a gyratory movement (a rotation) to "escape" from the path of rotation (Figs. 3.1, 3.2).

The closer sun and moon pull in the same direction, the stronger the resultant pull. This is easily understood by representing forces by vectors: the resultant is the diagonal of the parallelogram constructed by

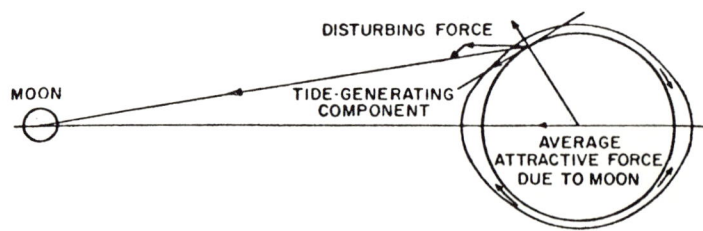

Fig. 3.1. Semi-diurnal tides. *(Source: Stone and Webster Engineering Corp.)*

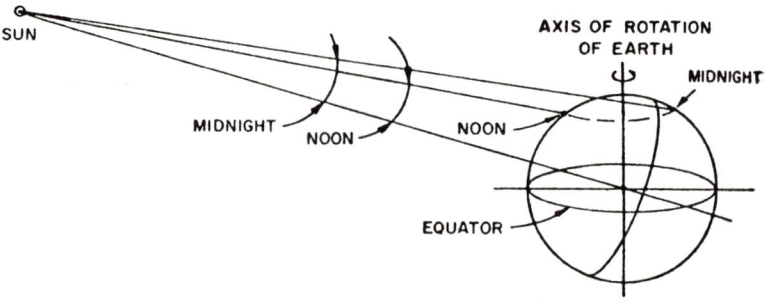

Fig. 3.2. Diurnal tides. *(Source: Stone and Webster Engineering Corp.)*

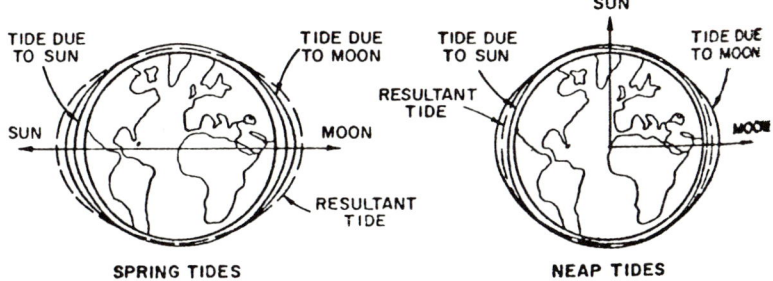

Fig. 3.3. Spring and neap tides. *(Source: Stone and Webster Engineering Corp.)*

using sun and moon force vectors, and it becomes longer (thus greater in strength) as the two gravitational forces become gradually parallel. (Fig. 3.3) The optimum condition is reached when the sun, the moon, and the earth all line up. This configuration occurs at new and full moon; then the "pulls" simply add up, the greatest ranges are observed, and spring tides occur (the term *spring* has nothing to do with the season of year). When the moon and sun oppose each other, as at the quarter moon, the least pull results, the force vectors are at right angles, and the resultant is the shortest. Then the smallest amplitudes are registered; these neap tides occur at the moon's first and last quarters. When the moon is at the point in its orbit nearest the earth, called *perigee,* the lunar semi-diurnal range is increased and the perigean tides occur. When the moon is farthest from the earth or at apogee, the smaller apogean tides occur. When the sun and the moon are in conjunction or opposition, the combined effect of the two masses produces higher tides. At conjunction, the moon is then between the earth and the sun, and the three bodies are all directly in line. At opposition, the moon is at the opposite side of the earth from the sun. The tides are equinoctical tides when the sun is near equinox; during this period, spring tides are higher than average. Furthermore, when the full or new moon and the sun have little or low declination and spring tides of greater than average range occur, particularly if the moon is also in perigee, tides are called equinoctical spring tides (Figs. 3.4, 3.5).

If the range is then, for instance, 80 cm, it may well only be 30 cm at quadrature, when the moon and the sun are nearly at a right angle relative to the earth. The moon is then at its first or last quarter, which occurs every two weeks (neap tide). The high tides do not occur at all

78 TIDAL ENERGY

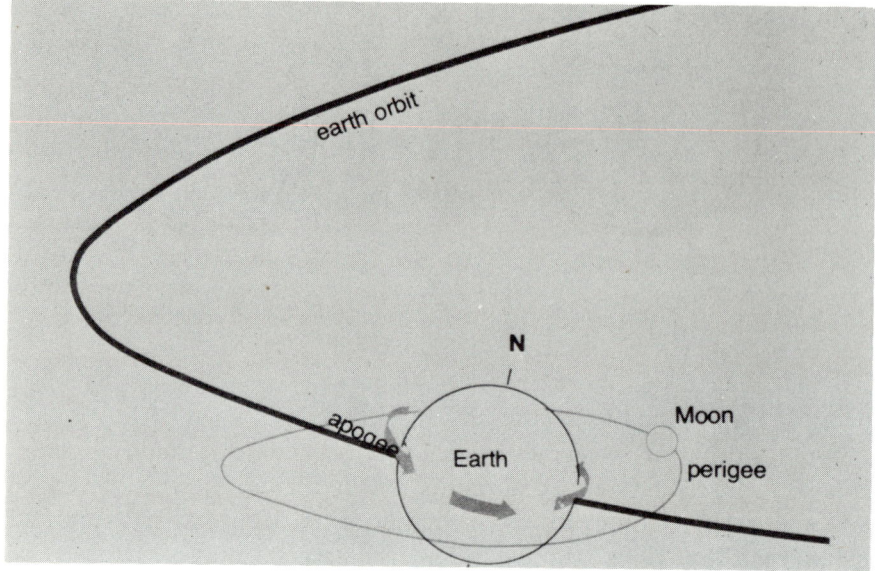

Fig. 3.4. Apogee and perigee of the moon.

when the moon is directly overhead, because of a time lack caused by irregularities on the floor of the ocean and the presence of continental masses that impede the even progress of the wave. Tides being characterized by a rhythmic rising and falling of sea level, they could actually be considered as waves. Tides have a very long period, the period of the lunar tide being approximately 12 hours and 25 minutes, and two lunar tides occur each lunar day (a lunar day is 24 hours and 50 minutes long). The moon's average distance to the earth is about 455,600 km, while the sun's average distance is 177,800 million km. Due to the motion of the earth and the moon, the arrival of the tide is delayed approximately 50 minutes each day.

Foucault's experiment with the pendulum may be recalled here. The direction of the swinging of a pendulum is actually a half cycle of the total movement; this is called a half-pendulum day. A half-sidereal day is about 11 hours and 58 minutes (approximately 12 hours) and represents a half-pendulum day for a pendulum mounted at the north or south pole.

The periodic rise and fall of the sea surface is referred to as the astronomical tide, because winds, earthquakes, and other forces coming from the land can produce long waves similar to tides resulting in periodic

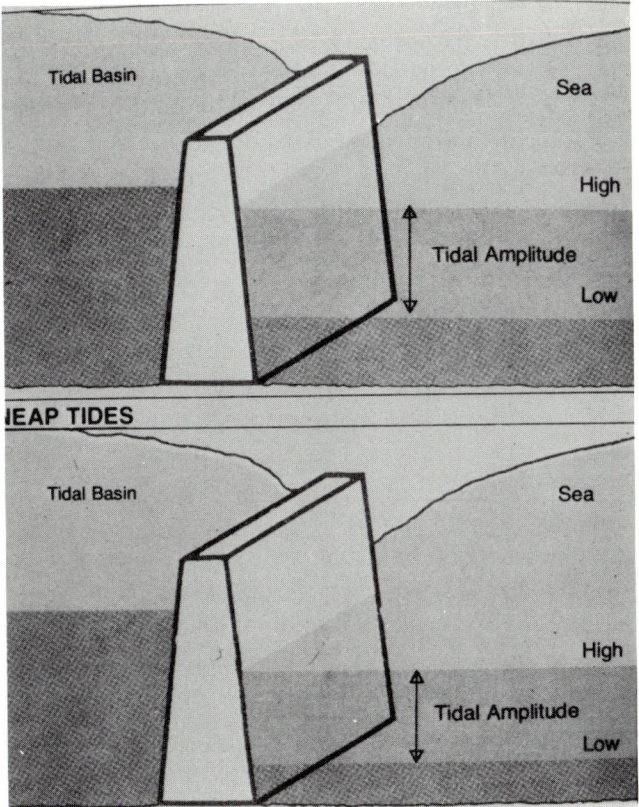

Fig. 3.5. Amplitudes at spring and neap tides.

changes of the sea level, yet are not due to the same causes as the astronomical tide. Tides were first studied in the Mediterranean as far back as the fourth century, although they are hardly noticeable there.

TIDAL AMPLITUDE[1]

Though the tide-producing forces are distributed over the earth in a regular manner, the size and shape of the ocean basins and the inter-

---

[1]Amplitude is used synonymously with range, as is customary in many languages; it is not half the range as used in some English language texts.

ference of land masses prevent the tides of the oceans from assuming a simple regular pattern. The way in which the waters in different parts of the oceans, as well as in the smaller waterways, respond to these known regular forces is dependent mostly upon the size, depth, and configuration of the basin or waterway. The rate of rise and fall of the tides is not uniform. From low water, the tide begins to rise slowly at first, but at an increasing rate until it is about halfway to high water. The rate of rise then decreases until high water is reached and the rise ceases. The falling tide behaves in a similar manner. The period at high or low water, during which there is no noticeable change of level, is called stand. The difference in level between consecutive high and low tides is the range or amplitude. The range of the tide varies in accordance with the intensity of the tide-producing force, though there may be a lag of a day or two between a particular astronomic cause and the tidal effect.

None of the oceans appears to be a single oscillating body; rather, each one is made up of a number of oscillating basins. As these basins are acted upon by the tide-producing forces, some respond more readily to daily or diurnal forces, others to semi-diurnal forces, and others almost equally to both. Hence, tides at a place are classified as of a given type according to the characteristics of the tidal pattern occurring at the place.

Tides are quite insignificant in the middle of the ocean. They increase considerably in importance along the coast, as coastal morphology can amplify the tide; for instance, near horizontal beaches and estuaries. Along two European rivers of an otherwise very regular regime, tides are felt as far inland up the Scheldt River as Ghent and, upstream, in the Seine River as far as Rouen. The "strength" of tides (the height of the tidal crest) varies considerably from place to place. In Nova Scotia, there are some 12 m of difference between high and low tides; in the Bristol Channel, the difference is 17 m, as in the Bay of Fundy. In the Channel Islands (Guernsey, Anglesey, and Jersey), the difference reaches 14 m (46 ft). At the historic site marking the boundary between Normandy and Brittany, the Mont St. Michel, the difference between high and low water is 14 m. Some other places have hardly any tide, such as Gulf Port (Texas) and Biloxi (Mississippi).

As a result of the rotation of the earth, there is also a diurnal inequality of semi-diurnal tidal amplitude. Spring tides have their greatest amplitude at half moon or at syzygy, when sun, moon, and earth line up in one plane and their differential forces of attraction are in phase. Neap tides, the lowest tides in a given new lunation, occur when sun and moon tractive forces are 90° out of phase (Fig. 3.5).

If there is a point at the surface of the earth, neither in line with the moon nor at right angles to that line, forces acting there have a resultant nearly parallel to the surface of the earth. All tidal movements depend on it; it moves particles along the earth's surface toward the line of the centers of the earth and the moon: this is the *tractive force*.

TIDE-GENERATING FORCES

So far we have mentioned three forces: gravitational pull from the sun and the moon and the rotational force of the earth. They are responsible for high and low tides, which succeed one another in the course of a lunar day. We speak also of a tidal day, defined as the interval between two upper transits of the moon over the meridian of a geographical site, taking 24 hours and 50 minutes or 24.84 solar hours, about 1.035 times as long as the mean solar day and causing a "delaying" of the tide from earth day to earth of 50 minutes (Figs. 3.6, 3.7, 3.8, 3.9).

However, superimposed on this apparently simple mechanism are several other rhythms. These influence the tidal range, not the tides'

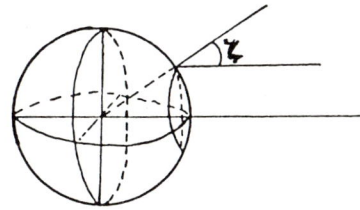

Fig. 3.6. Force of attraction at a given point of the earth.

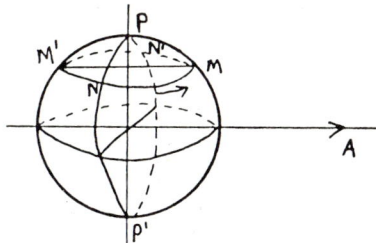

Fig. 3.7. Declination and rotation: influence on tide-generating forces.

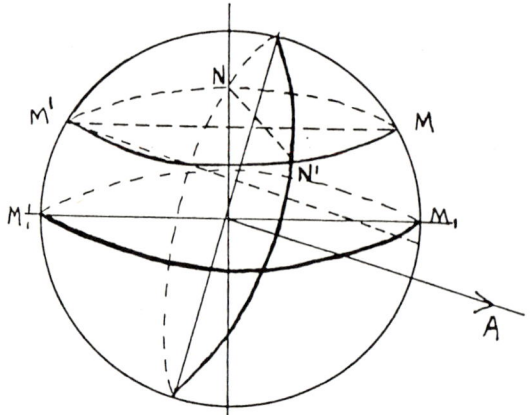

Fig. 3.8. Tide forces when declination is different from zero.

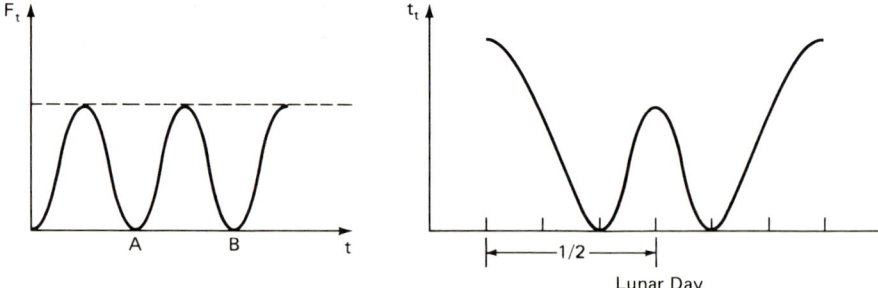

Fig. 3.9. Tide curves when declination is equal to zero and when different from zero.

basic rhythm; they include the earth's revolution and the moon's declination.

As mentioned, tides result from differences in the gravitational attraction of the moon and the sun upon different parts of the rotating earth. The gravity of the earth acts approximately toward the earth's center, holding the earth in the shape of a sphere. But the moon and sun provide tide-producing forces. The moon appears to revolve about the earth but actually the moon and the earth revolve about their common center of mass. They are held together by gravitational attraction and yet kept apart by an equal and opposite force. In this earth-moon system, the tide-producing force on the earth's hemisphere nearer the moon is in the direction of the moon's atrraction, or toward the moon. On the hemisphere opposite the moon, the tide-producing force is in the direction of the opposite force, or away from the moon.

The tide-producing forces tend to create high tides on the sides of the earth nearest to and farthest from the moon, with a low tide belt between them. As the earth rotates, a point on the earth passes through two high and two low areas each day if the moon is over the equator. There are similar forces due to the sun, and the total tide-producing force is the resultant of the two, though the sun's tide-producing force, as we will calculate later, is only 0.46 that of the moon. Only minute tidal effects are caused by other celestial bodies because they are too far away.

The combination of the relative positions of the celestial bodies and of tide components may bring about remarkable phenomena and cause exceptional amplitudes, although these are merely natural curiosities. Such an unusual tide occurred on March 28, 1967 and took particular dimensions on the French Channel coast. The preceding time the moon, the sun, and the earth lined up to cause such a tide was on January 1, 1900. Such tides can uncover as much as 10 km of normally water-covered continental shelf exposing not only sunken ships and scuttled men-of-war, but also ancient Roman and Gallic villages of the drowned former sea coast, tree trunks, old roads, and Neolithic tombstones.

The periodicity of forces ranges from 3.1 hours to 1,600 years. The maximum tractive force occurs once every 1,600 years, next due to occur in the year 3300.

THE EQUILIBRIUM TIDE

The theory of the equilibrium tide is due to Sir Isaac Newton, who, by imagining an ideal earth, free of inertia, entirely covered by water, succeeded in picturing tide-producing forces. However, actual action of these forces is quite different. Indeed, all points on the surface of the earth are not equidistant to the moon. The tide-raising capacities lie in the differential tidal force or the difference in the force of attraction at any given point of the surface of the earth and the attraction at the center of the earth (Fig. 3.6).

If M_m is the mass of the moon and M_e is the mass of the earth, d is the distance separating moon and earth, r is the radius of the earth, C

is the latitude of point X, and g is the acceleration due to gravity, then the tractive force can be expressed by the formula

$$F_x = \frac{3}{2} g \frac{M_m}{M_e} \frac{r^3}{d^3} \sin 2C,$$

where G, a universal constant, has been most recently determined as having the value

$$(6.673 \pm 0.003) \times 10^{-8} \frac{\text{cm}^3}{\text{g}^2/\text{sec}^2}$$

and g, the acceleration at sea level and 0° latitude is 978.03g $\frac{\text{cm}}{\text{sec}^2}$.

Concluding, the astronomical tide is caused by tangential forces, which are deviated, because of the earth's rotation, clockwise in the Northern hemisphere and counterclockwise in the Southern hemisphere.

FORCES OF ATTRACTION

The gravitational forces of attraction of the moon and the sun balance the centrifugal motion of the earth around their respective revolution centers. The magnitude of the differential forces of attraction can be computed from Newton's Law of Universal Gravitation. The maximum differential force on earth is

$$\Delta F = \frac{M}{(R \pm r)^2} - \frac{M}{R^2} \to \pm 2M \frac{r}{R^3},$$

with M the mass of the attracting body, R the distance from the earth's center of mass to the moon, and r the radius of the earth. Astronomical tides consist of forced oscillation whose periods are a function of the tide-generating forces. Their amplitude and phases are constrained by friction and by the depth of water. So while the astronomical tide should coincide with the moment the sun or the moon crosses the local meridian, friction and depth of water cause a delay of several hours before the arrival of the tide (Figs. 3.8, 3.9).

Mathematically, the tide-generating forces have an effect that can be explained as follows. Let F_1 be an attraction force exerted by a celestial body of mass A upon a molecule M of mass 1 at the surface of the earth. The distance from A to C, the center of the earth, is given by d. Then F_1, the absolute force, is

$$F_1 = k \frac{A}{L^2} \qquad [3\text{-}1]$$

if L, or AD, is the distance from A to the particle M at the earth's surface, and k is the gravity constant.

The relative force to the earth F_2 exerted upon the molecule is F_1, less the attraction to which M would be subjected from A if M were at the earth's center (C).

$$F_2 = k \frac{A}{d^2} \qquad [3\text{-}2]$$

The resultant is the vector $\overline{MB}$ whose direction is found by calculating the distance $\overline{CN}$.

$$\frac{NA}{BF} = \frac{MA}{MF} \qquad [3\text{-}3]$$

$$\frac{CA - CN}{k \dfrac{A}{CA^2}} = \frac{\overline{RD}}{k \dfrac{A}{AD^2}} \quad \text{or} \quad \frac{d - CN}{k \dfrac{A}{d^2}} = \frac{L}{k \dfrac{A}{L^2}}. \qquad [3\text{-}4]$$

Extracting CN,

$$CN = d - \frac{L^3}{d^2} = (d - L) \frac{d^2 + dL + L^2}{d^2}. \qquad [3\text{-}5]$$

However, because the distance from C to the earth's surface is comparatively very small in relation to the distance CA, d and L are about equal in value; thus,

$$\frac{d^2 + dL + L^2}{d^2} \cong \frac{d^2 + d^2 + d^2}{d^2} = 3. \qquad [3\text{-}6]$$

Hence:

$$CN = 3(d - L) \qquad [3\text{-}7]$$

In turn, if P is the intersection of the perpendicular dropped from M or CA, $AP \cong AM$. With $AM = L$,

$$CP = d - L \qquad [3\text{-}8a]$$
$$CN = 3\,\overline{CP}. \qquad [3\text{-}8b]$$

Based upon Eq. 3-3,

$$\frac{MA}{F_1} = \frac{NA}{F_2} = \frac{MN}{MB}. \qquad [3\text{-}9]$$

With $\overline{MP}$ perpendicular to $\overline{AC}$ and N chosen so that $CN = 3\,\overline{CP}$; the vector $\overline{MN}$ gives the direction of the generating force and is proportional to that force. This generating force is, in C, with r the earth's radius,

$$F_1 = k\frac{A}{L^2} \qquad F_2 = k\frac{A}{d^2} \qquad L_0 = d - r \qquad [3\text{-}10]$$

$$F_1 - F_2 = kA\left(\frac{1}{L_c^2} - \frac{1}{d^2}\right) = kA\left(\frac{1}{[d-r]^2} - \frac{1}{d^2}\right)$$
$$= k\frac{A}{d^2}\left(\frac{1}{\left[1 - \frac{r}{d}\right]^2} - 1\right). \qquad [3\text{-}11]$$

But $\dfrac{r}{d}$ is small; thus,

$$F_1 - F_2 = k\frac{A}{d^2} \times \frac{2r}{d} = \frac{2kAr}{d^3}. \qquad [3\text{-}12]$$

The force in D is with $L_D = d + r$:

$$F_1 - F_2 = kA\left(\frac{1}{[r+d]^2} - \frac{1}{d^2}\right) = -\frac{2kAr}{d^3}. \qquad [3\text{-}13]$$

The planet's attraction upon a molecule in C and in D is the same but the signs are opposite, and attraction forces in both locations thus tend to heighten the sea's surface.

The planet's attraction on the particle of mass and the gravity force (g) on the particle are then in a ratio of

$$R = \frac{\left|\dfrac{2kAr}{d^3}\right|}{g} = \frac{\dfrac{2kAr}{d^3}}{\dfrac{kM}{r^2}} = \frac{2Ar^3}{Md^3} \qquad [3\text{-}14]$$

since

$$g = \frac{kM}{r^2}. \qquad [3\text{-}15]$$

In points A and B, the celestial body's attraction is half that of the force in C or D and is directed toward the center of the earth,

$$\frac{kAr}{d^3} \qquad [3\text{-}16]$$

and is equivalent to an increase of the gravity force of $58.10^{1/6} \dfrac{\text{cm}}{\text{sec}^{-2}}$.

Let us apply these calculations to the moon as the celestial body.

Any point, any unit of mass on earth, describes an identical orbit, and thus the centripetal acceleration supplied by the gravitational attractions of the sun and the moon keeping the earth in its orbit must also be the same everywhere. The gravitational attractions, however, vary from geographical location to geographical location as points on earth vary in distance to sun and moon.

To keep the earth in its orbit, the mean gravitational attraction and the mean centripetal acceleration must equalize. While the latter is constant, the other is not. This difference creates the force that produces the tides. The force's magnitude is minuscule in comparison to the earth's gravity, but in most places on earth is partly parallel to the earth's surface, and that component of the tidal force is not counteracted by any other force and brings the waters to move.

The relationship between centripetal acceleration and gravitational attraction can be expressed mathematically. If g is gravity (or the gravitation constant), M_b is the mass of the celestial body (in this case the sun or the moon), d is the distance between the center of the earth and the center of the celestial body considered, r is the radius of the earth, F_t is the tide-producing force, and a_p, $\bar{a}$, and a_c are the local and average gravitational and centripetal accelerations respectively, then the equations become successively:

$$a_c = \bar{a} \frac{gM_b}{d^2}. \qquad [3\text{-}17]$$

If P is a point directly under the celestial body considered, here

$$a_P > \bar{a} \text{ and } a_P = \frac{gM_b}{(d-r)^2}. \qquad [3\text{-}18]$$

The difference between a_P and $\bar{a}$ is the tide-producing force (per unit of mass) F_t

$$F_t = a_P - \bar{a} = \frac{gM_b}{(d-r)^2} - \frac{gM_b}{d^2} = gM_b \left[\frac{1}{(d-r)^2} - \frac{1}{r^2} \right] \qquad [3\text{-}19]$$

$$F_t \cong gM_b \frac{2r}{d^3}.$$

The comparative effect of the moon and the sun upon the earth can be assessed as follows: if d_m and d_s are, respectively, the distance between moon and earth and sun and earth, and M_m and M_s are the corresponding masses, then:

$$M_s = 27.10^6 \times M_m$$
$$d_s = 389 \, d_m$$

and if $F_{t,s}$ and $F_{t,m}$ are the tide-producing forces of the sun and the moon, respectively, then

$$\frac{F_{t,s}}{F_{t,m}} = \frac{gM_s \frac{2r}{d_s^3}}{gM_m \frac{2r}{d_m^3}} = \frac{M_s}{M_m} \times \frac{d_m^3}{d_s^3} = \frac{27.10^6}{389^3} = 0.46 \qquad [3\text{-}20]$$

$$\frac{a_{c,s}}{a_{c,m}} = \frac{gM_s}{d_s^2} \times \frac{d_m^2}{gM_m} = \frac{M_s}{M_m} \times \frac{d_m^2}{d_s^2}$$

$$= 27.10^6 \times \frac{1}{389^2} = 178.5. \qquad [3\text{-}21]$$

Thus, the centripetal acceleration due to the sun is only about 180 times that due to the moon, even though the sun's mass is 27 million times that of the moon, and the tide-producing force of the sun is less than half than that of the moon. As for the generating force's direction, it is given by Proctor's rule. Considering a meridian containing plane AT, there is symmetry relative to AT and also to the plane of the illumination circle (see above, points C and D). In any point of the earth's surface, the vertical and horizontal components (F_v and F_h) of the generating force are

$$F_v = \frac{3kAr}{d^3} \left(\cos^2 \zeta - \tfrac{1}{3} \right) \qquad [3\text{-}22]$$

$$F_h = \frac{3kAr}{d^3} \sin \zeta \cos \zeta$$
$$= \tfrac{3}{2} g \frac{kAr}{d^3} \sin 2\zeta \left(\text{or } \tfrac{3}{2} g \frac{M_m}{M_e} \frac{r^3}{d^3} \sin 2\zeta \right) \qquad [3\text{-}23]$$

where ζ is the zenithal distance (zenith angle) from the planet to the point on the earth.

VARIATIONS OF THE GENERATING FORCES

Tide-generating forces vary in a given point as time elapses. This is caused by the rotation of the earth and the declination. If the planet is in the plane of the equator, the axis of the pole is perpendicular to the line joining the center of the earth to the planet. The point under study follows a parallel $M'NMN'$. When the point where the generating force is studied is in M or in M', the force is maximum, and the intensity of this force in M and M' is the same (Fig. 3.7).

When the point reaches N or N', both inside the illumination circle, the force is vertical and its horizontal component is nil. This horizontal component was at its maximum in M and in M'. Thus, in a solar or lunar day, the generating force passes through two maxima (in M and M') and two minima (in N or N').

If only the tangential component is considered, this component becomes nil when the point reaches N or N', hence the semi-diurnal periodicity of the generating force. On a graph of the variations of the horizontal force $OA = 12$ hours. Since the lunar day is 24 hours and 50 minutes long, the periodicity due to the moon is 12 hours and 25 minutes.

When the declination of the sun or the moon is not nil, the astre is not in the plane of the equator, and the resulting curve has a different aspect. The point where the variation of the generating force is studied follows the parallel $M'NMN'$; when it passes in N and N', situated upon the illumination circle, the generating force is vertical (Fig. 3.8).

When the point after its passage through N (where the tangential component is nil) proceeds toward M, this component increases, reaches its maximum in M, and becomes nil again when in N' to attain a second maximum in M. In some cases, the maximum corresponding to M' may be more important than the one observed for M (Fig. 3.9).

The time necessary for the moving point to travel the arcs $N'M'$ or $M'N$ is longer than the time needed to cover arcs NM or MN'. This gives to the variation curve of the horizontal component of the generating force a totally different aspect and results in an unequal diurnal periodicity. Apparently, the generating force is made up of a component with semi-diurnal periodicity, upon which is superimposed a component with a diurnal periodicity. For a point on the equator, there is no diurnal inequality, since the generating force is indeed the same in both M_1 and M'_1.

TYPES OF TIDES

The generating force can be decomposed in periodic terms of which it is the sum. First we will have to consider two semi-diurnal terms or periodic forces, respectively, the moon's attraction ($T_m = 12$ hr 42) and the sun's attraction ($T_s = 12$ hr). These components vary according to latitude; they are based upon a fictitious moon and sun. To these forces correspond the M_2 and S_2 "waves" or species.

The combinations of the relative positions of the sun, the moon, and the earth are quite numerous and are responsible for only an approximate repetition of tides during successive lunar months. About 20 severe astronomical frequencies, called tidal species, are identifiable. Among the principal semi-diurnal ones are the principal lunar (M_2, 12:42-hr period), principal solar (S_2, 12:00), larger lunar elliptic (N_2, 12:66), and luni solar (K_2, 11:97). The same names apply to diurnal tides: principal lunar (O_1, 25:82), principal solar (P_1, 24:07), and luni solar (K_1, 23:93). Long-period ones include lunar forthnightly (M_f, 327:86), lunar monthly (M_m, 661:30), and solar semi-annual (S_{sa}, 2191:43).

However, the sun and the moon do not remain in the equator's plane, nor is their distance to the earth constant; correction factors must thus be introduced. The planet's declination is not nil and a diurnal component must be brought in; the moon's declination has a mean period of about 13 days 66, and therefore two periodic terms have to be brought in whose sum is nil when the moon is in the equator's plane (declination is then nil) and maximum when the declination is maximum.

To these two terms correspond the K_1 and O_1 "waves" or species. Their angular speeds $\left(\omega = \dfrac{360°}{T}\right)$ must differ so that, in 13 days 66, one of the "waves" progresses 360° in relation to the other. This amounts to an angular speed difference of

$$\frac{360°}{13.66 \times 24} = 1°10'.$$

The mean speed is equal to the moon's diurnal movement,

$$\frac{360°}{24.84} = 14°49'.$$

Angular speeds and periods for K_1 and O_1 are then

$$\omega_{K_1} = 14°49' + 0.55 = 15°04'; \quad T_{K_1} = \frac{360}{15.04} = 23.93$$

$$\omega_{K_2} = 14°49' - 0.55 = 13°94'; \quad T_{O_1} = \frac{360}{13.94} = 25.82$$

The sun's period is about six months; the principal solar wave is P_1 ($T = 24$ hr 07) and the declination wave is the same as K_1. M_2 corresponds to a theoretical moon remaining equidistant from the earth. In reality, the moon travels in an ellipse, not a circumference, and an N_2 wave must be added to correct M_2 so that the effect of M_2 is increased when the moon is at its perigee and decreased when at its apogee. The moon's revolution is 27 days 55 long, hence to travel from apogee to perigee, one-half time elapses during which N_2 must lose 180° on M_2.

By calculations similar to the ones used for M_2, we find that $\omega_{N_2} = 28°44'$ and $T_{N_2} = 12$ hr 66. The distance to the sun being also variable, still another "wave" must be introduced: T_2 is the larger solar elliptic component, which has an angular speed and a period of $\omega_{T_2} = 29°95'$ and $T_{T_2} = 12$ hr 02.

THEORY OF TIDES

The matter at hand now is to examine how these forces can actually produce tides. Newton proposed a static or equilibrium theory, today merely of historical interest. Newton ventured the hypothesis that under the action of the generating forces the sea level takes instantaneously its equilibrium position. If we consider a single disturbing planet and a position of equilibrium, the earth (supposedly covered with water) is an ellipsoid. We know that the generating force corresponds to a decrease or an increase in the gravity pull, depending upon the respective position of a point at the surface of the earth. Basically, we should have tides that are identical, but they are not, and in some sites they even exceed 15 m. The static theory, hence, is not adaptable to the actually observed phenomenon.

If the ocean's surface were in all points in a perpendicular position to the resultant of the force of gravity and of the attractive forces of the sun and the moon and, if we had a hydrostatic equilibrium, then we would have an equilibrium tide. But this is not so. We do not have an "ideal earth" covered entirely with water, nor can we ignore the effects of inertia. While the theoretical equilibrium tide helps in understanding tide-producing forces, the reaction of oceans to these forces is very different from an equilibrium tide. Its value, however, still persists from a qualitative point of view. The great axis of the ellipsoid is always directed toward the perturbing planet. However, the tide which should show a semi-diurnal and a diurnal character, is, in some areas, only semi-diurnal. According to the static theory, there should be a large diurnal inequality and this is not the case. The ellipsoid resulting from solar attraction and the ellipsoid resulting from lunar attraction add up; according to the static theory, the maximum amplitude would occur every 14 days 18 hours at syzygy, with the minimum amplitude at quadrature. Yet, in the middle of the ocean, this occurs then syzygy later. No part of the static theory explains this; it fails to take inertia into account.

The sea level cannot instantaneously reach its equilibrium. Laplace was first in examining the tide mechanism from a dynamic point of view. A dynamic theory was then proposed based on harmonic analysis. He considered the following principals: 1) forced oscillations (under the influence of periodic forces, the oscillations of molecules are periodical and have the same period as the force); and 2) superimposition of small movements (the whole movement, as we have just seen, is due to a certain number of small forces and is the sum of the partial movements that each of these forces would produce individually).

Laplace decomposed the generating force in different partial forces, each corresponding to the effect of a fictitious planet, whose movement is simple.

Kelvin was first to show, in 1867, that the generating force may be entirely decomposed in a sum of rigorously periodic terms, such as M_2, K_2, and N_2.

Each component is of the form

$$C \cos \omega t$$

in which ω is the angular speed and C is a coefficient calculated in cm for each of the "waves." To each component corresponds a partial movement of equal period

$$KC \cos (\omega t - \varphi)$$

with K and φ empirical constants characteristic of the site considered expressed, respectively, in cm and degrees.

The sea level, for the site, at a given moment is then

$$K_1 C_1 \cos (\omega_1 t - \varphi_1) + K_2 C_2 \cos (\omega_2 t - \varphi_2) + \ldots + K_n C_n \cos (\omega_n t - \varphi_n) = \sum_{i=1}^{i=n} K_i C_i (\cos \omega_i t - \varphi_i). \quad [3\text{-}24]$$

Four major types of tide result from the combined effects of the tidal components.

In the *diurnal* type of tide, only a single high and a single low water occur each tidal day, thus within approximately 24 hours. This type occurs along the northern shore of the Gulf of Mexico, in the Java Sea, the Gulf of Tonkin (off the Vietnam-China coast), and in a few other localities such as Copenhagen.

In the *semi-diurnal* type, there are two high and two low waters each tidal day, with relatively small inequality in the high and low water heights. Tides on the Atlantic coast of the United States are representative of the semi-diurnal type.

In the *mixed* type, the tide is characterized by a large inequality in either the high water heights or the low water heights (or both). There are usually two high and two low waters each day, but occasionally the tide may become diurnal. Such tides are prevalent along the Pacific coast of the United States, in the Carribean, and in many other parts of the world.

In the *semi-diurnal tides with diurnal inequality,* two high tides and two low tides occur, but the heights of two successive tides or two successive low tides are unequal.

The literature mentions occasionally tropical and equatorial tides. These terms refer to the magnitude of the inequality and indicate that the moon is respectively over one of the tropics or the equator; in the first instance, the diurnal inequality is greatest, and in the second the moon is at its minimum declination and the inequality is minimal. There are, however, tides with longer periods than those mentioned above (approximately 12 and 24 hours). The M_f lunar long period tide is a fortnightly phenomenon whose period is 327.86 solar hours; M_m is a monthly lunar tide and S_{sa} is a solar semi-annual tide. There are in total about 20 constituent tides and the tide with the longest period, the chaldean period, nearly as long as the sares, extends over approximately 18 years and 8 months.

TIDES IN GEOMETRICALLY SHAPED BASINS AND ESTUARIES

Simple theory does not accurately describe tidal fluctuations because the earth is not entirely covered with water, water depth is not constant, the earth rotates producing a Coriolis deflection, and coastal shape, basin shape, and basin size exert considerable effect on the tide (Fig. 3.10).

Two progressive waves having the same period and wavelength but moving in opposite directions will result in a standing wave. When an enclosed body of water is made to oscillate, a particular period of oscillation, or resonance, depending on the body's dimensions is preferred. Standing waves are set up. In an enclosed basin, they are associated with a wavelength equal to twice the size of the basin. If l_c is the length of the basin and λ the wavelength, c is the velocity of the wave, and T_c is the period, and if the depth (d) is shallow:

$$c = \frac{\lambda}{T_c} = \frac{2l_c}{T_c}. \qquad [3\text{-}25]$$

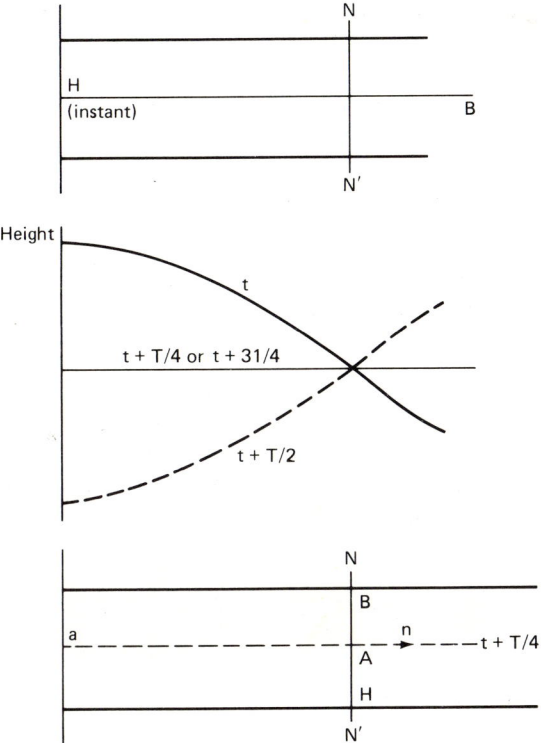

Fig. 3.10. Tides in geometrically shaped basins.

However, the velocity under such conditions is also

$$c = \sqrt{gd} \qquad [3\text{-}26]$$

because wave velocity is equal to

$$c = \sqrt{\frac{g\lambda}{2r} \cdot \tan_h \frac{2\pi d}{\lambda}} \qquad [3\text{-}27]$$

and, in deep water, $\tan_h \frac{2\pi d}{\lambda} \to 1$. Thus:

$$c \cong \sqrt{\frac{g\lambda}{2\pi}}. \qquad [3\text{-}28]$$

While in shallow water, d is small compared to

$$\tan_h \frac{2\pi d}{\lambda} \to \frac{2\pi d}{\lambda}; \text{ thus,} \qquad [3\text{-}29]$$

$$c \cong \sqrt{\frac{g\lambda}{2\pi} \cdot \frac{2\pi d}{\lambda}} = \sqrt{gd} \qquad [3\text{-}30]$$

and the resonant period, T_c, is given by

$$T_c = \frac{2l_c}{\sqrt{gd}}. \qquad [3\text{-}31]$$

In a one-end open basin, the standing wave has a wavelength (λ) four times the basin size (l_0), and

$$c = \sqrt{gd} = \frac{\lambda}{T_0} = \frac{4l_0}{T_0} \qquad [3\text{-}32]$$

and the resonant period T_0 is given by

$$T_0 = \frac{4l_0}{\sqrt{gd}}. \qquad [3\text{-}33]$$

If a number of periodic forces act together (with tidal forces), the body of water will respond most actively to the one closest to its resonant period. This explains semi-diurnal tides in some areas, diurnal ones in others, because of best response to forces with these periods. Tides being periodic, they may be considered as waves.

Tides rise and fall primarily because of the moon's gravitational attraction. The earth rotates; hence, twice a day tides occur. This rotation with regard to the moon is slightly out of phase with the solar day, and thus high tides are spaced 12 hours 24 minutes apart. The earth is also subject to the moon's gravitational attraction. Its angle of incidence to the sun and the moon changes, bringing about a diurnal tide superimposed on the semi-diurnal pattern. There are several other types of tides, but only spring and neap tides are of primordial interest here. As the moon revolves around the earth, the sun, the moon, and the earth occupy different relative positions; when all three celestial bodies are aligned, maximum effect is attained and spring tides occur, but minimum effect occurs when the angle between the three bodies is near 90° (neap tides).

The tidal wave is also influenced by the Coriolis and other geostrophic forces, and basin morphology, and rotates around an amphidromic point. Open ocean tides have an amplitude of about 50 cm, but coastal ocean tides often exceed several meters. This increase in height is caused by a combination of the convergence effect on the tidal wave energy of the coast, and semi-diurnal oscillation amplification by a coastal indentation of near-resonant length. Construction of the movements of an approaching tidal front by a narrow indentation is *convergence*. When length and depth of the indentation cause the tidal wave to travel a quarter of the wavelength in the approximately 12 hours 24 minutes, separating successive waves, *resonance* occurs; the quarter-wavelength for the Bay of Fundy is about 300 km, and locations considered for power plants installation there are, with 246–286 km, quite close to resonance.

The rise and fall of tides results in a dissipation of energy roughly estimated at 3.10^9 kW, of which 10^9 kW are spent in shallow seas. The total earth's potential water power approximates 3.10^9 kW.

Tidal power stations are usually planned in estuaries. When a tidal wave enters into an estuary, the speed of the wave's crest is greater than that of the trough, since the depth is greater at the crest than at the trough.

In a given point, flow is faster than ebb. Based upon the harmonic analysis of such a tide, we consider that there is superimposition upon the tide wave of another wave whose period is half that of the tide wave: the quarter diurnal lunar term M_4 corresponds to M_2.

A basin or a bay may show free oscillations whose periods are a function of only the geometric characteristics of the basin. These oscillations are called *seiches*. When the water level is submitted to a perturbing force such as, for instance, weather conditions, it continues to oscillate (even when the causing force has ended) with the period proper to the basin. These oscillations are thus different from those of the tides, which are forced oscillations. However, when the applied force has the same period as the very oscillation of the basin itself, it is a foregone conclusion that the amplitude of movements will be much larger and there is *resonance*.

It is thus possible to have resonance between an oceanic basin and one of the elementary generating forces of the tide, resulting in a considerable increase of the amplitude of the oscillation corresponding to this force. This is noticed, for instance, on some coasts of France where the tide is of the semi-diurnal type: the influence of the components of different periods is thus quite weak compared to the effect of semi-diurnal waves.

While a seiche is thus an oscillation due to such factors as atmospheric conditions (e.g., a storm), at the surface of the ocean or even a lake, the term is also applied to the "sloshing" of the water in a hemispheric shaped basin. Perhaps the high tides of the Bay of Fundy are due to a seiche. Indeed, the ocean floor off the Newfoundland coast is thus shaped that tides are thrown into resonance which, as explained before, is a reinforced oscillation because of the similarity of period of the applied force and the oscillation itself.

The Coriolis effect plays a primordial role. Considering a stationary wave in a basin with a nodal line NN', the amplitude of vertical motions on this line is nil and the horizontal currents are at maximum amplitude. Reciprocally, the horizontal current is nil at tides' low. The reversing of the current occurs at high and low tides that occur everywhere simultaneously. The Coriolis force given by the formula below is directed to the right of the velocity in the northern hemisphere:

$$C = 2\omega \sin \varphi$$

where ω is the angular speed of the earth's rotation and φ is the latitude.

Since the speed is periodic and of a period equal to that of the tide, the Coriolis force upon the particles also is periodic and will cause a transversal oscillatory movement, which will superimpose itself upon the stationary wave. There is a level increase to the right of the current and a decrease to the left.

Using as origin the time of the reversing of the current (high tide, left side of figure), at $\frac{T}{4}$ the current along NN' is maximum and directed to the right; the surface should be horizontal. The Coriolis effect causes accumulation of water to the right of the speed and only $\overline{ab}$ is at the mean level.

At the intersection A of NN' and $\overline{ab}$, there are no vertical movements. High sea occurs at time $\frac{T}{2}$ on Ab, $\frac{3T}{4}$ on An and T on Aa.

Cotidal lines—lines joining points of simultaneous high tide—turn counterclockwise in the northern hemisphere. Point A is the *amphidromic point;* it is a nodal point caused by the Coriolis effect deflection of the motions in such a manner that the antinodes rotate about the central point: no nodal line but a nodal point, no simple sloshing back and forth as in a normal standing wave. The cotidal lines radiate out from the amphidromic point.

While along the cotidal lines, high tide occurs simultaneously, equal phase in tide exists, and co-range lines are about concentric around the amphidromic point and join points of equal tide range, and that range increases as the distance to the point increases.

The great amplitudes of the Bay of Fundy, the Rance, the Severn, occur at the end of an inlet resonant to a tidal period.

TIDAL CURRENTS

Horizontal movement of the water is "current." Tidal current is the periodic horizontal flow of water accompanying the rise and fall of the tide, and results from the same cause. The current experienced at any time is usually a combination of tidal and nontidal currents.

Offshore, where the direction of flow is not restricted by any barriers, the tidal current is rotary; that is, it flows in continuously changing direction through all points of the compass during the tidal period. The tendency for the rotation in direction has its origin in the deflecting force of the earth's rotation, and unless modified by local conditions, the change is clockwise in the northern hemisphere and counterclockwise in the southern hemisphere. The speed usually varies throughout the tidal cycle, passing through two maxima in approximately opposite directions, and two maxima about halfway between the two maxima in time and direction.

In rivers or straits, or where the direction of flow is more or less restricted to certain channels, the tidal current is reversing; that is, it flows alternately in approximately opposite directions with an instant or short period of little or no current (called slack water) at each reversal of the current. During the flow in each direction, the speed varies from zero at the time of slack water to a maximum (called strength of flood or ebb) about midway between the slacks. The movement toward shore or upstream is the flood; the movement away from shore or downstream is the ebb.

Clearly, during a tide the speed vector makes a full 360° turn and speed varies. Tidal current and tide itself are thus two aspects of one and the same phenomenon and both can be studied by harmonic analysis. This leads to the following conclusions: 1) the maximum speed of the current is reached at high tide and at low tide, but the directions are opposite; 2) at mean tide, away from the coast, when the current inversion occurs, speed is equal to zero; and 3) we observe a flow current from the start of the mean tide rises to the mean tide retreat, and an ebb current from the retreating (falling) tide's mid-point to the midpoint of the rising tide.

Treating tides as waves, a progressive tidal wave will have a shallow wave horizontal orbital velocity expressed by

$$U_s = \frac{A\sigma}{kh} \cos(kx + \sigma t) \qquad [3\text{-}34]$$

where

A is the amplitude of the wave
σ is the angular velocity of a particle undergoing circular motion as the wave passes by
k is the wave number
h is the depth of the water in which the wave is traveling
x is the distance from a point of origin
t is the time from a particular instant.

From the viewpoint of harnessing tidal energy, the principal consideration is that tidal currents are appreciable in relatively shallow water near continents; elsewhere, they are rather weak. For instance, off San Francisco, the maximum speed is 1 knot (1.85 km/hr). However, when particular geometry (the shape of embayment, river, or estuary) comes into play, water flow is impeded by bottom and sides and speeds of 9–19 km/hr may be encountered. Occasionally, such a rate of flow may vary according to depth; while incoming or flood tide usually has a uniform speed, the ebb or oceanbound flow may be faster near the surface than in depth. This latter situation develops when the two currents (namely, the river flow of fresh water and the returning salt water flow) are about equal and are layered, fresh water on top. Friction against the bottom reduces speed; this effect may be mathematically expressed by $v = az^m$, in which v is the speed, z the distance to the bottom, and a and m constants. The tidal bore, an abrupt solitary wave, is a very rapid rise of the tide and the advancing water presents suddenly a front of considerable height. The tide wave, not to be confused with tidal wave or tsunami, is the ridge of water raised by tidal action and resulting in tides around the world.

When estuaries are shallow, funnel shaped, and tidal amplitude is great, the high water is pushed inside the inlet faster than the low water because of the greater depth at flood time, resulting in a flood movement of non-uniform speed. The upper mass of water may overtake the lower mass of water and the abrupt front is created; the upper water crest eventually falls forward as the flood tide continues its advance. Several regional names designate bores; most common are egre, eager, pororoca, and mascaret. Among the most impressive bores are those of the Severn River (Wales), a tidal power station selected site, the Trent (England), the Amazon (Brazil), the Chien-Tank River (North China), the Messina Straits (Sicily), the Seine, Orne, and Gironde (France), the Hugli River (India), the Petitcodiac (New Brunswick), and the Bay of Fundy (Canada)—also selected as a potential tidal power station site. Bores can exceed three meters in height and progress at speeds from 18 to 28 km/hr. The Amazon bore may reach 7.5 m. They are caused by friction action in long narrow estuaries tending to reduce tidal effect. When instead the estuary is wide open and edge shaped, the tidal amplitude tends to rise to the head area; we will encounter such circumstances when discussing the Cook Inlet (Alaska) tidal power project (Fig. 3.11).

A bore is thus a very rapid rise of the tide in which the advancing water presents an abrupt front of considerable height. When estuaries are shallow and the tidal range is large, the high water is propagated upstream faster than the low water because of the greater depth in the channel at high water. If the high water overtakes the low water, an abrupt front is presented with the high-water crest finally falling forward as the tide continues to advance. The French call the bore "le mascaret." (Fig. 3.13)

The Coriolis effect produces, in the northern hemisphere, the counterclockwise rotation, and, in the southern hemisphere, the clockwise rotation, of the tidal crest in an estuary or bay, thereby creating—when such an inlet is broad—wave-like tides and rotating tidal currents advancing progressively around the edges. Le Floch (1958) holds that in rotating currents there is superimposition of "true" tidal currents and "outpour" currents. The latter are due to the emptying and filling of the

Fig. 3.11. The tidal bore at Moncton, New Brunswick. *(Source: New Brunswick Department of Tourism)*

coastal area and are perpendicular to the coast. They can be quite important along an indented coast and in a narrow-inlet gulf flow at speeds of 14 km/hr (Fig. 3.14).

In estuaries, changes of direction occur slightly after the low tide slack and after the high tide slack. Ebb current refers to the current which starts after the low tide and the flood current begins after high tide. Currents can be studied using harmonic analysis and for rotary currents (rotating currents) the components can be analyzed in a two right-angle axes system. A "currents rose" can be drawn using speed vectors with a common point of origin, and a curb drawn connecting their terminal points.

The ebb current is the movement of the water away from shore and stream downwards; the flood current is the movement towards the shore and upstream. The apogean tidal currents are currents of decreased speed occurring when the moon is near apogee. When the moon is near

THE TIDES 103

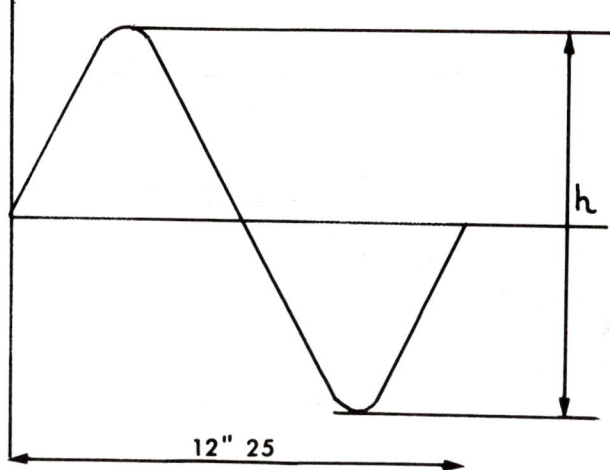

Fig. 3.12A. Tidal curve. The curve of a tide is approximately sinusoidal.

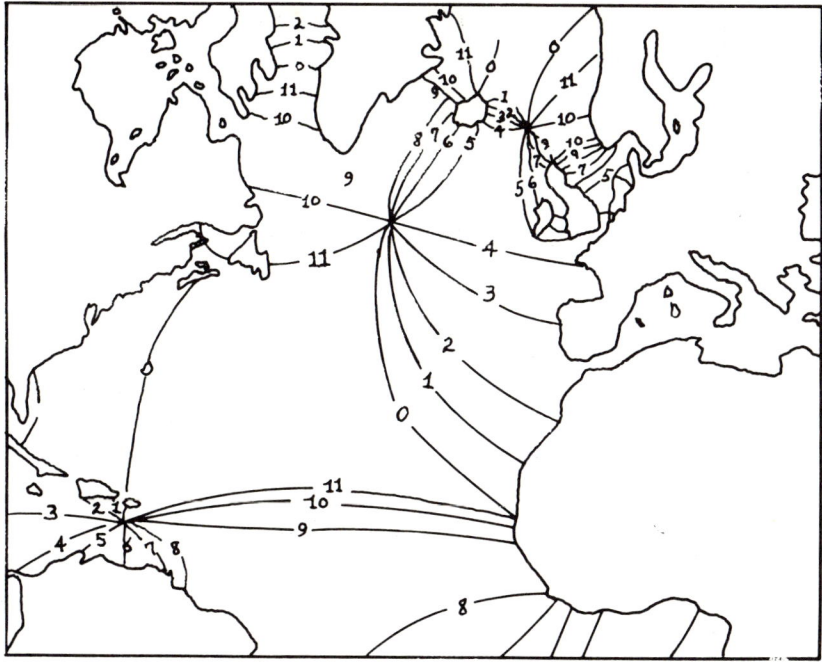

Fig. 3.12B. Amphidromic points. Graph for such points in the North Atlantic Ocean. The numbers indicate the cotidal lines.

104   TIDAL ENERGY

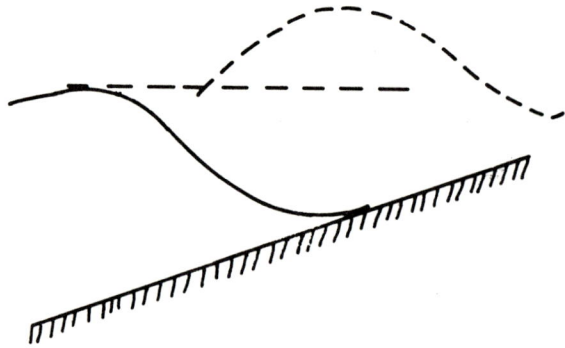

Fig. 3.13. Tides in estuaries. The wave front becomes steeper as the wave crest travels faster than the trough (with height at crest larger than height at trough).

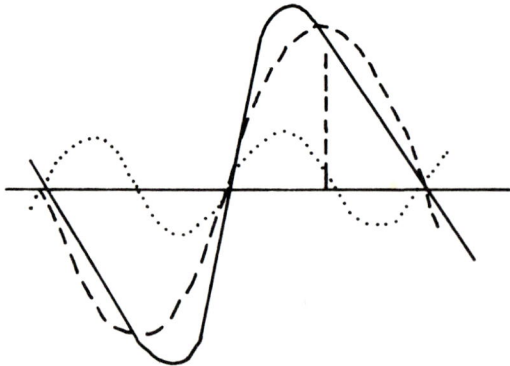

Fig. 3.14. Tides in estuaries. There is superimposition upon the tide wave of a wave whose period is only half that of the tide wave (harmonic analysis).

perigee, the currents flow faster; that is when the moon is nearest to the earth. Tidal currents are thus alternating or rotating (gyratory); in an alternating current, which is of interest for tidal power harnessing, the speed vector, at the point of observation, is at all times situated on a same straight line, and current carries alternately in one direction, then in the opposite one. A "current inversion" occurs when speed drops to naught and the current changes direction; this inversion takes place close to coasts at the time of high and low tide, but farther away from coasts it occurs at mid-tide or better mean tide, the halfway mark between high and low tide. Tidal currents may be of the semi-diurnal,

diurnal, or mixed type, corresponding largely to the type of tide at the place, but often with a stronger semi-diurnal tendency. At most places, however, the type is mixed to a greater or lesser degree.

TIDAL ENERGY

Let us first consider the case of a broad basin such as the North Sea. The semi-diurnal tide enters the North Sea from the Atlantic Ocean as a progressive wave. In the Netherlands and in Belgium, there are tidal amplitudes of between 3 and 4 m, but along the Norwegian coast they often are less than 25 cm. Along those western coasts, ranges do not exceed 5 m.

The water mass is deflected due to the Coriolis effect to the right upon entering the North Sea and pushes the mass of water against the Norwegian and Danish coasts. But amplitudes are much greater along the Scottish and English littorals. Why? Moving in from the Atlantic Ocean, the progressive wave has more energy but loses some of it steadily as depth is reduced; the reflected wave has substantially less energy than the inbound one coming from the north. In addition, because the Coriolis force tilts the wave as it moves through the basin, this Kelvin wave develops in the broad channel.

In the case of the Bay of Fundy, we have an inlet with a very wide opening into the Atlantic Ocean that splits into two narrow basins, Minas Basin and Chignecto Bay. Toward its northern end, waters are shallow, the inlet narrows, and resonance occurs, since the tidal period and that of free oscillation in the bay are about the same. The M_2 component has a quarter-wavelength of 345 km and basin length has a median value of 312 km, close enough to explain an amplitude amplification of 3.

Amplitudes increase as we progress up the inlet and they are larger on the south shore because a progressive component is added to the standing wave. The range reaches 17 m at perigee.

As we have seen, tidal amplification in some basins results in increases of range and thus of head usable for power generation. There are hundreds of sites where such conditions exist, though a limited number is seriously considered due to economic and physical considerations.

The energy that is dissipated through friction can be harnessed before this occurs. According to Wicks (1980), the worldwide tidal energy flux is the equivalent of 1.5×10^6 MW and represents an equivalent water head of 3.12 m; Lawton (1974) believes that much of the 13,000 MW of the world's tidal energy resources can be exploited where estuaries and embayments have a sufficient head or tidal range. Saunders (1975) estimates that world tidal energy capable of utilization probably totals 350 TW-hr.

To visualize the practicability of harnessing the energy of the tidal phenomena, we must keep in mind that the incremental amount of energy per cycle obtainable from tidal flow of water is given by

$$dE = gR(DSdR) \qquad [3\text{-}35]$$

with R the tidal range, D the water density, g the acceleration due to gravity, and S the area of the enclosed basin. Assuming the area independent of the range, integration of Eq. 3-35 gives

$$E = DgS \cdot \frac{R^2}{2}. \qquad [3\text{-}36a]$$

With double effect, power being generated at ebb and flow, the maximum energy is

$$E_{max} = DgR^2S. \qquad [3\text{-}36b]$$

The total annual (approximately 700 cycles) energy in kW-hr that can be generated in a single basin has been estimated as shown in Eq. 3-37, with a double effect unit.

$$E\frac{\text{kW-hr}}{\text{yr}} = 0.017 \, R^2S \qquad [3\text{-}37]$$

Thus, available energy varies with the surface area of the tidal basin enclosed (S) and with the square of the tidal range (R^2). Consequently, sites with tidal ranges of 7.60 m and 78 km² of basin, 10 billion kW-hr/yr could be generated.

With a tidal cycle of a semi-diurnal tide being 12.4 hr, the average power generated would be

$$\overline{P} = (\tfrac{1}{2})6 \times 2D \times g \times SR^2. \qquad [3.38]$$

However, only about 30% can in fact be retrieved, and Bernshtein (1965) calculated the annual energy output in kW-hr considering a basin of 1 km² at

$$E_a = 5 \times g \times 10^5 \, SR^2, \qquad [3\text{-}39]$$

the installed capacity in kW capable of handling the largest tides at $N_i = 311 \, AR^2$ and the maximum discharge through the barrage (in m³/sec) at $Q_{max} = 57 \, AR$.

Gibrat (1953), Lewis (1963), and Wickert (1956) provided a power formula applicable to emptying, filling, and pumping in a tidal power scheme. Using the following symbols, the equation takes the form

$$E_e = \int_{z=0}^{H} \gamma f(z) z \, dz, \qquad [3\text{-}40]$$

where

$f(z)$ = area of the basin as a function of water depth
γ = unit weight of seawater
V_b = volume of basin (capacity)
V_p = volume pumped
E_c = work necessary for a complete tidal cycle
E_e = work necessary to empty the basin
E_f = work necessary to fill the basin
E_p = work necessary to pump
$+A$ = normal high water
$-A$ = normal low water
$+C$ = a point above normal high water
H = tide height

Velocities of tidal flow, no matter how high the tide, are seldom high, hence utilization of the tidal kinetic energy is not profitable. Therefore, only the potential energy will be considered. In the theoretical situation, sea and basin are separated by a dam.

With

$$E_e = \int_{z=0}^{H} \gamma f(z) z \, dz, \qquad [3\text{-}40]$$

to fill the basin it will require

$$E_f = \int_{z=0}^{H} \gamma f(z)(H-z)dz \qquad [3\text{-}41]$$

and, for a complete tidal cycle,

$$E_c = E_e + E_f \qquad [3\text{-}42]$$

$$E_c = \int_{z=0}^{H} \gamma f(z) z \, dz + \int_{z=0}^{H} \gamma f(z)(H-z) dz$$

$$E_c = \gamma H \int_{z=0}^{H} f(z) dz. \qquad [3\text{-}43]$$

The volume of the basin is

$$V_b = \int_{z=0}^{H} f(z) dz \qquad [3\text{-}44]$$

and hence Eq. 3-42 can be written by introducing Eq. 3-44.

$$E_c = \gamma H V_b \qquad [3\text{-}45]$$

The total volume changeover is

$$0 \to V_b \text{ to } 0 \to 2V_b$$

and thus the average operating head is $\dfrac{H}{2}$.

To lower the level to level B, by pumping, a certain amount of energy is needed; it is given by

$$E_{p_1} = \int_{z=0}^{-B} \gamma f(z) z \, dz. \qquad [3\text{-}46]$$

The energy provided by the flood tide is

$$E_{f_1} = \int_{z=-B}^{H} \gamma f(z)(H-z) dz. \qquad [3\text{-}47]$$

If we increase the level in the retaining basin to the level $+C$, we need energy to pump equal to

$$E_{p_2} = \int_{z=H}^{+C} \gamma f(z)(z - H) dz. \qquad [3\text{-}48]$$

The energy provided by the ebb tide is

$$E_{e_1} = \int_{z=0}^{+C} -\gamma f(z)(z - H) dz. \qquad [3\text{-}49]$$

And, consequently, the net energy production for a complete tidal cycle is

$$E_c = E_{f_1} + E_{e_1} - (E_{p_1} + E_{p_2}) \qquad [3\text{-}50]$$

$$E_c = H \int_{z=-B}^{C} f(z) dz. \qquad [3\text{-}51]$$

Introducing in Eq. 3.51 the value of V_b given by Eq. 3-44,

$$E_c = \gamma H (V_b + V_{p_1} + V_{p_2}). \qquad [3\text{-}52]$$

By subtracting Eq. 3-35 from Eq. 3-52, we find the difference between operations involving and not involving pumping:

$$E_{gain} = E_{c_4} - E_{c_3} \qquad [3\text{-}53]$$
$$E_{gain} = \gamma H(V_b + V_{p_1} + V_{p_2}) - (V_b) \qquad [3\text{-}54]$$
$$E_{gain} = \gamma H(V_{p_1} + V_{p_2}) \qquad [3\text{-}55]$$

It should be noticed that Eqs. 3-50 and 3-53 are independent of basin shape and tide symmetry. Each basin, moreover, has a proper natural frequency according to size and proportions, which makes it possible to take advantage of resonance and thus increase the tidal range during one cycle. In a basin of length L and depth d, this frequency is

$$T = \frac{2L}{\sqrt{gd}}. \qquad [3\text{-}56]$$

A tide range increase can run as high as 10%.

Most schemes which are proposed involve the generation of a peaking supply of power, consistent with the nature of tides. Power would be delivered in large blocks one or two hours a day; it has been calculated, for instance, that two sites in the Bay of Fundy could provide a 5,000-MW block of power.

4
The Electricity Generation

The tidal power scheme involves a barrage with gates and sluices, a powerhouse, and one or more basins. (Fig. 4.1)

BARRAGE

We have used synonymously the terms barrage and dam. Barrage has been suggested as a more accurate term for tidal power schemes because it has only to withstand heads a fraction of the structure's height, and stability problems are far more modest. However, the literature does not always make the distinction, even though heads are small with tidal power cutoffs.

Tidal power barrages have to resist waves whose shock can be severe and where pressure changes sides continuously. Side slopes cannot be steep. Most future construction tends to be inspired by the Soviet Kislaya plant and intends to use *in situ* found earth or rock-fill cofferdamming, used by the French, which adds extremely high costs to the construction price. Closure of the barrage in fast tidal waters has been studied in connection with several plans, among others for the Severn River and the Kimberley coast, and engineers are confident that they can cope with the problem.

Since barrage length adds also to the price tag of the plant, short barrages are preferred even if basin size may have to be smaller as a result of site choice. Up to a height of 20 m, cost remains proportional to length as it is not changed by the need to build a dam wall to withstand high hydrostatic pressure. When the elevation exceeds the 20-m limit, costs increase faster with length. Most tidal power plants do not have heads exceeding 20 m.

The barrage needs to provide channels for the turbines in prestressed or reinforced concrete. To build these channels, a temporary cofferdam

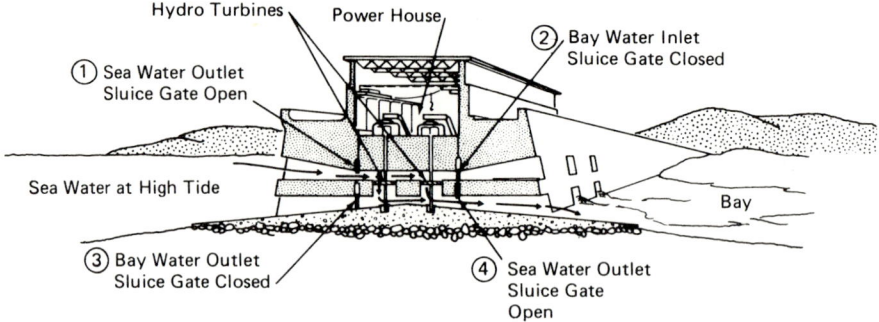

Fig. 4.1. How a tidal power plant works.

is necessary, but it is now possible to build them on land, float them to the site, and sink them into place.

Flatness is required for the sea bottom; sandy bottom usually necessitates piling. There, where sand or rock can bear the weight of the structure to be implanted, the bottom can be prepared, the structure placed on it and then anchored. Prefabricated concrete blocks can be used as the core for large barrages and voids filled with rocks or concrete, remaining holes with sand, and the entire construction then asphalted.

Construction of a barrage usually will influence the tidal amplitude (see *Environmental Impact*). Indeed, such a construction modifies the effective length of the embayment or basin and its shape as well, particularly if the scheme involves supplementary spur dams, or brings about relocation or disappearance of natural obstructions as is foreseen for the Severn plant and has occurred in the Rance estuary.

The constructions influence the resonance of the bay, and most bays are less than the resonant length of the tidal wave. If resonance is reduced, the range will decrease; if measures are taken to augment the resonance, tidal amplitude may be increased.

LOCKS AND GATES

Tidal power basins have to be filled and emptied. Gates are opened regularly and frequently, but heads vary in height and on the side where they occur, which is not the case with conventional river projects. The gates must be opened and closed rapidly and this operation should use a minimum of power. Leakage, all authors agree, is tolerable for gates and barrage. Ice problems have been solved in the Soviet plant and apparently are no longer an impediment to build plants in inclement areas. Since we are dealing with seawater, corrosion problems are acute, but as shown for the Rance River and for the Soviet scheme, they have been very successfully solved by cathodic protection, and where not possible, by paint. (Fig. 4.2)

Gate structures can be floated as modular units into place. However, in the Passamaquoddy project, for instance, weight of structure would cause compaction of clay-sand bottom, and settlement would be substantial; this factor will have to be taken into consideration if a plant is built.

Though, in existing plants, vertical lift gates have been used, technology is about ready to substitute a series of flaps that operate by water pressure. Flap gates are gates that are positioned so as to allow water into the holding basin and require no mechanical means of operation.

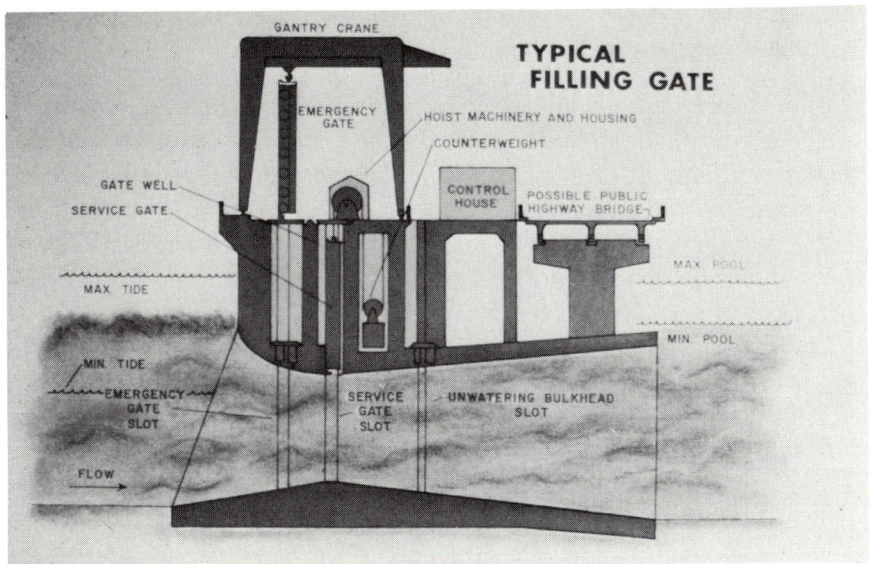

Fig. 4.2. Typical filling gate, Passamaquoddy project. *(Source: U.S. Corps of Engineers)*

If used, they are positioned, in the case of modular construction, in the caissons. A caisson is then floated into place. They have to be built so that they will be adequate for the maximum tidal amplitude. Top hung on a gate-hoisting beam, a gate would transfer its hydraulic load to the concrete structure. If operation is to be rapid and efficient, gates must open under the minimum differential head; this favors as flat as possible the tidal basin face of the sluice.

The flap gates allow flow only in the direction of sea to basin. Hence, the basin level rises well above sea level as ebb flow area is far less than flood flow area.

POWERHOUSE

Because small heads only are available, large size turbines are needed; hence, the powerhouse is also a large structure. Both the French and Soviet operating plants use the bulb type of turbine. Of the propeller type, with reversible blades, bulbs have horizontal shafts coupled to a single generator. The cost per installed kilowatt drops with turbine size, and perhaps larger turbines might be installed in a future major tidal power plant. (Fig. 4.3)

The Bulb Group (Rance Example)

A bulb type turbine is an axial flow turbine. The bulb set, resembling in appearance a small submarine, is made up of an ogive shaped steel shell containing an alternator and a Kaplan turbine. It is placed in a horizontal hydraulic duct and entirely surrounded by water; a shaft provides communication with the engine room of the power plant. The set functions as a turbine and as a pump, and regulates the flow in both directions of flow, tide to reservoir, and reservoir to tide. (Fig. 4.4)

The alternator, 5 m in diameter, is directly coupled to the turbine and turns at a pressure of 2 kg/cm^2 absolute housed in air; its power is 10,000 W for cos 1. The turbine is a 5.35-m-diameter Kaplan wheel with four mobile blades and guide vanes; it has a 10,000-W power and a speed of 94 rotations per minute. Starter drive speed is 380 rotations per minute. (Fig. 4.5)

The group, functioning as a turbine, in the direct sense basin to sea, absorbs a discharge of 110 million cm^3/sec and 200 million cm^3/sec for drops of 11 and 3 m, respectively, furnishing power of 10 and 3.2 MW.

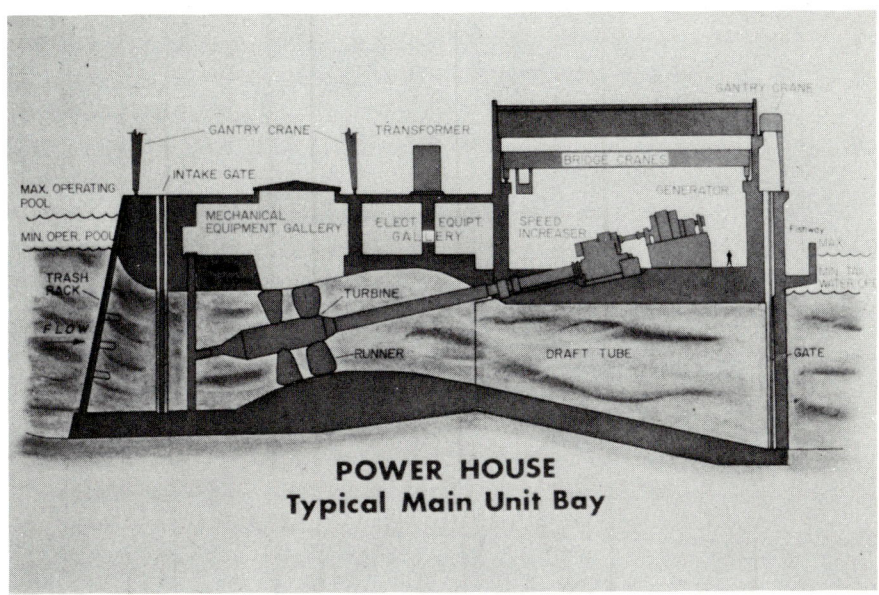

Fig. 4.3. Powerhouse, Passamaquoddy project. Typical main unit bay. *(Source: U.S. Corps of Engineers)*

In reverse, thus the sea to basin direction, similar drops furnish 10 and 2 MW and discharges of 130 and 135 million cm^3/sec are absorbed. Functioning as a pump in a direct sense (thus, sea to basin), the maximum pumped discharge is respectively 225 and 170 million cm^3/sec for drops of 1 and 3 m. (Fig. 4.6)

The largest bulb units in the world are installed at the Rock Island Hydrostation on the Columbia River (Washington). They have a 7.6-m diameter and can produce 51,300 kW each at a head of 12 m.

Rim-Type Turbines

Different types of turbines are under study; usually mentioned are inclined shaft turbines, rim type turbines, or straight flow turbines, where the generator is attached peripherally on the turbine blades, an arrangement that couples two turbines of conventional type to one generator, and a hydraulic system in which up to six turbines are coupled

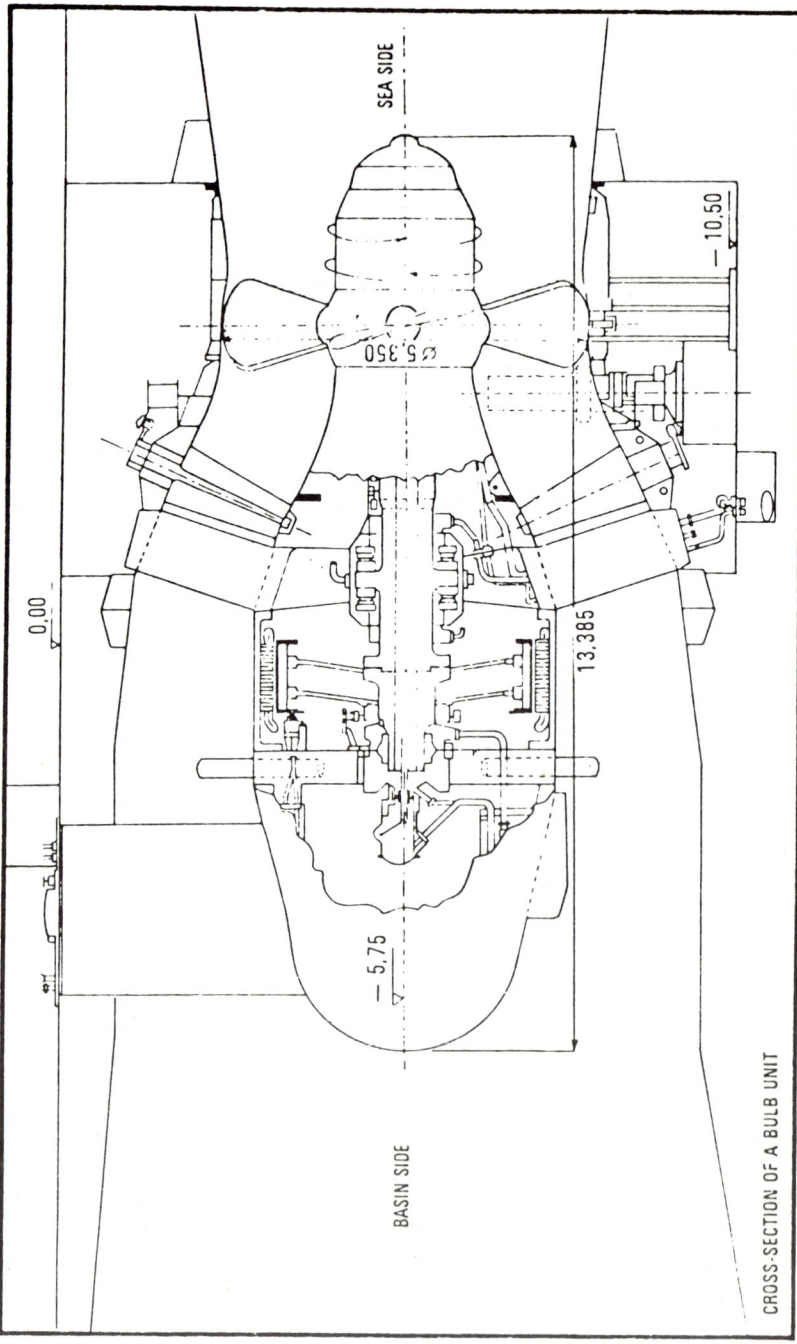

Fig. 4.4. Cross-section of a bulb unit. (Lebarbier, *Nav. Eng. J.*, April 1975)

THE ELECTRICITY GENERATION 117

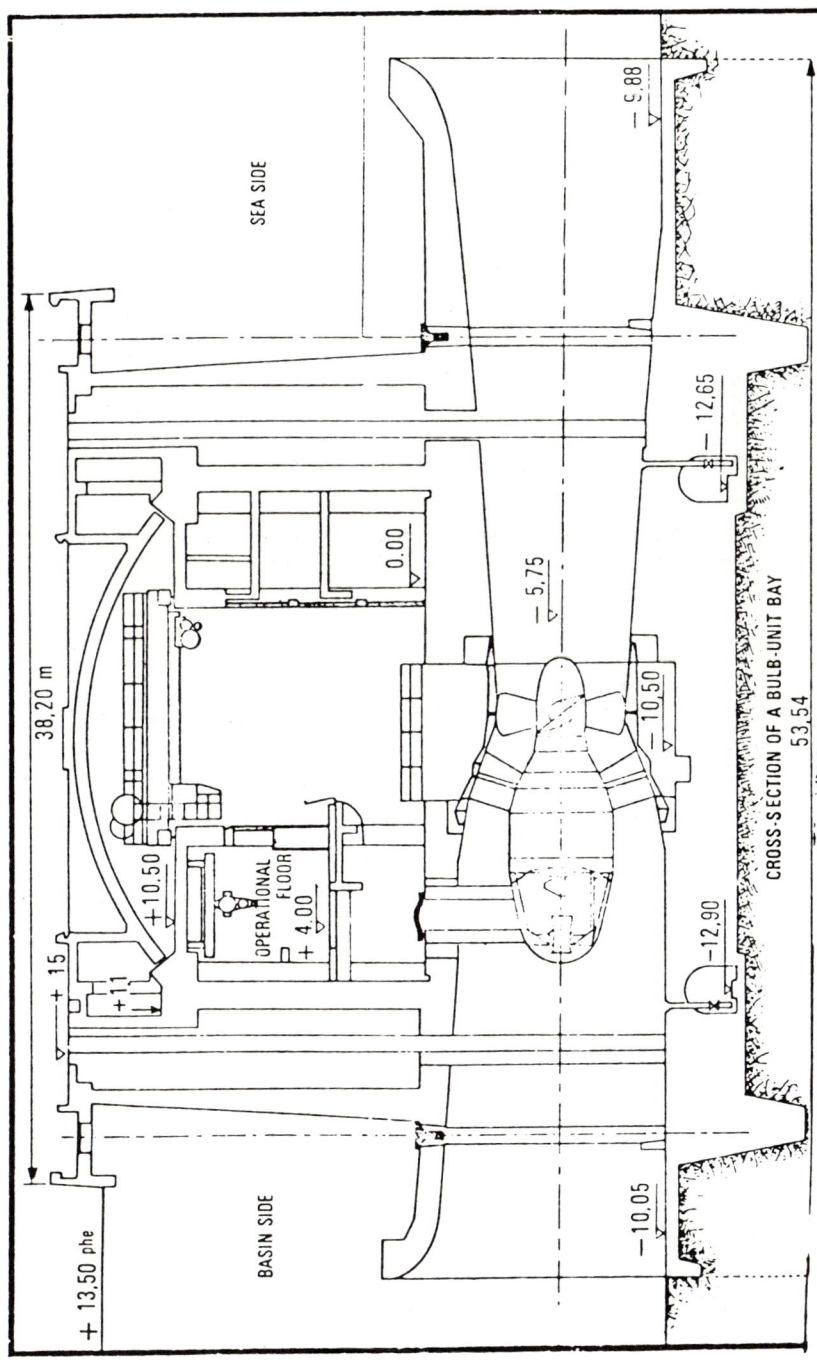

Fig. 4.5. Cross-section of the Rance River plant power station. (Lebarbier, *Nav. Eng. J.*, April 1975)

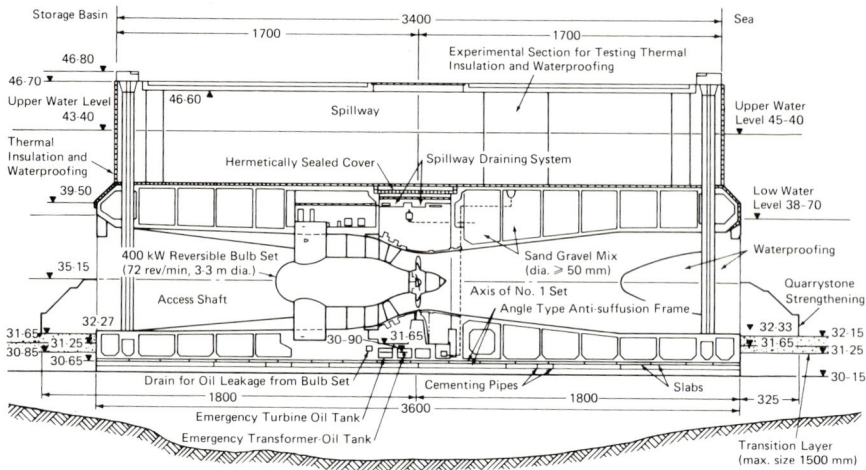

Fig. 4.6. Cross-section through the prefabricated-caisson powerhouse and reversible bulb set of the Kislaya scheme. (Bernstein, *Water Power*, **26**, 5, 172–177, 1974)

to hydrostatic pumps used to drive a Pelton wheel, which, in turn, drives a high-speed generator. (Fig. 4.7)

Constans (1978), discussing the rim type of turbine, in which the rotor surrounds the turbine runner as a rim carried by the runner blades, pointed out that "the main problem with such a design is the seals between the stationary parts and the rotating rim. It is questionable whether it will be possible to make variable pitch runners with this design, in which case the productivity will be lower than for a corresponding bulb turbine with variable pitch."

Engineers who favor straight flow generators against the bulb turbogenerators point out the lower inertia characteristic of the bulb type, claiming this could lead to problems during power system disturbances. The designers of a new type of straight flow units put forth savings in civil works and in generator and auxiliary electrical equipment because larger unit capacities than with the bulb-type unit, for the same head, would be possible. Such a 7.6-m-diameter unit is currently being tested at the mouth of the Annapolis River (Nova Scotia); if conclusive, insertion of such turbines would perhaps cut capital costs by up to 10%.

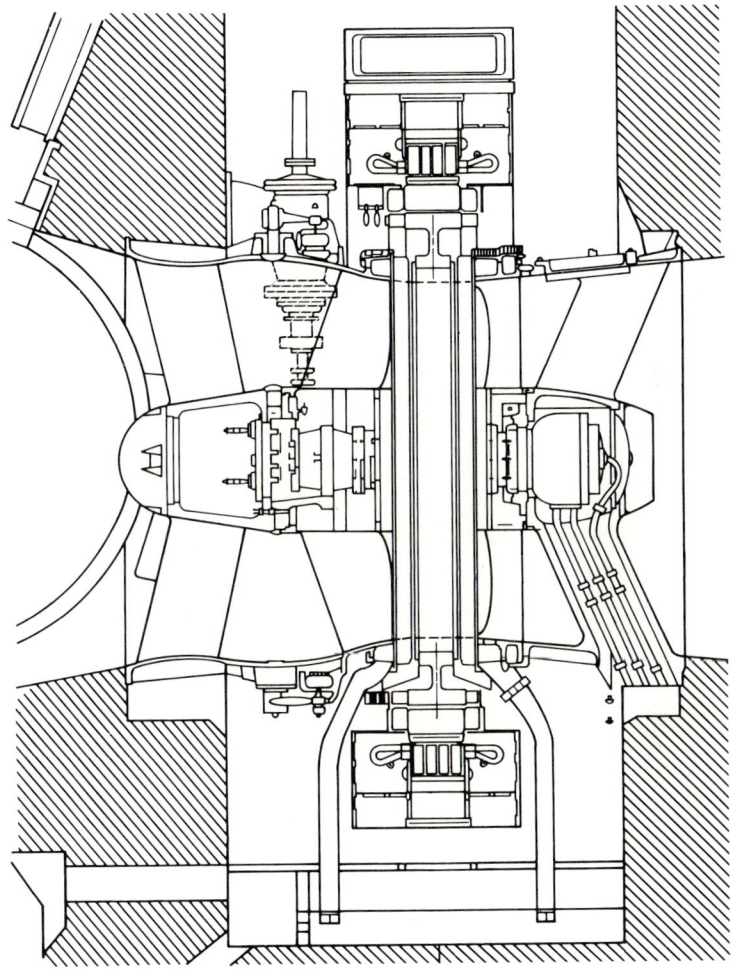

Fig. 4.7. Cross-section of a typical rim-type generator. (Miller, "Die Straflo Turbine, die technische Realisation von Harza's Idĕĕn." Zurich: Straflo Group, 1975)

BASINS

Basin, pool, and reservoir are all synonymous terms. In a basin, the water is retained. Basins may be coastal indentations, embayments, bodies of water between islands and continents, or estuaries. A tidal power scheme can be a single or a multiple basin project. The single basin is the simplest arrangement and currently is the most talked about for any possible plant, as it is the most economical.

Single Basin Schemes

Several alternatives exist: generation may take place in one direction only, or in both, and pumped storage may be included in the operation. (Fig. 4-8)

One-Way Operation. The basins fill at flood tide and power is generated at ebb flow. Turbines and spillways are in the barrage separating pool and sea. Only fixed-blade turbines are necessary. However, up to 65% of the energy needs retiming. The head is created when the basin is filled and the tide is low. Power is intermittent since turbines are started only when the head has been created. The supply is mostly off-peak. (Figs. 4.8, 4.10, 4.11)

Ebb and Flood Operation. (Figs. 4.9, 4.11). If reversible-blade turbines are installed, then power can be generated as the basin fills at flood tide and when it empties at ebb tide. The head is on the seaward

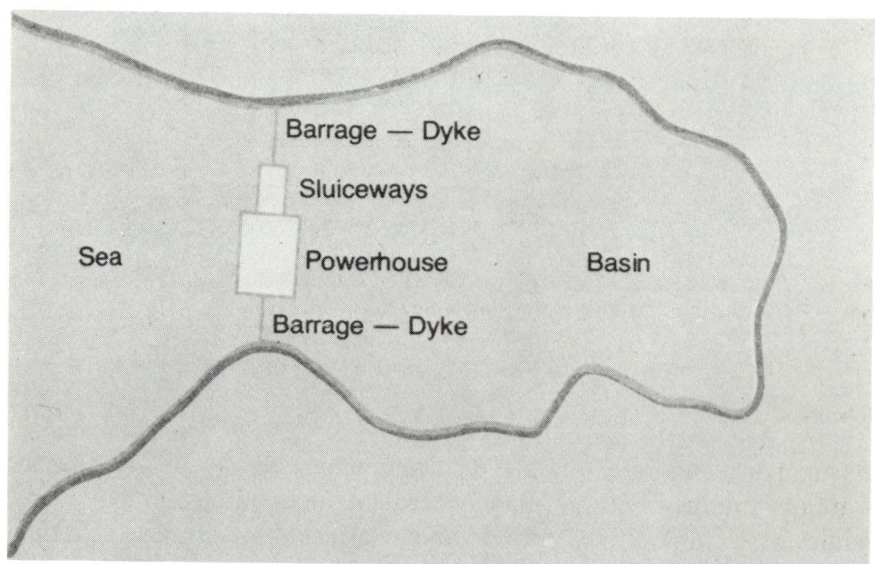

Fig. 4.8. Single basin scheme.

THE ELECTRICITY GENERATION 121

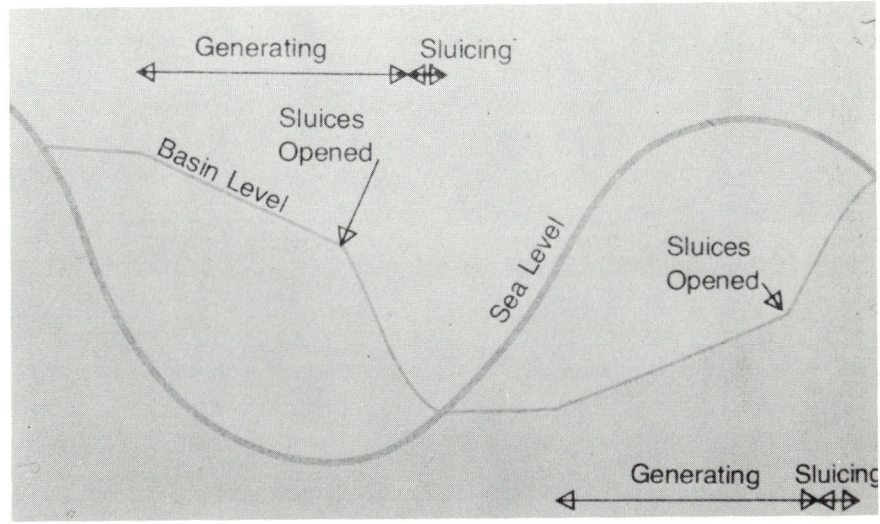

Fig. 4.9. Single basin, ebb and flood operation.

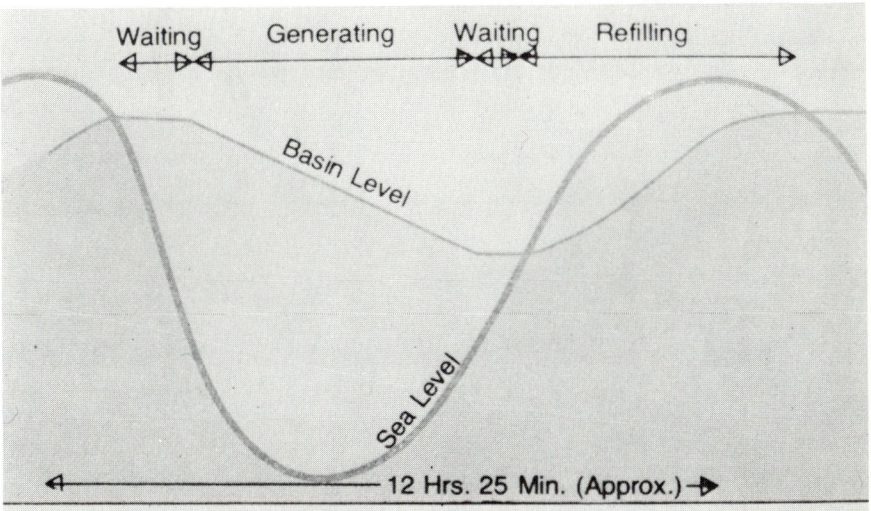

Fig. 4.10. Single basin single effect operation.

122 TIDAL ENERGY

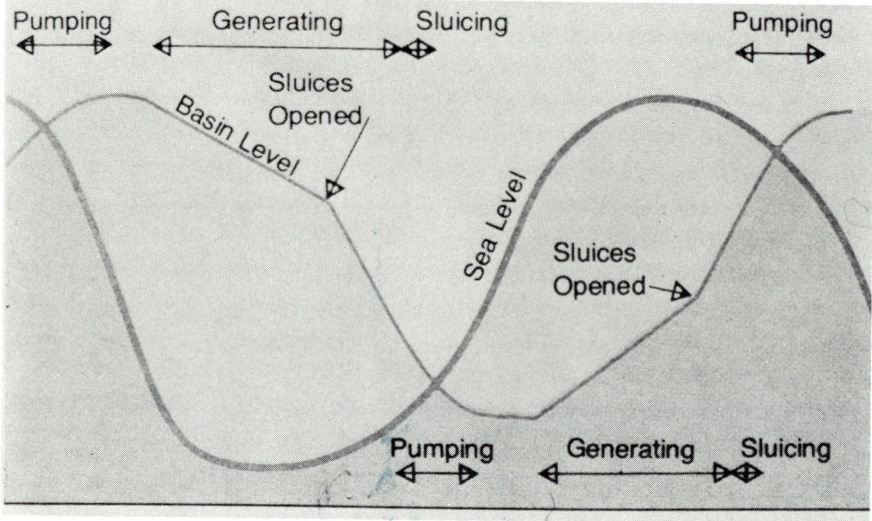

Fig. 4.11A. Single basin double effect without and with pumping.

S = Sluicing T = Turbining

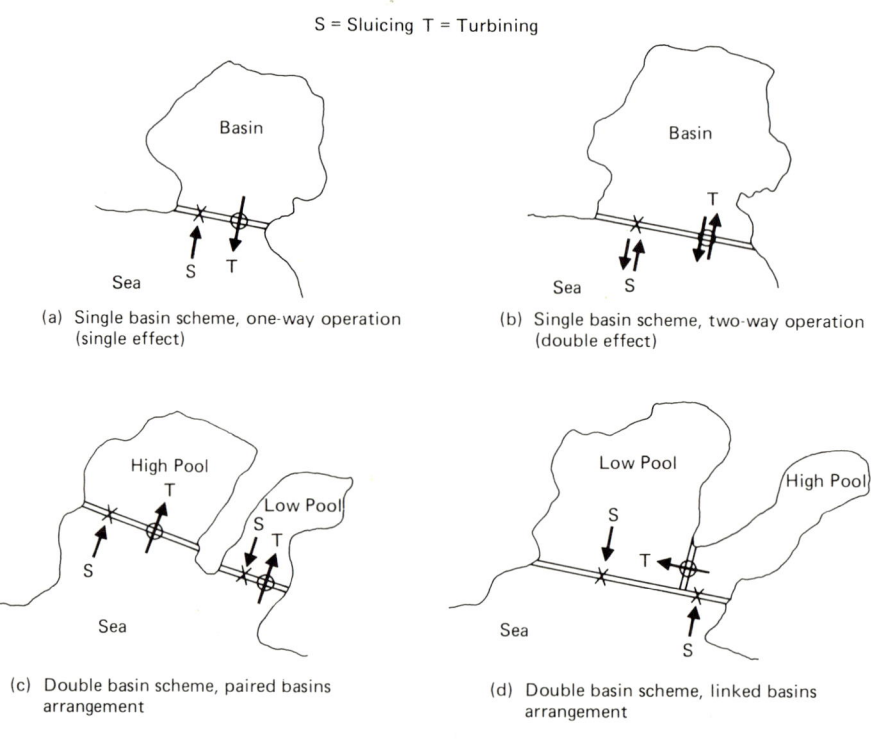

(a) Single basin scheme, one-way operation (single effect)

(b) Single basin scheme, two-way operation (double effect)

(c) Double basin scheme, paired basins arrangement

(d) Double basin scheme, linked basins arrangement

Fig. 4.11B. Various basin schemes.

side as the high tide rolls in, on the basin side as low tide rolls out. Power fluctuates, as is the case with the one-way mode of generation but only 50% of the energy requires retiming. (Fig. 4.11)

Ebb and Flow Operation with Pump Turbines. In this system, generation is similar to that in the preceding scheme but, in addition, the turbines may be used as pumps to increase or lower the reservoir level. Obviously, some energy is lost to activate the pumps. Power can be produced continuously and on demand independently of the tide state.

High-Head Pumped Storage. It is possible to store energy without a two-way operation. The pumped storage scheme requires only simple turbines. Continuous, not tide-connected, maximum efficiency production is achieved. However, some authors consider this a highly expensive scheme and doubt that the additional capital investment is worth the extra power. (Fig. 4.12A)

Multiple Basin Schemes

Most projects envision only the use of two basins. Again, several alternatives are available. (Fig. 4.11B)

Reverting to studies published as many as 30 years ago and schemes then proposed, one can differentiate between multiple-basin schemes that are paired or linked. In the linked basin approach, the powerhouse is set in the barrage between the high and the low pool and it can operate continuously. The flow is controlled from sea to both basins. Continuous production is less important when the tidal power scheme is part of a system fed by different types of power sources; production is lower than that of one of the basins used on its own. In addition to the powerhouse dam, dykes and basin-level control gates are needed. Rather expensive mechanical interfacing is required.

Paired Basins (Fig. 4-11B.c). When opting for the paired basin system, the adjacent tidal basins are not mechanically linked, but an electric link is created through coordination of production between a high and low pool: the high pool would generate at ebb tide, the low pool on flood tide.

With the paired basin system version, supplementary tidal basins can be added. Energy production is additive. Two or more basins with about the same installed capacity, when coordinated, can provide at lesser unit cost, greater dependable peak capability, than the sum of such capabilities from those basins independently operated.

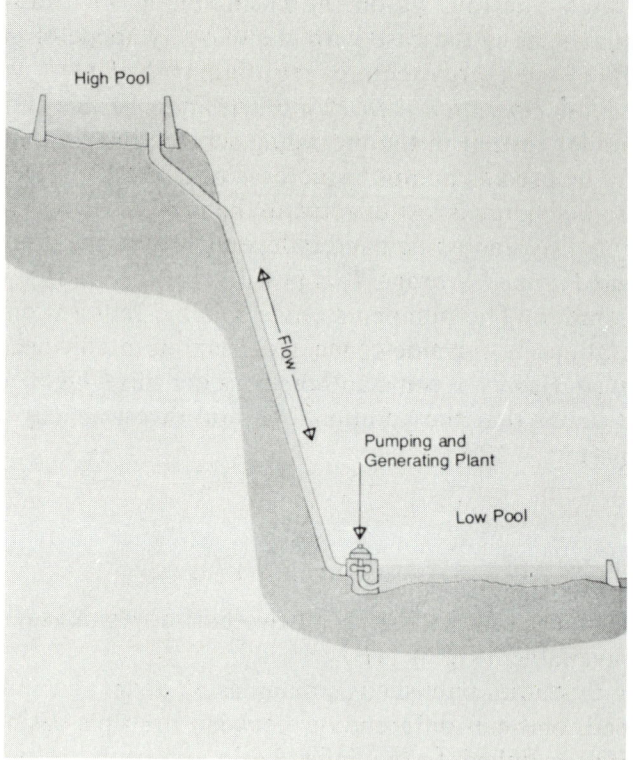

Fig. 4.12A. Pumped storage.

Linked Basin Schemes (Fig. 4-11B.d)

Double Pool. Simple turbines will suffice. Continuous power production is possible by placing the powerhouse between the two basins in the barrage separating them. One basin is filled at flood flow and the other is emptied at ebb flow. There is an upper basin, with its spillway, and a lower basin, also with its own spillway. Water flows from the upper to the lower pool. Turbines can thus work continuously, but head and output vary. Half the available energy is sacrificed and two sets of sluices are required.

Double Pool with Pumping in Two Sea-Pool Barrages. This scheme permits pumping high pool up and low pool down using the off-peak available power. An energy gain can be achieved when sea-pool head is smaller than pool-pool head.

Double Pool with Pumping in Pool to Pool Barrage, Separate Pumps. The pumps are a separate system. The off-peak period power can be absorbed. In the double pool with pumping in a pool to pool barrage with a pump-turbine system, the operation is similar to that in the preceding one, but the turbines are used as pumps as in the single pool two-way with pumped storage scheme described above. Off-peak power is absorbed.

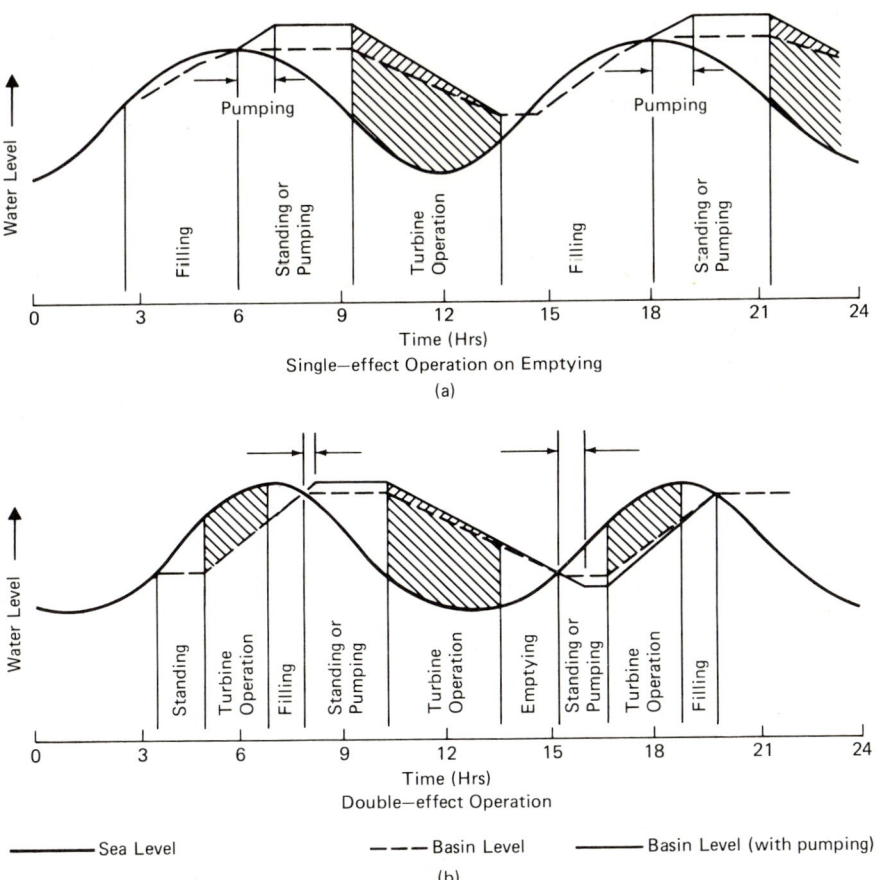

Fig. 4.13. Water level vs. time under different operations.

126 TIDAL ENERGY

Multiple Basins and Tide-Boosted Pumped Storage. The difference with the preceding systems is that the basins may be of different sizes and be in different sites. Actually, the tidal energy is not the "heart" of the scheme; rather, it is used to increase the output of another plant with pumped storage. This system can be built with several variations. (Fig. 4.13)

Hydrostatic Pump and Pelton Wheel

Instead of being used to provide electricity, tides can be used to drive a hydrostatic pump. In this case, a simple propeller turbine is linked to a hydrostatic pump, and tidal energy is converted to a flow of high-pressure oil, which, in turn, drives a Pelton wheel coupled to an alternator. No electrical equipment is located in the marine environment, but the scheme is less efficient.

The Plastic Barrier

A rather recent development at the U.S. Department of Energy is its examination of a small tidal range system; a major interest of the system is that it would make tidal power harnessable in many geographical sites heretofore excluded as unsuitable. Another advantage of the new proposal developed by Alexander Gorlov (1979), of Northeastern University, is that it involves lightweight construction rather than rigid dams, using a thin plastic barrier hermetically anchored to bottom and bay sides. A cable spanning the bay would support the barrier and hold it above the water level; the cable would be attached to several floats. The submerged part of the barrier subjected to a water pressure that is equal to the difference in water levels across the dam of about 0.2 atmosphere.

Tidal energy would be converted into power through use of compressed air. A large piston of an air motor is driven by the energy derived from the flow of water from a higher to a lower elevation; two chambers are used. Electricity can be generated directly or compressed air can be stored for later release and power generation. By heating, the air output could be increased. The use of a gas turbine for conversion will be investigated.

Construction costs are considerably reduced with such a scheme. Dams, whether built on site or floated in as caissons, are a major expense and so is the powerhouse, and, in addition, no special sluice gates are necessary.

No site has been selected to construct a pilot plant. Boston harbor had been considered, but currently, locations along the Maine coast are being assessed. Environmental impact is probably minimal, although there is a risk of creating a pool of stagnant water behind the barrier; however, the barrier could be removed and pulled to one side and navigation would be permitted, as well as "cleaning out" of the impounded water.

As shown in his not yet released report to the U.S. Department of Energy, Gorlov establishes that his hydropneumatic system can provide a high-velocity air jet even for very small water heads; the kinetic energy of the air jet is quite sufficient for a stable operation of industrial-size air turbines. The dimensions of the turbine blades are quite reasonable notwithstanding the low input air pressure, because the air plenum works in a closed chamber and inevitably transfers all the energy provided by the falling water to the turbine blades. (Table 4-1)

Precise cost figures cannot be provided without a detailed design, but a rough estimate can be given of the capital investment based upon costs of powerhouse, water dam, and air chamber. An economic analysis was made by Gorlov in his 1980 annual report to the U.S. Department of Energy and by the Anderson-Nichols & Co. report to Northeastern University (1980); both reports took the proposed Passamaquoddy Tribe Half Moon Cover Project, in Maine, as a prototype. Some of the calculations are shown in Table 4-2. The unit cost in U.S. $/kW is 880; comparing this amount to the average capital investment costs of various alternative methods of energy conversion, one finds, in $/kW, 760

Table 4-1. Air Turbine Characteristics. *Source: A. N. Gorlov, Northeastern University (Boston).*

TURBINE POWER (MW)	AIR PRESSURE (GAGE) (atm)	AIR MASS FLOW RATE (kg/sec)	BLADE LENGTH (m)	GAS VELOCITY (m/sec)	MINIMUM BLADE LENGTH/DIAMETER (ratio)
9.57	0.2	702.3	1.00	164.2	2.5
16.9	0.3	856.7	1.00	195.9	2.5
25	0.2	1,799	1.60	164.2	2.5
25	0.3	1,264	1.22	195.9	2.5
25	0.4	966.3	1.01	220.1	2.5
25	0.5	834.8	0.872	241.1	2.5

Table 4-2. Passamaquoddy Tribe Half Moon Cove Tidal Power Project. *Source:* A. M. Gorlov, Northeastern University (Boston).

	POWERHOUSE	BARRIERS*	AIR CHAMBER**	TOTAL
Construction cost (U.S. $ × 10^6)	1.45	0.78	6.2	8.43
Capacity (MW)				9.47
Unit cost (U.S. $/kW)				880

*The dam structure used for calculation is a flexible barrier.
**Second alternative as used in Anderson-Nichols & Co., Inc., *Report to Northeastern University (1980).*

for conventional oil thermal power plants, or about 14% less; but the cost for a coal thermal power plant is $/kW 900, thus about 2.4% higher than for the Gorlov tidal scheme. The differential is even higher for a nuclear power plant, which would show a cost of $/kW of 1,800, or more than twice the cost of the tidal power plant. The table illustrates two alternatives: with barriers and with chambers, the project thus seems economically viable. A more accurate comparison of conventional and tidal installations requires an in-depth feasibility study.

The number of natural hydropower sources that can be tapped for energy production can be increased spectacularly, since numerous sites are currently not exploitable in view of the ultra-low-heads prevailing there. In fact, not only tidal basins, but some rivers, brooks, and even irrigation canals with heads of less than 2 m could be used. Furthermore, development of low-pressure air turbine technology on an industrial scale would also strengthen the competitive position of hydropneumatic power plants.

STORAGE

Storage is necessary when alternative electricity production schemes are not, or cannot be, connected with the electric grid. Tidal power plants are not an exception to this rule; there is a strong case for associating them with storage to provide for the varying needs of the consumer, also taking into account the eventual presence close-by of other plants. The tidal power plant has, by its very nature, many of the components required of a pumped-storage scheme. It has also the advantage that, with such a scheme, added tidal power becomes a firm daytime source of power. (Fig. 4.12A)

This is important because the principal saving provided by tidal energy is that of fuel, and probably also of pollution abatement costs.

The fuel saved depends on the types of plant whose load factor is reduced and how such plants perform with the addition of tidal power or without it. Evidently, this argument applies only to highly developed countries where there are numerous other plants, and far less to developing countries.

The alternative to storage, when energy is available (but power is not needed because of timing), is not to generate, but the potential thus left untapped is not recoverable at a later time. It is pretty much a case of "store it or you will never have it when needed and no potential is available." Hence, the extra capital investment may well be justified. (Fig. 4.13)

Gibson and Wilson determined, in 1979, that in an average hydrological year, the combination of a tidal power scheme involving the Minas Basin in Canada and a Dickey-Lincoln hydroelectric scheme (U.S.) would provide a 9% increase in the value of tidal energy, while for the Cumberland Basin (Canada) this would even be 64%. The additional monetary benefit, however, they reminded us, is fundamentally dependent upon the character of the electrical system, and on the pumping capacity available. Furthermore, the additional benefits of the retiming of off-peak tidal energy will climb up to the point when installed pumping capacity approximately equals installed tidal capacity.

For tidal power being produced in "mechanical form," only batteries, compressed air, and hydraulic storage can be considered. Flywheel could be used, but its technology is not yet ready.

Flywheel Storage. By connecting a flywheel to a motor/generator, it is spinned up to high speed and absorbs energy. This energy can be returned to the motor-generator by decreasing the flywheel's speed. However, only small quantities can be stored, and then are limited to short periods of time.

Batteries Storage. Perhaps most flexible and practical are electrochemical batteries. These batteries are the most common, but research is being conducted with less expensive and lighter types. Also presently available are aqueous solution nickel and nickel-cadmium types, but they are much more costly. Again, this type of storage is suited to small plants, particularly self-contained ones.

Hydraulic Storage. Hydraulic (or pumped-water) storage is most frequently mentioned in connection with tidal power plants. Two reservoirs, at different elevations, are linked by pumps and turbines and their motors and generators. They constitute the accumulator. In the

130 TIDAL ENERGY

Rance River scheme—and elsewhere, too—a reversible pump-turbine, connected to a motor-generator, achieves the same result. Off-peak energy, which could not be used, is put to work to pump water from the lower into the upper reservoir, which then can be run through the turbine and generator and provide energy when needed at a later time.

Power can be provided immediately on demand; storage stabilizes an otherwise intermittent source and constitutes an efficiency improvement. Reservoirs must be found or built, of course, and a certain amount of energy is lost, as shown in the chapter on the Rance plant, to activate and work the pump. (Figs. 4.14, 4.15)

Fig. 4.14. Rance River plant. Wheels with a diameter of 5.35 m have four movable blades which face the sea.

Fig. 4.15. Rance River plant. Close-up view of bulb unit. *(Source: Electricité de France)*

This is the only scheme presently in operation.

Compressed Air Storage. Air is compressed during off-peak periods and thus stored underground. It can then be called upon to work gas turbines that provide electricity. Tidal energy can be used directly to drive air turbocompressors. (Fig. 4.16)

Existence of such storage facilities near Canadian schemes has been mentioned as a supplementary enticement for tidal power plant construction near some sites. This system needs a rather high-pressure compressor so that the facility will be economically viable. Experts do not

132 TIDAL ENERGY

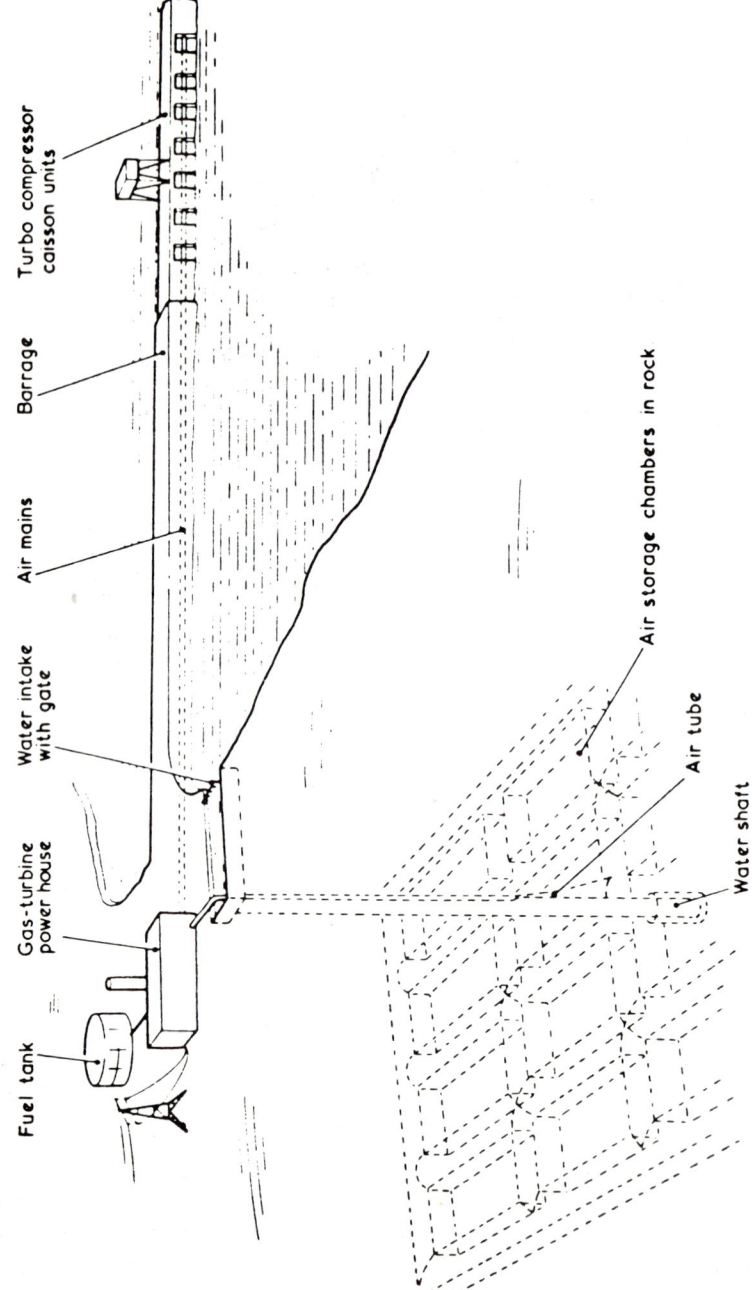

Fig. 4.16. Principal components of tidal/air storage/gas turbine electricity generation scheme. (Wilson, *Underwater J.*, August 1973, p. 184)

THE ELECTRICITY GENERATION 133

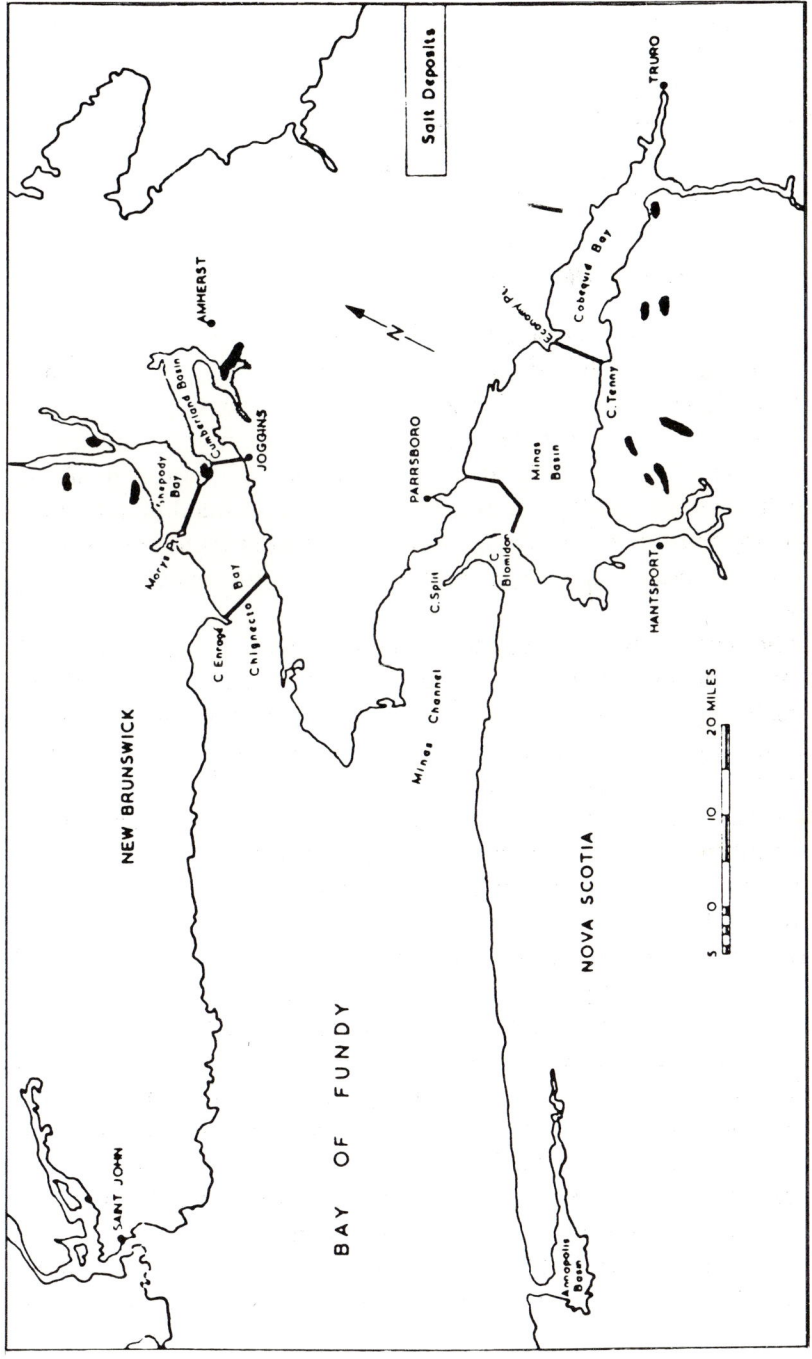

Fig. 4.17. Map of the Bay of Fundy showing salt deposits and barrage sites. (Wilson, *Underwater J.*, August 1973, p. 185)

consider this type of storage as efficient as pumped storage and it requires, furthermore, consumption of oil or gas (burned in the compressed air; 1,000 kcal for each kW-hr). (Fig. 4.17)

Water displacement is one of the ways to provide constant pressure air supply; if instead pressure air storage is to be used, then a specially designed turbine and controller will have to be designed.

CORROSION PROTECTION

Seawater is not a simple solution of electrolytes, but a living medium which undergoes modifications brought about by the metabolism of organisms. These local modifications are due to vegetals (oxygen production and an increase in pH), to animals (oxygen consumption and carbonic gas production), to bacteria (formation of hydrogen sulfide). Hence, laboratory results can differ from results attained in actual field tryouts. Indeed, to the problems actually resulting from corrosion are added those caused by algae and barnacles, for instance through fouling.

Vinylic paints, martensitic, and austenitic steens (made, respectively, of chrome, nickel, copper and chrome, nickel, and molybdenum), and a cuproaluminum bronze with low manganese content, were found to be the most satisfactory materials. The use of titanium in critical areas, such as for turbine bleeding, appears to offer interesting possibilities in overcoming sand erosion and corrosion problems in turbine design. The major limiting factor in the early 1930's in not constructing the U.S. plant in Maine was the intermediate life of stressed components in the seawater of the area.

Corrosion control investigations of the Rance River plant were made at submerged, intertidal, and atmospheric test sites on the shore and in the estuary. A wide variety of coatings were assessed, and so was the design of a cathodic protection system and the influence of seawater on materials in the power race. Anti-fouling coatings applied over vinyls showed good performance, lasting some three years. For the cathodic protection system of the final installation, coating systems involved sandblasting to white metal, one coat of wash primer, three coats of zinc chromate primer, and two coats of copper oxide.

During construction, rubber tape was put on surfaces suspected of damage. Coatings were made in different colors so as to be easily checked.

It was found during operation that aluminum-bronze items stood up well to corrosion, whether or not provided with cathodic protection because of the anti-fouling properties of this material. Such components must not, of course, be painted. It was also found that ordinary steel almost always needed cathodic protection, painting being insufficient after several months in contact with water. Cathodic protection is also necessary for 17% chrome/4% nickel steel to prevent local corrosion in still water caused by seaweed and shellfish.

Cathodic protection was to be used initially only for the sets combining different metals but was later extended, with great success, to the large dam gates in 1968 and to the lock gates in 1970. Only the tops of the lock gates, the bulkhead gates, and the metallic parts of the superstructure are scheduled for a six-coat repainting, which means, in fact, all parts which cannot be provided with cathodic protection because they are either continually or intermittently above water. Thus, in the Rance River plant, corrosion has been overcome by means of cathodic protection, which aggregates 10 kW.

Plastic was substituted for most of the steel and aluminum-bronze pipework. No problems occurred with aluminum-bronze bolts on stainless steel items, nor with cadmium-plated steel bolts when they were painted and provided with cathodic protection.

Corrosion-resistance of concrete and reinforcement bars was excellent; blast furnace cement containing 75% slag-rich concrete mixes with low sand content was used. Reinforcing bars received at least 5-cm coatings.

According to Lebarbier (1975), excellent behavior of the plant over the first eight years of existence was recorded in regard to impermeability, cracking, scour, and corrosion.

REFERENCES

Anderson-Nichols & Co., Inc., 1980. Conceptual design of tidal air chamber, *Report to Northeastern University.*
Cohen, A., 1980. Energy equation in the analysis of air chamber operation, *Report to Northeastern University.*
Congressional Budget Office, 1980. *The World Oil Market in the 1980s: Implications for the United States* (May).
Constans, J., 1978. *Energy.* Monaco: Eurocéan.
Gibson, R. A. and Wilson, E. M., 1979. Tidal energy integration using pumped-storage, *ASCE, J. Pow. Div.*

Gorlov, A. M., 1978. Apparatus for harnessing tidal power, U.S. Patent 4, 103, 490 (August).

Gorlov, A. M., 1979. Some new conceptions in the approach to harnessing tidal energy, *Proc. 2nd Miami International Conference on Alternative Energy Sources*, pp. 1171–1795.

Gorlov, A. M., 1980. A novel approach to exploitation of tidal energy, *Annual Report to the U.S. Department of Energy*.

Lebarbier, C. H., 1975. Power from tides: the Rance tidal power station, *Nav. Eng. J.* **83**, *No. 2*:57–71.

U.S. Army Corps of Engineers, 1980. *National Hydroelectric Power Study* **XV**.

5
Economics of Tidal Power

A tidal power plant has been repeatedly identified as similar to a river hydropower station. Some have suggested, however, that although both use water power, beyond that there is no resemblance—and, actually, even the water is not the same! Fentzloff (1972) wrote that the failure to recognize that hydraulic, machine, and structure concepts are different would lead to uneconomical tidal power plants. The river plant is made possible by the downward-acting terrestrial gravitation causing the water to flow on the earth's surface, while a tidal power plant is based on the upward-acting lunar gravitation causing the water to oscillate in embayments, gulfs, and estuaries.

As a tidal power plant operates, the water level in the pool or pools fluctuates, which means that the head varies continuously. Power and energy that can be developed are governed by interrelated factors: head, basin area, sluiceway capacity, generating units capacity, and type of operation. Reviewing these factors, we observe that the head varies, sluiceways are not selected for maximum discharge but for energy optimization or dependable peak time power, and installed generating capacity depends not on available flow but on economic considerations.

If equivalence with the output of conventional "firm power" energy sources is sought, the tidal power plant loses out because here energy is delivered intermittently in spans of time, particularly if a single-basin scheme is under consideration. The latest Fundy Bay schemes are based on a system approach involving studies of alternative generation expansion programs, including the incorporation of tidal development into the power network. (Fig. 5.1)

In economic assessments, consideration must be given to *capital* investment. This factor plays a major role in site selection since it involves transportation cost differentials, availability *in situ* of material from which to construct barrage and, if needed, auxiliary "dead" dykes, type of geological foundations, accessibility for construction, and tem-

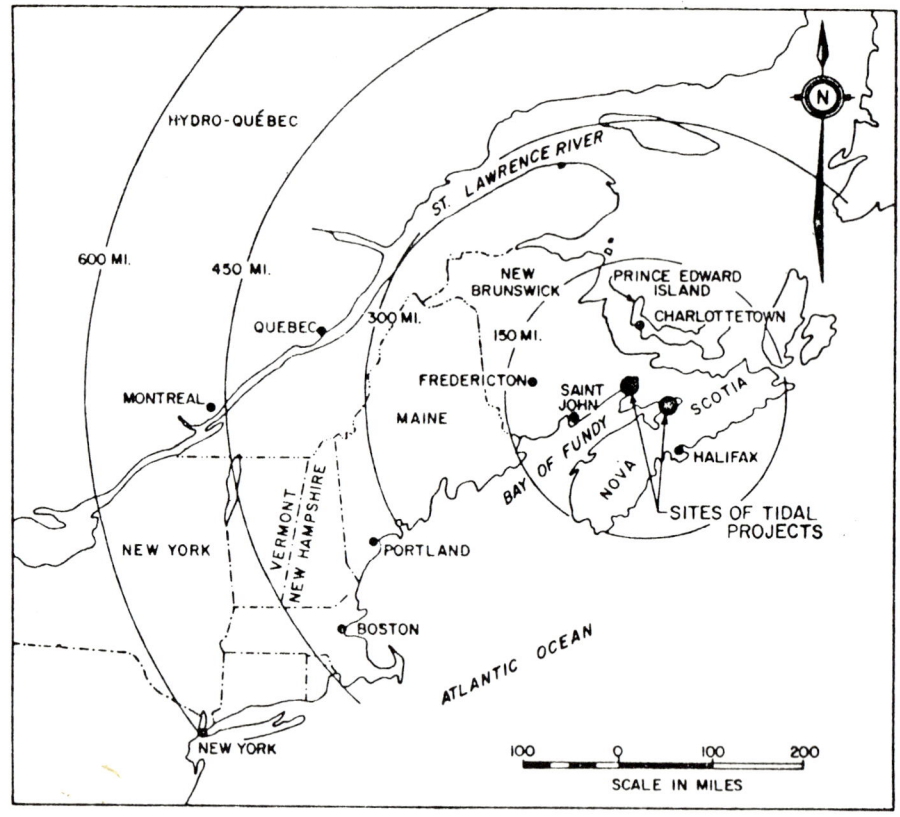

Fig. 5.1. Canadian projects. Relation to contiguous markets. (R. H. Clark)

porary housing for workers. Next are *annual* costs; they must include amortization of capital, interest on outstanding capital borrowed, maintenance expenses, and costs for renewals and operation. An inflation escalation clause is quite important, since construction time foreseen for some projects exceeds ten years. (Table 5-1; Figs. 5.2, 5.3)

Some factors are often overlooked in comparative cost studies, though they are mentioned more frequently [Constans (1978), Shaw (1977), Clark (1978), and Lawton (1972)]. They favor the tidal power plant. First, there is the economic life of the scheme. A conventional thermal plant is usually given an economic life of 35 years; a nuclear plant, 25 years; a hydroelectric river scheme (if well constructed), 50, or perhaps 75, years; and a tidal power plant, 75–100 years. In other words, a fossil

Table 5-1. Comparison of Capital and Annual Costs of Generating Equipment for Gas, Oil, Nuclear, and Tidal Power Plants.

FACILITY	UNIT SIZE (IN MW)	CAPITAL COSTS (IN APPROX. 1976 U.S. $ MILLIONS)	ANNUAL FIXED CHARGES (5½% REAL INTEREST)	
Gas turbines	100	15	9.45	1.4
First unit oil, thermal	475	132.9	8.90	11.8
First unit nuclear	635	482.3	8.97	43.3
Canada	1250	748.2	8.63	64.5
First unit NEPOOL	1150	743.8	12.24	89.9
Including taxes	1750	992.2	12.24	121.4
Pumped storage	200	55	6.231	3.4
Tidal power, Cumberland site (Canada); usable plant capacity	1085	1077	6.231	67.5

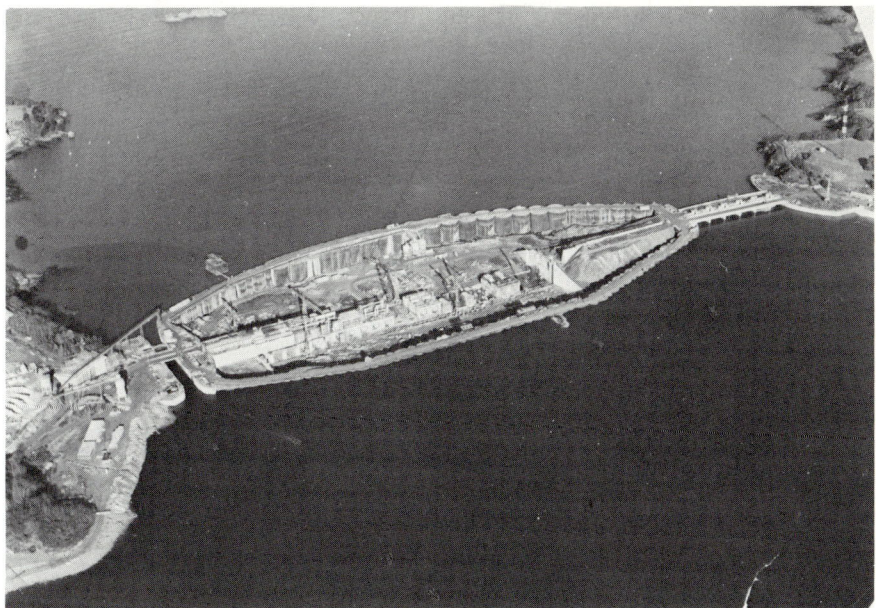

Fig. 5.2. Construction of plant inside a cofferdam. Rance River plant. *(Source: Electricité de France)*

Fig. 5.3. Construction "in the dry" on land. Kislaya plant.

fuel thermal plant would have to be replaced 1½ times, and a nuclear thermal plant twice, to equal the length of service equal to that of the tidal power plant. Initial capital costs for tidal plants are high but less frequently incurred and thus not subject as often to the inflationary spiral of replacement. Then, the tidal plant has no related fuel cost so that once operating loss is recovered, it will produce cheaper electricity, the more so if fossil fuel prices keep climbing to the predicted $86/barrel in the mid-1990's (from $23.50 in mid-1979 and $40 at the end of 1979). For instance, in the case of the feasibility study of Canada's Cumberland Basin project, the cost of oil used for calculations of "equivalent cost" of a kilowatt was less than $15 for the span 1985–1990. This appears now totally unrealistic; double, triple, or perhaps quadruple that amount, in constant dollars, appears far more likely, a factor militating actively in favor of tidal power, not only in Canada, but especially in oil-poor countries. Finally, other energy sources require continuous environmental pollution monitoring and abatement, a costly factor not affecting pollution near-free tidal power. To briefly illustrate this last point, it may be underscored that, in calculating the competitive price of producing electricity with a coal-fired plant, the cost of sulfur dioxide pollution is not taken into consideration, nor is that of oil spills and tanker collisions when speaking about oil-fired thermal plants, nor is the expense included of decommissioning a nuclear plant and its waste products disposal added to the cost of electricity generated by "nuclear power."

Yet these are staggering costs. We must now cope with an acid rain resulting from the sulfur dioxide combination with water vapor in the air, and with water pollution in the ocean. Some 780 million liters of oil were "lost" in 1978, and the Bay of Campeche (Mexico) runaway well and the supertankers collision alone account for (in a single month of 1979) 1,840 billion liters.

Last but not least is nuclear reactor failure. Contemporary nuclear plants, designed for baseload generation, will suffer reactor "poison out" if their power drops below 60–70% of maximum continuous rating. Then, 26 hr are needed before restarting so that waste products (e.g., xenon) have decayed. Failure could result from repetition of such occurrences, with ensuing reactor shut-down and expensive repairs.

VALUE OF THE POWER

It becomes necessary, when selecting the site from an economic point of view, to calculate for each potential location cost of energy and dependable peak, and to assess the markets that can be serviced. Several very promising sites are not being developed because they cannot absorb the production and other consumers are too far away. For instance, the Angoon Indian project has little chance of implementation because the village cannot use all the power. In Canada, Bay of Fundy power would have to be (at first) in part "exported" from the Maritimes to Quebec and New England.

In most cases, transmission is necessary, and hence the capital costs relative to construction of lines. Lawton calculated in 1972 that such costs, including interest during construction, would add $63 million for a facility in Canada, another $124 million to reach a load center in the U.S., and $124 million for each additional line. With operation and maintenance, between $8.80 and $9.00 per kW is added if tidal power is utilized.

The conclusions reached in 1977 by the Federal and Provincial governments in Canada were based on new marine construction technology and hydraulic equipment, particularly the rise of floated-in caissons and on-land construction of powerhouse, gates (etc.), which reduce substantially capital investment. This is significant because of the large investments required: capital amortization accounts for about 90% of the unit cost. A 1½ percentage point increase in interest rate affects by 10% the present value of development cost. (Fig. 5.2, 5.3)

Warnock and Tanner (1978) calculated optimal levels of installation of generation and sluiceway capacity regarding Bay of Fundy schemes providing the minimum unit cost of energy. It involved an iterative analysis of regime conditions, energy output, and cost of development. They found that installed capacity and output can be varied significantly with minimal impact on at-site cost of energy. A plant at the most seriously considered location, the Cumberland Basin, with a net capacity of 1085 MW and an average annual output of 3,423 GW-hr, would cost, in $ 1976, $1.197 billion, with an extra $37 million for transmission links, thus totaling $1.234 billion. Projected in-service (by 1990) cost—inclusive of inflation escalation and interest charges—is $3.12 billion. (Fig. 5.4)

For projects using renewable resources, life-cycle cost analysis can be properly substituted for benefit-to-cost analysis. The method groups expenses occurring at different stages during the estimated lifetime: levelized power cost, tabulating yearly operation costs under an assumed yearly inflation rate (affecting any plant), and under an assumed yearly increase of fossil fuel price. Thus, for each year of lifetime, a benefit-to-cost ratio is calculated; this is a prognosis of the economic viability based on assumed yearly operation costs. This method introduces the probable fossil fuel price increase in the comparison.

ECONOMIC FEASIBILITY

Capital investment, level of nuclear energy penetration or installation in the system, and the cost of fuel strongly influence any analysis of economic feasibility for a tidal power scheme. Between 80 and 90% of

ECONOMICS OF TIDAL POWER 143

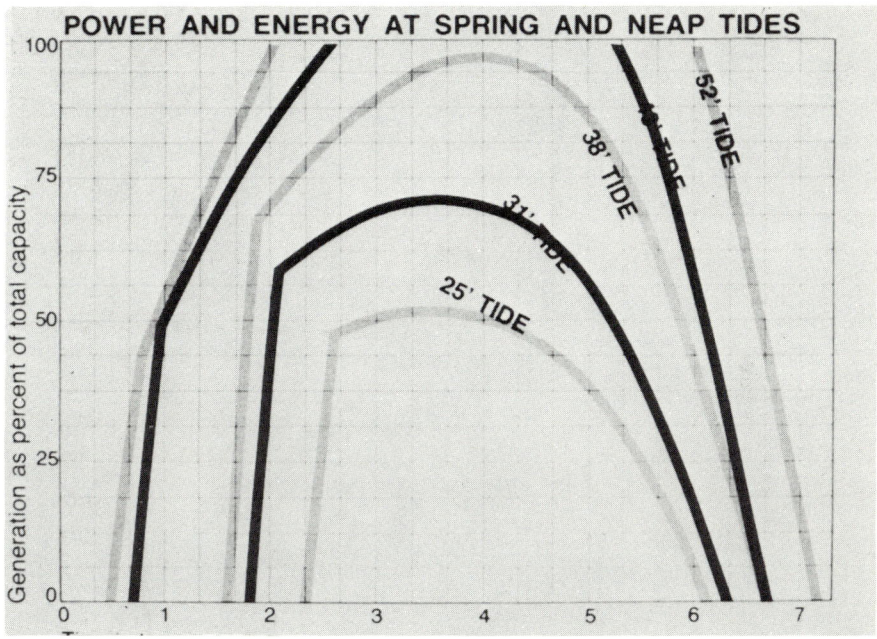

Fig. 5.4. Total energy demand.

benefits derived from inclusion of a tidal power plant in an electric system are attributable to the cost of fuel saved. The remainder results from considerations of amount of alternative generation capacity displaceable by tidal power, while maintaining a reliable and comparable system. If nuclear expansion proceeds unhampered, tidal power development will probably be adversely affected, but if it is restricted in any way, the impact on tidal power would be noticeable. (Fig. 5.5)

The primary role for tidal power is to replace fossil and nuclear generated energy. It would not displace nuclear energy, in the Maritimes Integrated System, but would reduce oil needs by half, save 3 million barrels (477 million liters), 330,000 tons of coal, and 90.8 tons of uranium. Savings are very high when fossil fuels (oil, coal, and natural gas) are replaced, but quite modest when nuclear or hydroenergy are "displaced." Based on the recent Canadian reassessment report, the Cumberland basin proposed station (1085 MW) would save on a percentage basis over a period of 30 years the approximate amounts shown in Table 5-2. (Fig. 5.6)

144 TIDAL ENERGY

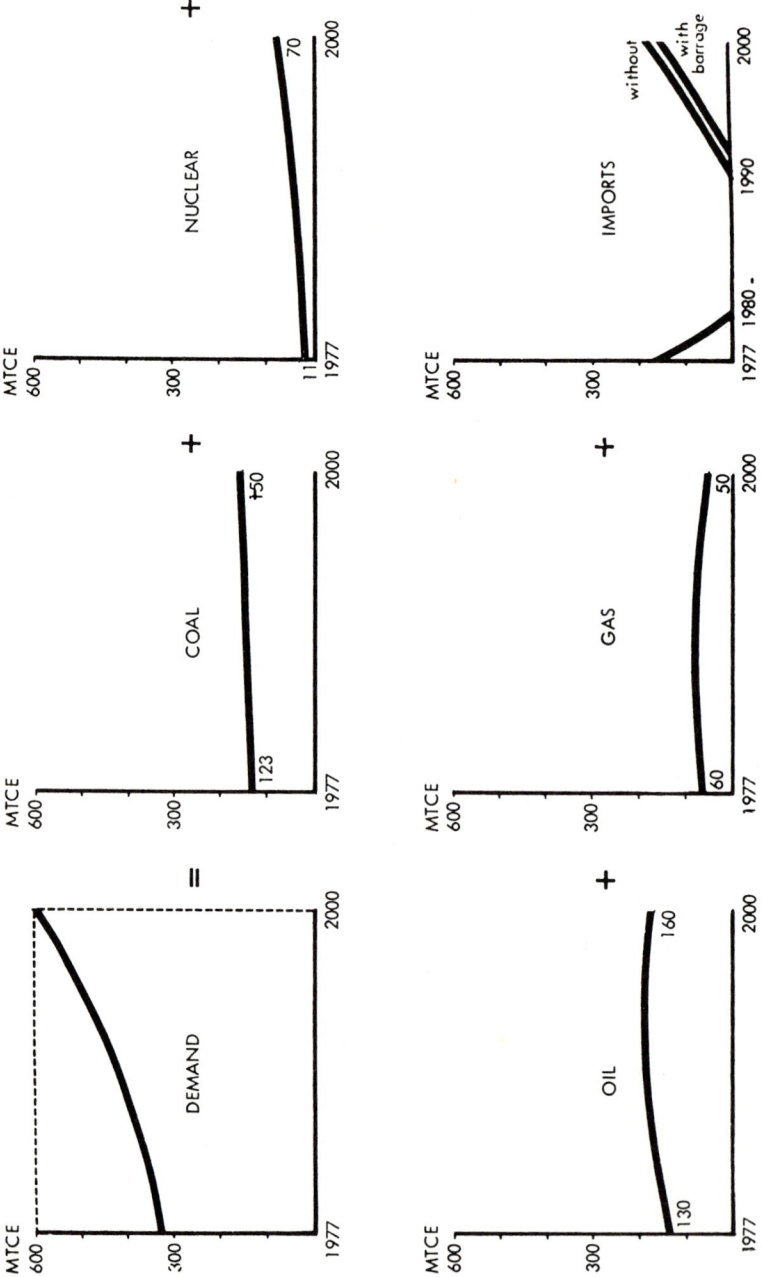

Fig. 5.5. Australia. Total annual energy demand (1977–2000) in million tons coal equivalent.

ECONOMICS OF TIDAL POWER   145

Table 5-2 Energy saving expressed in percent of raw tidal energy generation. (Value of tidal energy expressed in terms of thermal energy displaced in interconnected systems.)

ENERGY SOURCE	YEAR				
	1990	1995	2000	2005	2010
Oil	100	97	99	100	100
Coal	35	41	29	35	34
Nuclear	2	15	12	20	25
Utilized energy		2	1		

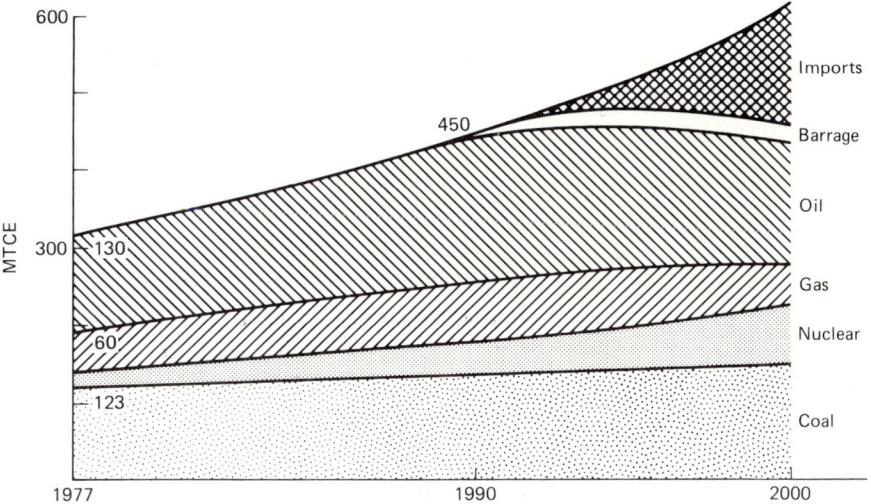

Fig. 5.6. Australia. Total annual energy demand (1977–2000) in million tons coal equivalent. Reduction of imports through tidal power harnessing.

With 1% escalation of oil prices till 1990, then 2%, interest of 4¾%, load growth of about 7%, export proceeds of 50% of exported power value, the Cumberland site shows a benefit-to-cost ratio of 1.2 (unadjusted in 1979 is 0.93), with a break-even period of between 30 and 35 years. (Table 5-3)

In the analysis of the costs involved in various "Quoddy" projects, accurate claims were made (and this is true even at present) that a thermal plant can produce cheaper power. In fact, the conventionally

Table 5-3. Economic Evaluation of Benefits and Costs, in 1976 Canadian $ (Millions), 5½% Discount Rate, with Strong Nuclear Introduction in MIS, for a 250-MW Retimed Storage (Value Subtracted), MIS-NEPOOL Tie, 500 MW, Cumberland Basin Site.

Maritimes integrated system, gross value from tidal energy	950.7
New England pool (NEPOOL), system gross value from tidal energy	48.6
Total gross value from tidal energy	999.5
Maritimes integrated system, internal transmission cost	6.0
MIS-NEPOOL, transmission cost	22.2
Total transmission cost	28.2
Net (unadjusted) benefits from tidal power	999.5 − 28.2 = 971.3
Present worth of energy (in GWL)	47,732
Levelized value of energy (in mills/kW-hr)	20.3
Total capital cost of tidal power plant	1,197
Present worth in 1985, unadjusted, of annual charges for years 1990–2045	1040
Annual charge at 6.231% (see Table 5-1)	75
At-site cost of energy (in mills/kW-hr)	21.8
Benefit-to-cost ratio	971 ÷ 1,040 = 0.93

produced electricity cost in 1965, in New England, about 7 mills/kW-hr. However, tidal power opponents were comparing constant generation (baseload power) with peaking power. Thermal units could not then, nor now, produce peaking power for 7 mills/kW-hr. Holyoke Water Power Company's Mount Tom highly efficient plant then produced power at 7.16 mills/kW-hr on the basis of a 22.8 hr/day of operation. A tidal power facility can be operated any number of hours a day, no matter how few; the cost of a kW-hr produced by a thermal plant used for peaking power rises considerably: for Mount Tom, it would have been 10.67 mills if used 12 hr/day, 32.91 for a 3-hr day, 56.4 mills for a 2-hr day; for the Quoddy tidal plant, it would have been 11.3 mills. Similar results would have been obtained for the Connecticut Yankee Atomic Steam plant.

CONCLUSION

Though initial capital outlay and early years of operation are very high for a tidal power plant, a break-even plateau is reached at the time that a same-age nuclear facility has already had to be replaced, and a fossil fuel plant is due for total replacement. Rising costs of fuels, particularly oil, will make tidal power competitive soon, and the oil's declining supplies ought to encourage tapping tide energy. In many economic analyses, comparison bases have often failed to take into account factors proper to tidal power and favorable to its development. Retiming of tidal energy (e.g., by pumped storage) turns unneeded energy into a valuable reserve that can be called upon at times of peak demand. Clark (1978) emphasized in connection with the Cumberland Basin project in Canada that were nuclear generation restricted to only 50% of the peak, this would result in more than a 35% increase of the gross benefits of tidal power there.

Table 5-4. Absorption Capability of the Maritimes Integrated System for Energy from a Tidal Power Station of 800-MW Capacity. *Source: Bay of Fundy Tidal Power Plant Review Board.*

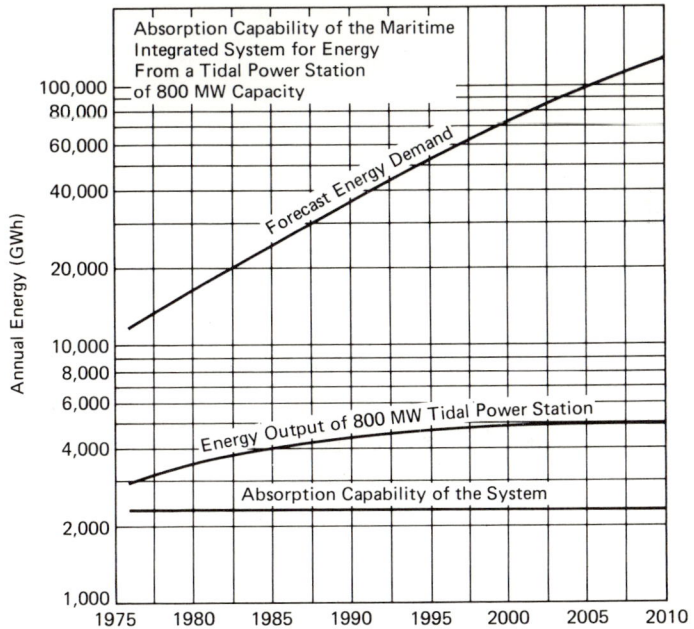

The system could absorb little more than half the output of a small site now, but is expected to be able to use well over 80% by 1990. This estimate is based upon single-effect operation. In double-effect operation, with pumping employed according to need, a higher percentage of the energy would be absorbed. In either case, retiming only the tidal energy which does not fit the demand cycle, could result in full utilization.

It is perhaps not being overly bold to forecast that the benefit-to-cost ratios of tidal power plants, even if they contribute only modestly to current and future energy supply needs, improve steadily, significantly, even rapidly.

REFERENCES

Clark, R. H., 1979. *Fundy Tidal Power: A Retrospective View.* Unpublished paper.
Clark, R. H., 1978. The economics of Fundy tidal power, *Proc. Intl. Conf. Wave and Tidal Power.*
Karas, A. N., 1977. System planning for Bay of Fundy tidal power developments, *Trans. IEEE*
Lawton, F. L., 1972. Economics of tidal power, in Gray and Garhus (Eds.), *Tidal Power.* New York: Plenum, pp. 105–129.
Tidal Power Review Board, 1977. *Reassessment of Fundy Tidal Power.* Ottawa: Bay of Fundy Tidal Power Review Board, pp. 24–40.

6
Tidal Power Around the World

Some hundred sites suitable for the construction of tidal-power plants exist in the world, some of which could be coupled for still greater efficiency and productivity. India, Australia, and New Zealand examined harnessing possibilities and so did Brazil, Argentina, Germany, Scotland, Wales, and Spain. Actual moves toward implementation of such projects were only made in the Soviet Union, France, Korea, China, the United States, Canada, and Argentina. Yet, most projects remained dormant until the Electricité de France actually went ahead with the Rance River project.

Tidal range is of primary concern when considering a site. The incremental amount of energy (E) per cycle that can be obtained from the tidal flow of water is given by Eq. 6-1,

$$dE = gR(D \times S \times dR) \qquad [6\text{-}1]$$

where R is the tidal range, D the water density, S the area of the enclosed basin, and g gravity's acceleration. Considering S independent of R,

$$\int dE = E = D \times g \times S \times \frac{R^2}{2}. \qquad [6\text{-}2]$$

In a double effect scheme (generation at ebb and flood tides),

$$E_{\max} = D \times g \times R^2 \times S. \qquad [6\text{-}3]$$

Hence, the total annual (700 cycles) energy in kW-hr that can be generated in a single basin with a double-effect action unit is

$$E_{(kW\text{-}hr/yr)} = 0.017\ R^2 S. \qquad [6\text{-}4]$$

E varies with the square of R and the first power of S.

This means that a site with a range R of about 6 m requires only an enclosed area of one-fourth of that needed by a site with amplitudes of only 3 m. Capital requirements for dam and generator increase with smaller ranges. A site where tidal ranges would be about 7.6 m could theoretically generate some 10 kW-hr for each 0.09 m². To generate 10 billion kW-hr/yr, 78 km² of impounded water would be needed. Hence, sites with amplitudes that are small are considered uneconomical, although some cases are being reexamined due to the development of low-head turbines.

If perhaps the largest number of sites considered and projects studied concern France and the British Isles, more obscure, and more modest, schemes are not lacking. Among these is the Spanish project of Vigo Bay, on the Atlantic Coast, just north of the estuary of the Minho River that separates Spain and Northern Portugal. But there has been no shortage of projects in North America, including one in Baja California, Mexico. (Fig. 6.1)

In England, Scotland, and Wales, some 32 locations could possibly be suited for the establishment of tidal power plants, with 12 locations where tidal amplitude exceeds 10 m. Best known sites here are the Bristol Channel and the Solway Firth. The Channel Islands have tides exceeding 10 m in Jersey (St. Helier) and Guernsey (St. Pierre). Thirty-eight potential locations can be listed for France, of which 23 have tidal amplitudes of over 10 m.

In Asia, suitable areas exist in the Sea of Okhotsk (U.S.S.R.) and, somewhat farther south, the Seoul or Kan River. These have differences in tide levels of over 13 m. Spectacular tide ranges occur on the coast of China near Shanghai, and at Amoy (or Szeming), facing the Saikoe River. Rangoon (Burma) on the Gulf of Martaban is close to the Irrawaddy delta, and here, as near Kandla (India), on the Gulf of Cutch, great tidal amplitudes prevail. In the Gulf of Cambay (India), the range reaches 12 m. In Africa, ranges are 5.5 m at Porto Gole (Guinea-Bissau), a possible location.

In Australia, sites abound along the Northwest coast, and additional

TIDAL POWER AROUND THE WORLD 151

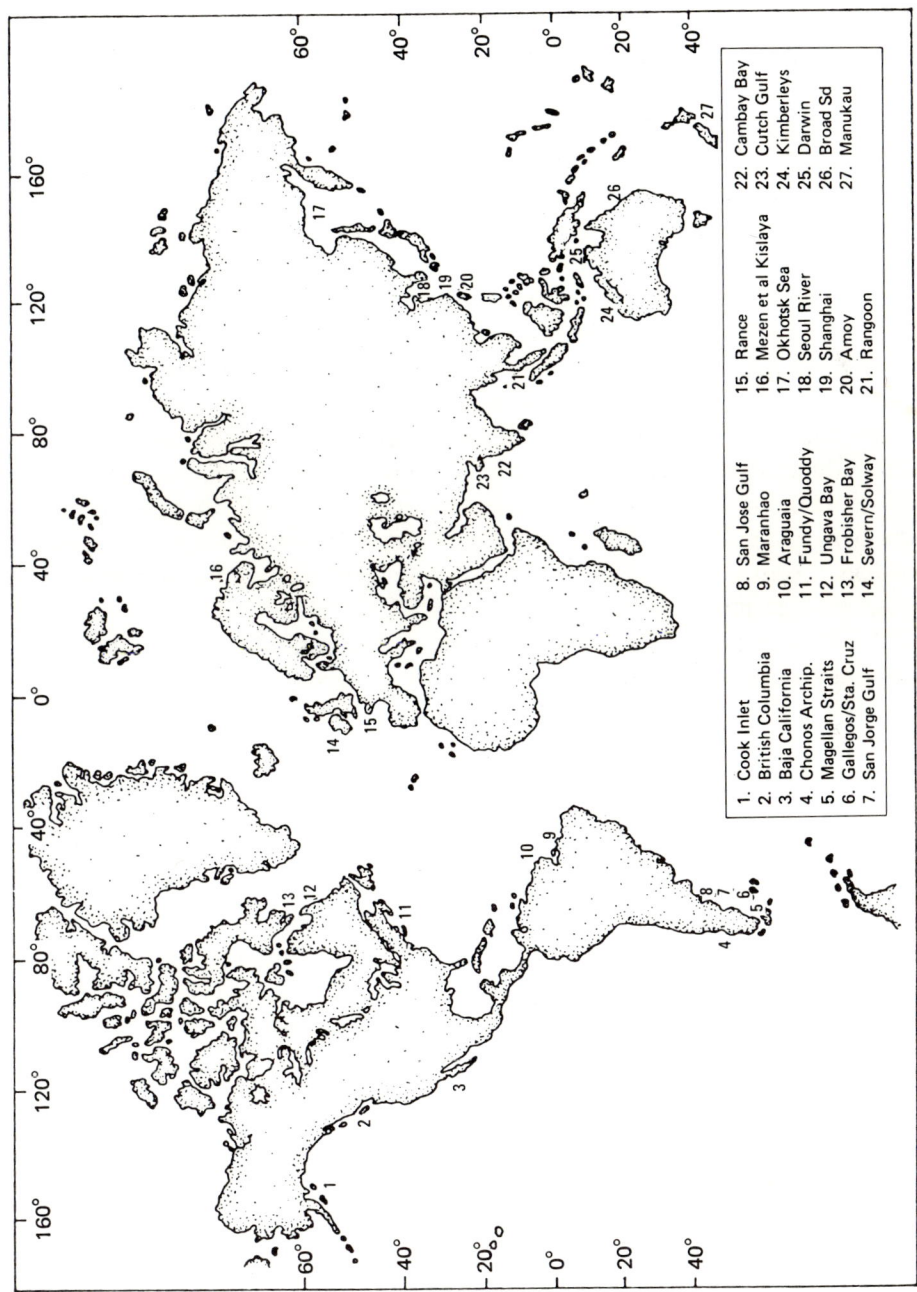

Fig. 6.1. Major tidal power plant sites: world distribution.

ones exist at Darwin, at the extremity of the Van Diemen Gulf, and on the Broad Sound (East Coast). Tides exceed 10 m.

In the Americas, the most often mentioned locations are on the Bay of Fundy, Passamaquoddy Bay, and Golfo San Jose (Argentina). However, several sites could be considered near Anchorage (Alaska) on Cook Inlet, which receives the Susitna River; in addition, five general locations, of which three have tides exceeding 12 m, exist in Canada, along the coast of British Columbia, and on the Ungara and Frobisher bays; one in Mexico, with tides exceeding 12 m (Baja California); two in Brazil, respectively, on the Araguaia-Tocantins River debouching in the Rio de Para, south of Belem (Para), and near São Luís in the State of Maranhão; and five more in Argentina, of which three have tides of over 10 m. Here, besides San Jose, the best known sites are on Santa Cruz Bay, an inlet of the Atlantic Ocean that receives the Santa Cruz and Chico Rivers, on Gallegos Bay, opposite the Falkland Islands, which receives the Gallegos River, also on the Atlantic Ocean, and in the Magellan Straits. The Chonos archipelago, on the southwest coast of Chile, south of the Gulf of Corcovado, has also been mentioned as a possible location. (Fig. 6.2)

All these sites have tidal ranges exceeding 5 m, and many have far greater amplitudes and are combined with easily dammed inlets or estuaries. However, many are, unfortunately, also located at distances of more than 175 km from areas of great electric demand.

AMERICA SOUTH OF THE RIO GRANDE

At least five sites have been studied in Argentina. Conservative average annual energy outputs have been estimated for Puerto Gallegos, (7.50–14-m range) at 2,000 GW-hr, for the Gulf of San Julian (6–9-m range) at 400 GW-hr, for the Deseado estuary (3.5 m) at 700 GW-hr, for the Santa Cruz River (7.5 m) at 4,000 GW-hr, and for the Gulf of San Jose, in the province of Chubut (6 m) at 9,000 GW-hr.

The Gulf of San Jorge, a wide indentation of the Patagonian littoral between the Cape of Two Bays in the north (Dos Bahias) and Three Points Cape (Tres Puntas) in the south, has tides of exceptional amplitudes reaching 12 m. The harbor of Comodoro Rivadavia is located at the head or the gulf. (Fig. 6.3)

Farther up north, similarly to the situation on both sides of France's Cotentin Peninsula, tides are out of phase on either side of the Valdes

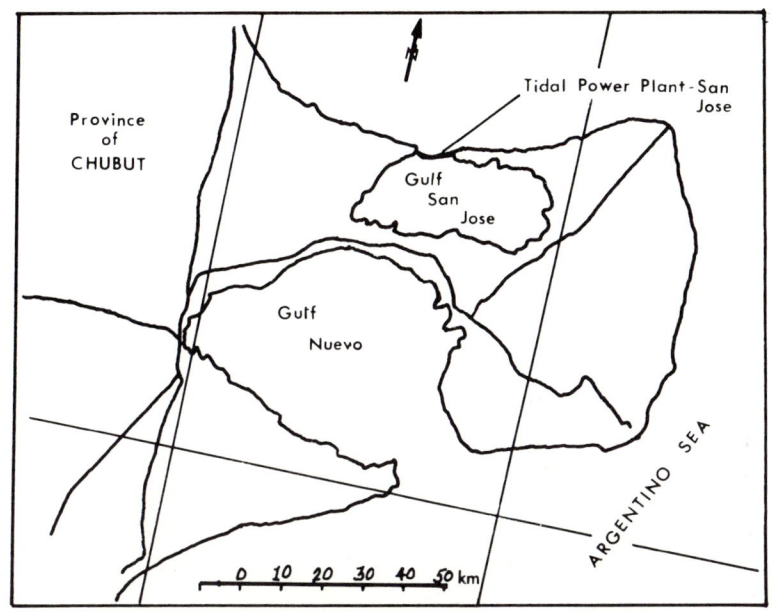

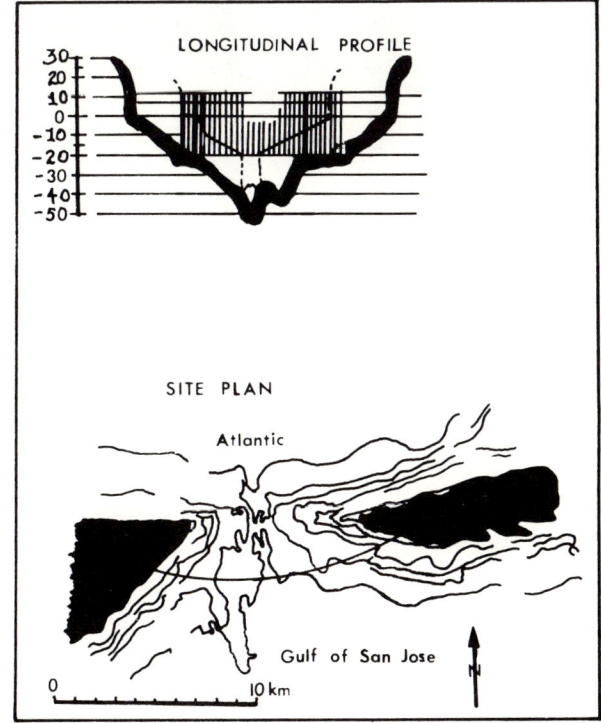

Fig. 6.2. San Jose tidal power plant. (Adapted from Fentzloff)

Fig. 6.3. Cabo Tres Puntas: potential tidal power plant site.

Peninsula; here, when it is high tide in the Nuevo Gulf, the tide is low in the San Jose Gulf, a few kilometers away.

The San Jose canal, on the Valdes Peninsula, would connect the San Jose Gulf in the North and the Nuevo Gulf in the South.

San Jose lies between 42° and 43° of latitude south. The coastal morphology causes the peculiar propagation of the tidal current. The Valdes Peninsula projects considerably into the Atlantic Ocean and the tide's advance from west to east is slowed down along the Patagonian littoral because of a submarine elevation. It requires four hours for the tide to round the peninsula, and the tides are delayed a half-period between the northern coast of the isthmus, or San Jose Gulf, and the southern coast, or Nuevo Gulf. A level difference of 16 m results and this drop could be used to produce electricity by digging a canal through the isthmus.

Local topography and geology favor two locations for the construction of the canal: a western route from the San Jose Gulf between Espada Point and Blanca Point to the Bajo del Piche depression on Nuevo Gulf; an eastern route, about 4½ km to the East of the preceding one, from San Jose Gulf to Nuevo Gulf, passing between the small Ladera island and Logaritmo Point near the Iriarte hacienda. The eastern route is 7.58 km long and lies lower than the 6.2 km-long western route; the latter, however, would not deprive the Iriarte hacienda of its water supply.

Structurally and petrologically, the isthmus shows remarkable uniformity. The sediments are predominantly tertiary and constitute a compact and impermeable strata, yet easily removable.

The power plant could be constructed at the canal's end on the San Jose Gulf. The 1959 feasibility study showed no unsurmountable technical difficulty in constructing the tidal power plant or any problems not already encountered for the planning of the Rance River plant. The only new challenges were the width of the canal and the number and dimension of the units.

A basin of 2,020 km^2, the San Jose Gulf has tidal amplitudes in the range of 3.5–7.8 m, and a width at the outlet of 7 km. If a single-cycle generation plant were to be built, an 8-km-long structure would only be required, against a 16.8-km barrage for dual-cycle generation. H. E. Fentzloff (1972), of the German concern Hochtief A.G. of Essen, recommended a powerhouse equipped with tubular turbines with rim generators: the maximum feasible runner diameter for such turbines fell in 1972 between 7.5 and 9 m. However, these sizes might be modified if a pumped storage scheme is inserted in the plans.

As in the case of Cobscook in Maine, the fill to be used would be material extracted from near-shore and estuary bank areas. Foundations near the shore would rest on the solid tertiary deposits of the sea bed. The plant would be concave toward the ocean.

The San Jose single-cycle plant would provide 4,965 MW of installed capacity versus a dual cycle capacity of 6,820 MW. Yearly output would be, respectively, 9,500 and 75,000 GW-hr. Some estimates, figuring on 400 turbines, foresee a production of 12 billion kW-hr/yr; such capacity would nearly be the quadruple of that of the Rance River plant.

Comparing existing and possible tidal power plants, and using Fentzloff's proposed "energy criterion" formula, or the ratio of yearly output to dam length expressed in GW-hr/km, the San Jose project leads most others: the single-cycle scheme has a ratio of 98, the dual cycle of 46.5, against 74 for Knik Arm (Alaska), 42 for the Bristol Channel (Great Britain), 37 for Minas Basin (Canada), and 28 for the Rance River (France).

As in the Rance plant, turbines could operate in both directions in order to attain maximum rentability. Using the French coefficients of rentability, the most favorable scheme, according to Sogreah (1959), would require turbines with a diameter of 6.90 m, with 60 rotations per minute; the group should be made up of 50 units with an individual power of 12,000 kW. With an 8,000-m^2 section for the canal, a favorable coefficient of 7.91–8.65 is attained.

The Chocon River plant was completed a few years ago. Thought was given to combining the hydroelectric and tidal power schemes, similar to the proposed linking of the Passamaquoddy project with the Rankin Rapids River plant on North America's east coast. (Fig. 6.4)

The San Jose plant could satisfy, by itself, a daily demand of 18,000 kW-hr/yr. Combined with Chocon, the total climbs to 45,000 kW-hr/yr. Furthermore, Chocon could compensate for, when needed, the San Jose production as peak demand occurs at tidal low productivity. According to Sogreah's 1959 report, the San Jose power plant would become rentable; it offered three solutions, depending upon the use of machines of diverse origins. However, some considerations are now outdated.

Already in 1928, voices were raised to "fetch energy in the gulfs of Patagonia." Several sites were considered, but the San Jose project retained the maximum support. The original plan to build a barrage across the bay is of such magnitude that it was shelved as too ambitious, too difficult, and too costly. On the other hand, the digging of a canal, only 2 km long, cutting the isthmus and thus connecting the two gulfs, appears quite feasible and able to provide, for a discharge of 6,000 m^3/sec, about one billion kW-hr. (Fig. 6.5)

The more ambitious Argentine plant (San Jorge Gulf) would have generating equipment of 376 units, which would deal with a discharge of 90,000–150,000 m^3/sec and produce 10,000 million kW-hr.

In Brazil, the Araguaia-Tocantins Rivers are suitable for tidal power development; so are other locations near São Luís, a northeast seaport, where manufacturers are established. In Itagui, amplitudes are 5 m and, in São Luís, they reach 8 m. (Fig. 6.6)

Though favorable, sites in the Straits of Magellan are unlikely to be used.

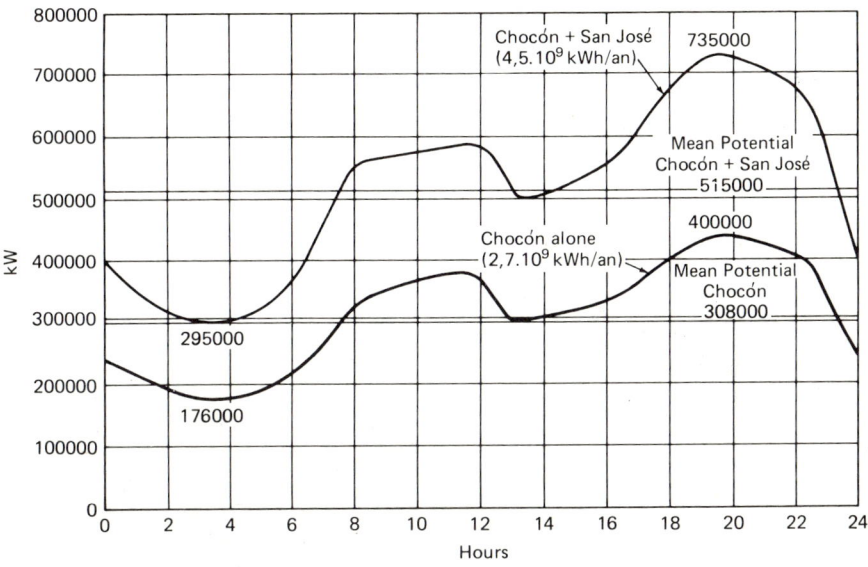

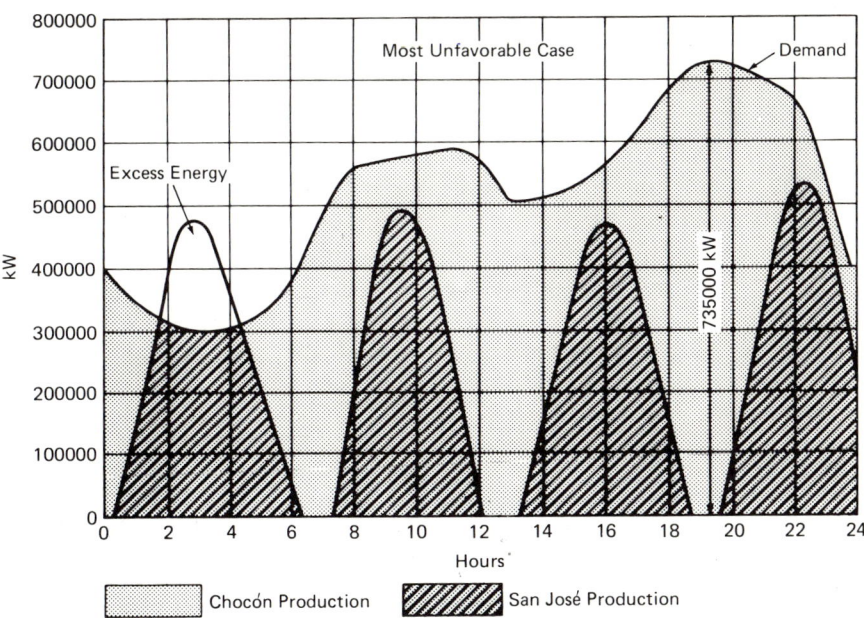

Fig. 6.4. Compensation graph San Jose-Chocon.

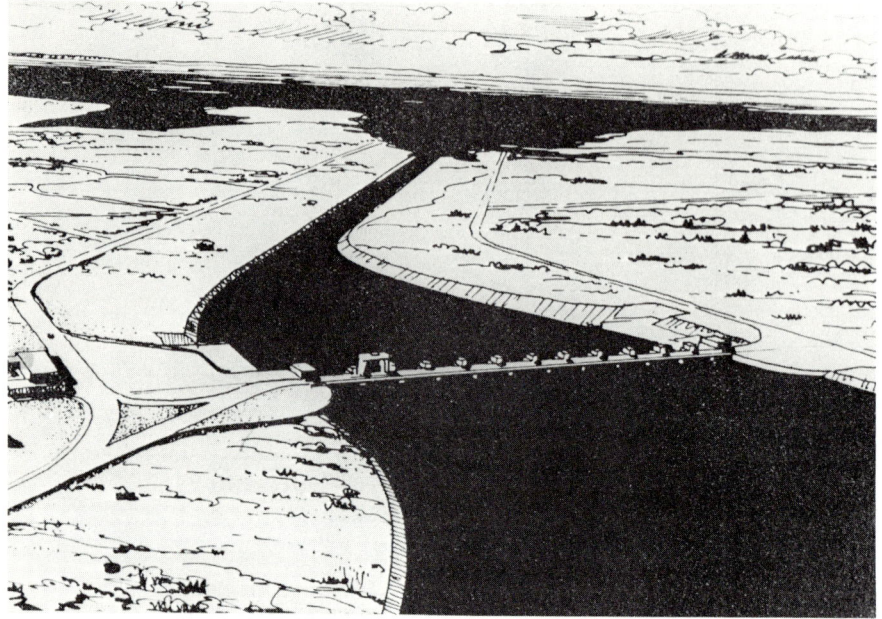

Fig. 6.5. Artist's concept of dam across Limay River for the Chocon project (Argentine).

In the north, tides vary by 4.85 m on Naos Island in Panama, but Baja California, on Mexico's Sea of Cortez coast, received the most attention. Here, tidal ranges reach 10 m in some locations. Yet Baja California has been repeatedly ruled out because it lies astride the San Jacinto Fault, part of the fracture zone that includes California's San Andreas Fault system and is subject to intense seismic activity.

NORTH AMERICA

Alaska has several promising sites, some of which have been the subject of studies. Less attention has been given to potential locations in British Columbia. Most has been directed toward the Atlantic coast, with projects under consideration at Passamaquoddy and in the Bay of Fundy. A particular revival of interest has occurred in recent years for a Fundy project.

Fig. 6.6. Tidal power sites in South America.

AUSTRALIA AND "DOWN UNDER"

According to a survey conducted some years ago by Fagnoni (Sogreah, 1963), who participated in the Rance plant construction, the Collier Bay site in Northwestern Australia, some 250 km north of the port of Broome, alone would provide many times the volume of power furnished by the French power station. Fifty sites have been examined along some 1,700 km of coastline and half of them were retained for detailed study to determine their potential. (Fig. 6.7)

The Kimberley region of Western Australia contains most of Australia's potentially harnessable tidal energy. Tidal range runs 9–12 m and numerous coastal indentations are suitable for barrage construction. Australian generation capacity was 20,000 MW in 1975 and electricity production reached 70 TW-hr. The four most attractive sites for a tidal power plant are Secure Bay, Walcott Inlet, George Water and St. George's Basin. Tidal power and energy have been estimated by Saunders (1975) and compared to that of the Rance. (Table 6-1)

A cursory assessment of average annual energy output foresees 3,950 GW-hr, with tides of 12 m, at Walcott Inlet, 3,500 in the St. George Basin, 2,400 for George Water, and, with amplitudes of 7 m at Secure Bay, about 1,650 GW-hr. Port Headland, where the tidal range is 5.18 m might also be suitable. In all, 25 of the 50 locations show maximal promise. Among these, Collier Bay, with amplitudes of 7.5–12 m, exceeds by far the Rance River in potential. Port Darwin (amplitude 5.12 m), on the edge of the Van Diemen Gulf, has been repeatedly mentioned for a possible plant location and, on the east coast of Australia, sites on Broad Sound.

Gigantic problems dealing with design, supply, and actual construction of plants "in the sea" face the Australians, but they can be overcome. Furthermore, in the experts' opinion, once the initial plant would be established, its scope could be rather easily expanded to satisfy increasing demand.

The only land-based report dates back to 1921 (Easton), and a 1950 thesis (Raynor) dealt with the potential of tidal power. The French Sogreah report recommended, in 1965, a Secure Bay scheme with a power station, sluice structures and a rock-fill dam at the Collier Bay end of The Funnel, the entrance to the bay's outer basin. As in the Rance River, natural islands were to be incorporated; two such islets are situated at the north of The Funnel. The power station, 570 m long, would house 30 bulb-type turbines, 7.35 m in diameter, and generating at ebb tide only; the dam would be 700 m long, and there would be two gated sluiceways of 100 and 500 m, respectively. Engineering problems were amounts of rock excavation and cofferdamming, construction time involved (9–10 yr) and the irregular Z-pattern of the natural tidal flow. (Fig. 6.8)

TIDAL POWER AROUND THE WORLD 161

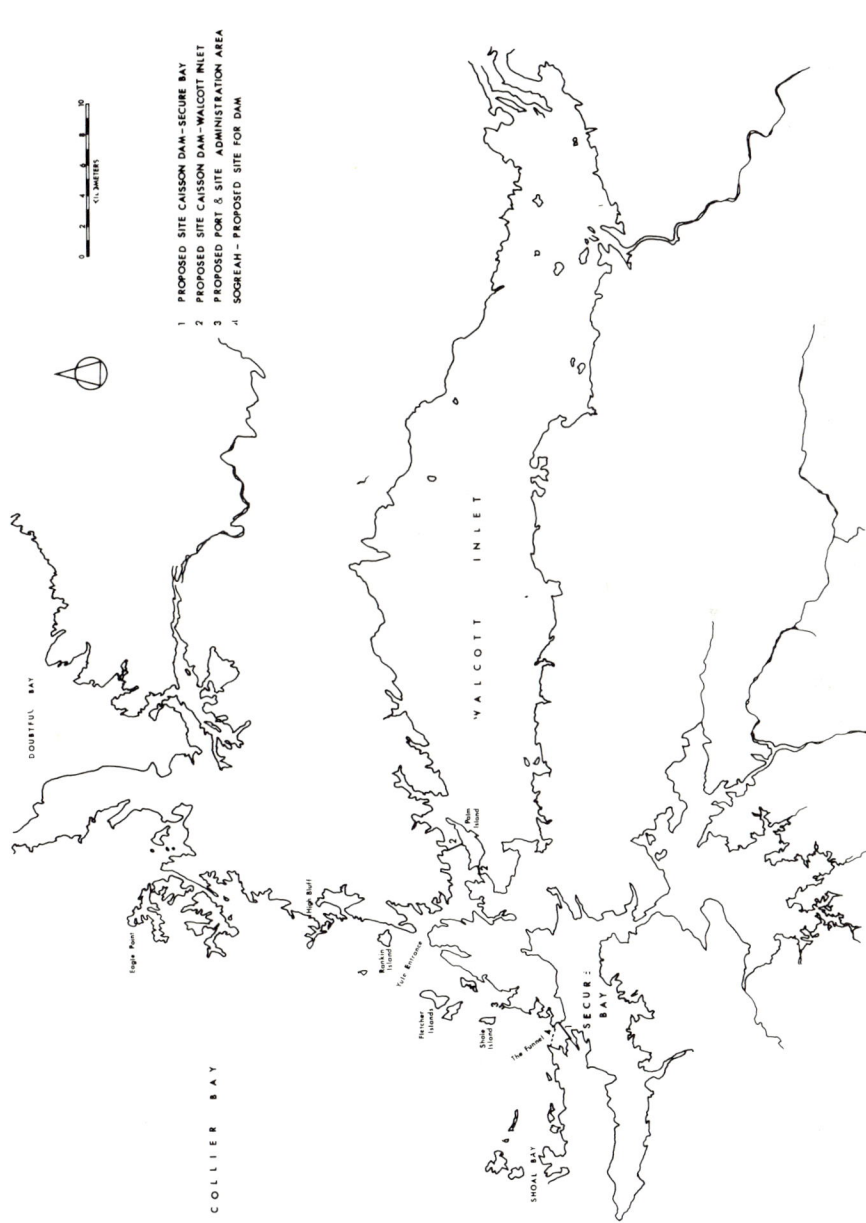

Fig. 6.7A. Secure Bay and Walcott Inlet, Australia. (Maunsell, *Kimberley Tidal Power*)

162 TIDAL ENERGY

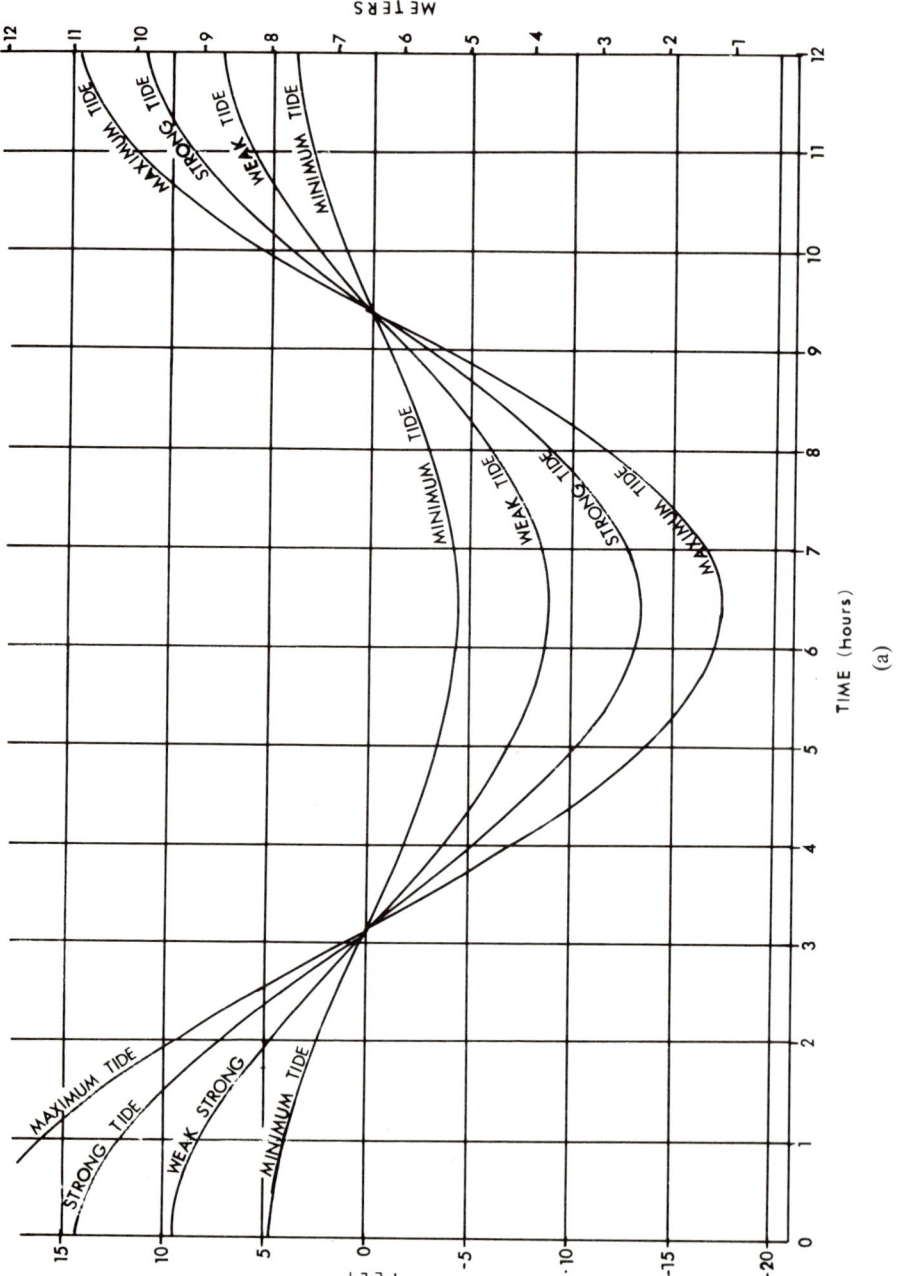

(a)

TIDAL POWER AROUND THE WORLD   163

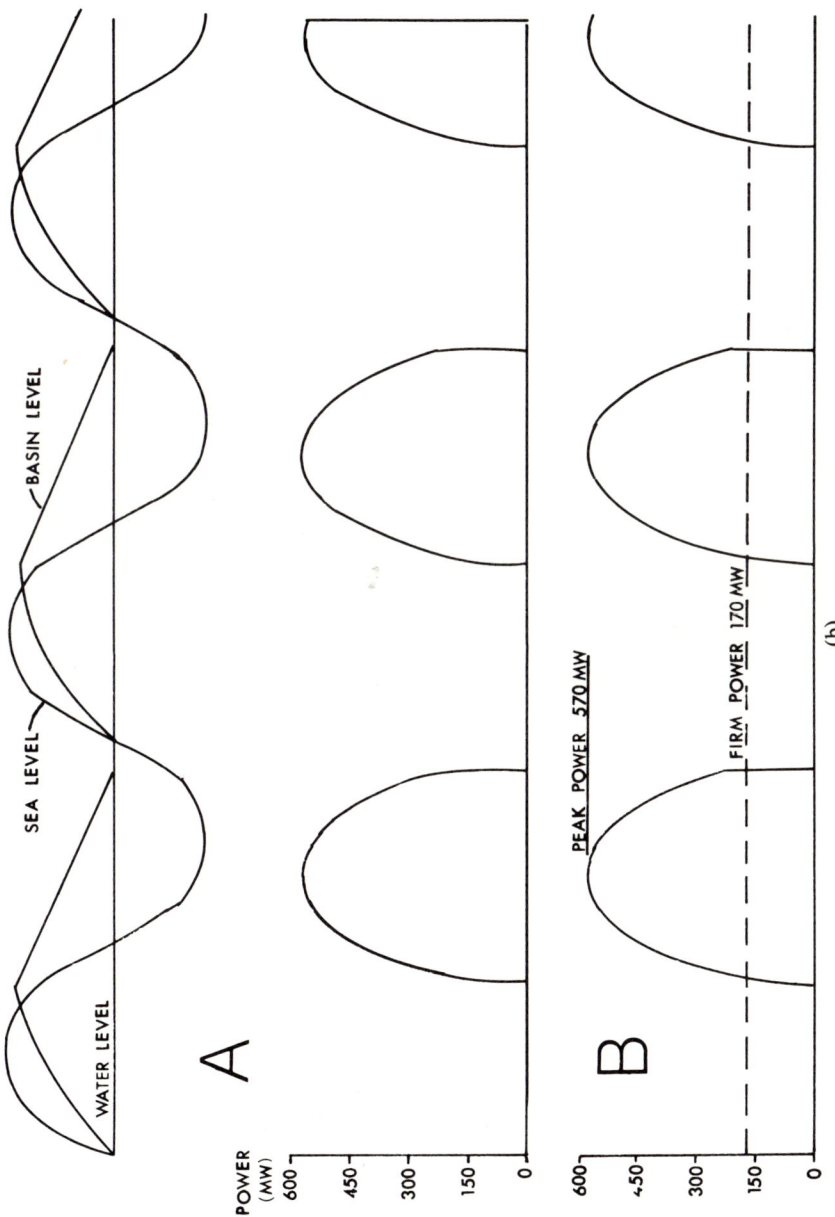

Fig. 6.7B. Secure Bay: (a) actual tidal range; (b) power and energy from proposed scheme. (Maunsell, *Kimberley Tidal Power*)

Table 6-1. Tidal Power and Energy of Four Australian Sites Compared to Sites in France, Canada, and the United States.

SITE	MAXIMUM POWER (MW)	ANNUAL ENERGY (TW-hr)
Secure Bay	570	1.650
Walcott Inlet	1,254	3.950
George Water	800	2.400
St. George's Basin	1,000	3.500
Rance River (France)	240	0.544
Cumberland Basin (Canada)	1,147	0.342
Passamaquoddy (M_3 or M_4)	180	0.590
Alaska (A_1)	750	2.870

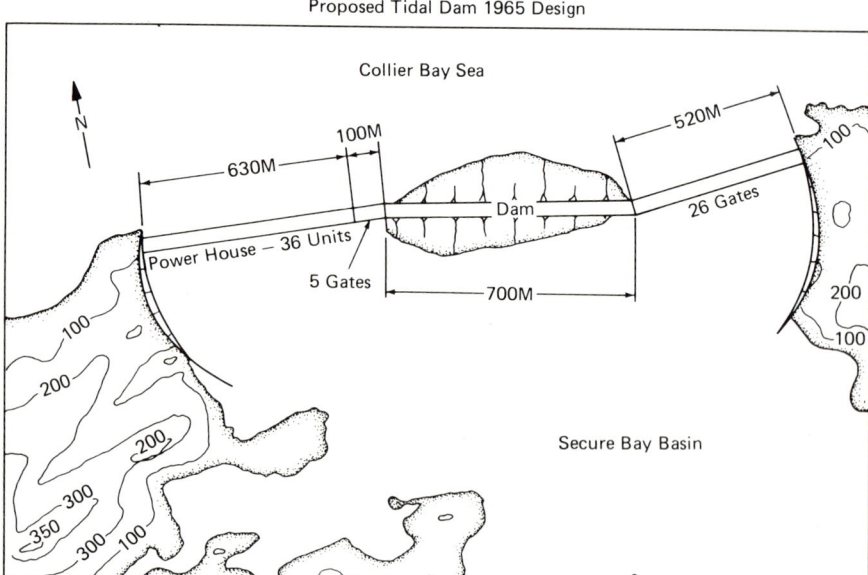

Fig. 6.8. Australia. Kimberley tidal power plant. Proposed tidal dam, 1965 design. (Maunsell, *Kimberley Tidal Power*)

Since the French Sogreah study of 1964, considerable changes can be taken into account. Capital costs, due to improved technology, are now far less, and oil prices are far more. The time span needed for construction can be reduced by half. Unfortunately, other expenses are now higher (e.g., electrical and mechanical equipment). Nevertheless, the contemporary cost is 25% lower (although the Sogreah estimate was in 1964 $). The price of a kilowatt of tide-generated electricity remains at least three times as high as that from fossil fuel.

Among the preferred sites are Secure Bay and Walcott Inlet. Here, pumped storage and a double-basin scheme appeared most attractive. The Yule Entrance to Walcott Bay and The Funnel at Secure Bay could be dammed and seawater could enter through the first bay and exit through Secure Bay. Water level in Walcott Bay would be held at a constant level above that of Secure Bay. Bulb-type turbines are small and have a small unit capacity. Studies are now pursued, whether large turbines, fewer in number, could be substituted and successfully floated in place. It must further be weighted whether the added cost of a double-basin scheme, which provides more power and more energy capacity but sacrifices some of it to achieve a firm output level, is in the long run economically profitable. (Fig. 6.9)

The scheme proposed in the Maunsell *et al.* 1976 study is based on the use of caissons as was done in the U.S.S.R.; foreseen are a 465-m power station, 30 bulb turbines of 7.65 m in diameter, two sluice sections each 152 m long, and flanking dams of reinforced concrete and rock-fill. Flap gates, allowing flood flow into, but not ebb flow out of, the tidal basin, would let the average basin level rise well above mean sea level, ebb flow area then being far less than flood flow area. Differential head in ebb tide direction would be about 10 m under some tidal conditions after flap gate installation, while differential head in flood tide direction would remain about 2 m. This feature would help when closing off the superstructure of the tidal power station and sluices. (Figs. 6.10, 6.11)

Of all the Australasian areas studied by Lewis (1963) and covering besides Australia, Tasmania, Indonesia, New Guinea, and New Zealand, nothing appeared to even remotely compare with the Kimberleys. For instance, the tidal amplitude at Waitemata is 2.7 m, north of the isthmus where Auckland is located, and 4.2 m at Manukau, south of the isthmus; but, in the St. George Basin (Australia) at the Prince Regent River mouth, tidal differences reach about 11.7 m and at least another 10 sites have equal tidal level differences. (Fig. 6.12)

166 TIDAL ENERGY

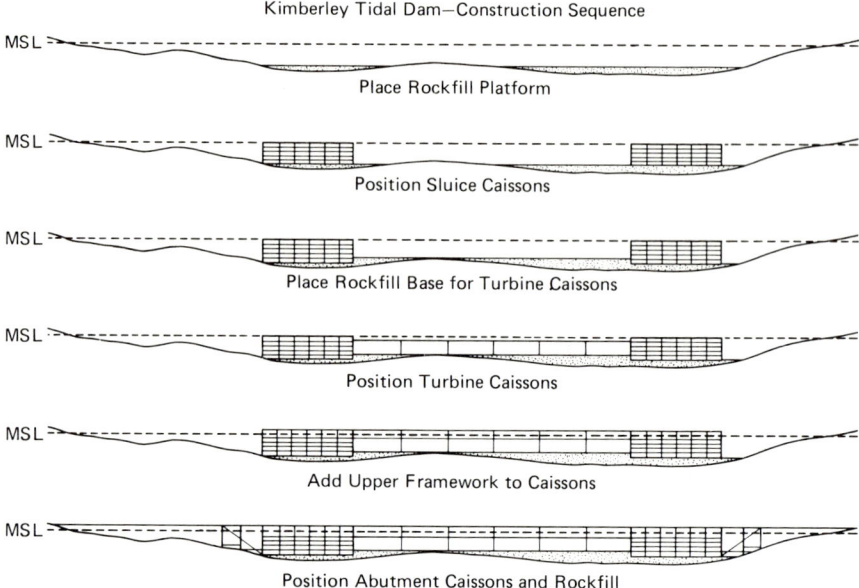

Fig. 6.9. Australia. 1976 proposal for Kimberley tidal power plant. Dam construction sequence. (Maunsell, *Kimberley Tidal Power*)

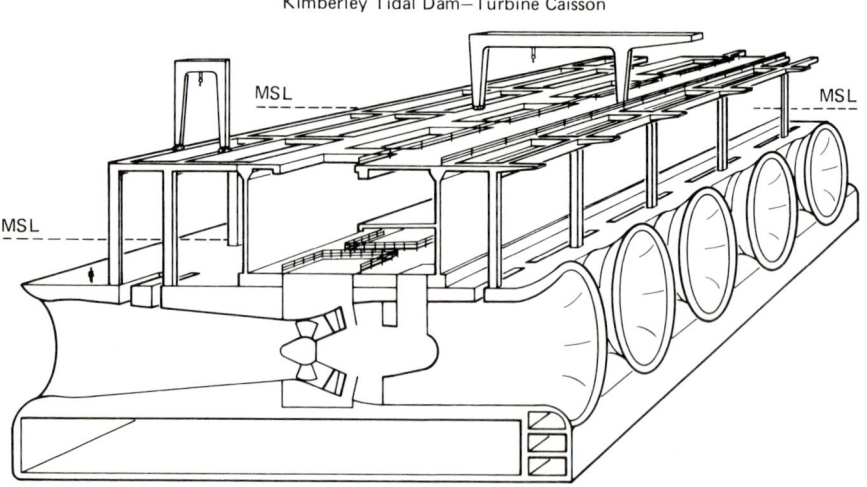

Fig. 6.10. Kimberley tidal power plant. Turbine caisson. (Maunsell, *Kimberley Tidal Power*)

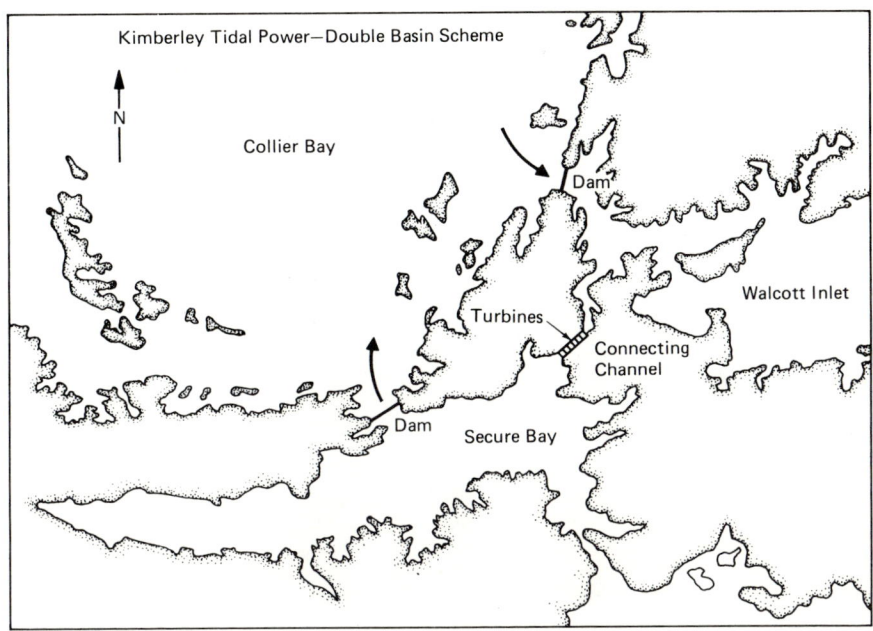

Fig. 6.11. Kimberley tidal power. Double basin scheme. (Saunders, *Kimberley Tidal Power*)

One can safely foresee the development of heavy industry (e.g., steel, aluminum, copper, lead, and zinc) in Northwest Australia if one or more power plants were constructed here. All sites are not equally favorable from an economics viewpoint, but according to Lewis, the most encouraging facet of the Australian schemes is that the original loan could be very rapidly amortized by a sinking fund. Thereafter, the cost of producing power would most probably remain below that of nuclear-produced power.

The best prospects for tidal power development would be implantation of mineral processing industries using electric smelting or other refinery techniques.

While Australia's present power production is about 7 million kW, the tidal power resources along its Northwest coast between La Grange Bay, just south of Broome, and Darwin Harbour, amount to some 300 million kW. By world standards, the possibilities are not only spectacular from a production point of view, they are also uniquely favorable tidal-wise and topographically.

Transmission costs, an important consideration because the power

Fig. 6.12. Government ship *Kangaroo* at Broome, Western Australia. *(Source: Australian National Tourist Office)*

could not be used up locally, can be reduced by higher loads, high load factors, and voltages exceeding 500 kV.

Nevertheless, construction of any plant has been shelved because of the remoteness of the region and the absence of a sizable population, hence of a market for the electricity eventually produced. And thus, what is probably the biggest single hydroelectric power potential in Southeast Asia, remains still untapped.

ASIA

Besides sites in Soviet territory locations under consideration include Korea, China, and Burma in Eastern and Southeast Asia. In Korea, barrages have been envisioned on the Seoul River in South Korea and Yangkakto in North Korea, where tidal range is 7.85 m.

Tides reach for instance 4.87 m at Pingyang and top 8.77 m at Jinsen (Chemulpo), originally a small fishing village which expanded during the Japanese occupation, now a fine harbor on the west coast of Korea.

On Korea's east coast, the maximal tidal amplitude exceeds 10 m and such high ranges are particularly frequent along a 160-km stretch of coast between the north shore of Kanghwa Island and the north end of Chonsu Bay. Very favorable conditions for tidal power harnessing are reported by Won near Inchon Bay where large tidal ranges (minimum 6 m), shallow waters, and suitable bedrock are all present. A plant of 400 MW could be built there. At least ten good sites have been identified. They could provide a capacity in excess of 1,300,000 kW. The Korean government has scheduled construction of a 400 MW plant in Inchon Bay to be completed in 1985, at an estimated cost of $300 million. Studies are conducted for a facility at Asan Bay (320 MW), a second at Cheonsu Bay (460 MW), and still another at Garorim Bay (330 MW). A five-year span is considered normal between in-depth study of site and scheme design, and construction completion. (Fig. 6.13)

Korea had examined potential sites in Inchon and along Kyonggi bays as early as 1930; in 1957, ten west coast bays had been suggested as suitable locations; and, in 1974, Sogreah, who made feasibility studies in South America and Australia, selected from the Asan, Garorim, and Cheonsu bays, Asan Bay as the best.

Along Korea's west coast, semi-diurnal tides, with some diurnal inequality a day or two after maximal lunar declination, increase in amplitude from Cheonsu Bay to Inchon; the high tide lag between these two locations is about an hour and a half. Interestingly, at Inchon, highest tides always occur around 6 AM and 18:00 hours and, inversely, the lowest tides are always at noon and about midnight. This phenomenon affects the ability of a scheme to supply power at times of peak demand. The sites lie at distances 35–150 km from the capital city of Seoul on the littoral band, where keen competition for space between a wide range of economic endeavors is developing.

Preliminary reports tend to favor a single-basin, single-effect plant, but possibly with adjustable blades to allow eventual use of the turbines as pumps. A caisson-type construction plan was used in rating several sites, and after a first round of eliminations, ten sites were retained. They were ranked by cost of energy, based upon the ratio of capital cost to total annual energy production, multiplied by an annual fixed rate of 8.14%. Inner and Outer Asan Bay, with ranges of 6.1 m, placed first and second (46.0 and 54.5 mills/MW-hr, respectively), with Incheon and Garorim bays a close third and fourth (54.6 and 58.6 mills).

170 TIDAL ENERGY

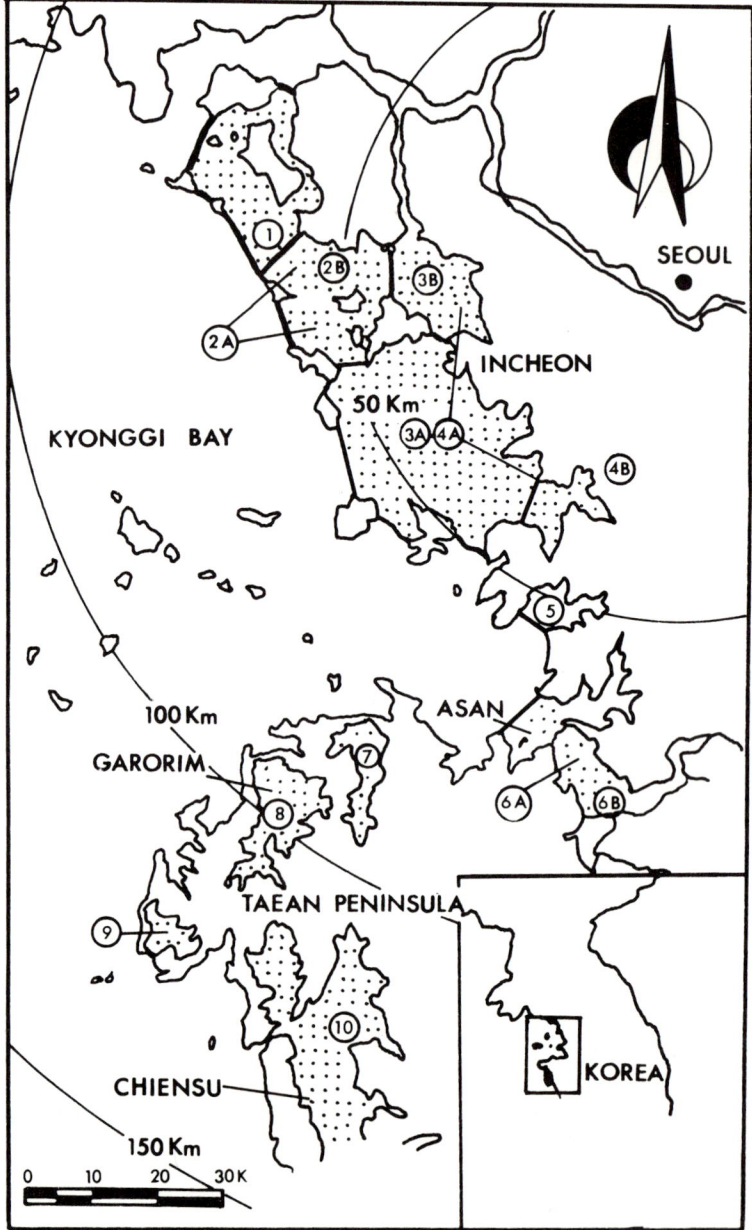

Fig. 6.13. Korea. Tidal power plants location map.

Basic calculations were made using a 15-MW fixed-blade bulb-type turbine and venturi-type sluices. In the ensuing economic analysis, Asian sites show benefits to cost ratios of 0.94 and 0.83 and savings per year of respectively 2,201,430 barrels of oil (350,000 kl or approximately 300,275 metric tons) and 3,648,084 barrels (580,000 kl or about 497,600 metric tons). However, the average value of "displaced energy" was only estimated to be 21 mills/kW-hr for fuel cost, because a price of $13/barrel had been used. With oil now well on its way to cost three times that much, the average value of "displaced energy" is much higher.

The preliminary study had concluded that none of the sites was economically viable "at a discount rate of 6% and a differential escalation in the price of oil of 2% from the year 1990 to 2010." Multipurpose development schemes, however, could provide that viability. So can the price of oil. As a part of its five-year plan (1982–1987), Korea is proceeding with a full feasibility and environmental impact study that might lead to a start of construction in 1984 and operations in 1986.

No plans exist, apparently, to build a tidal power plant on the Gulf of Martaban near Rangoon, but the Chinese are reputed to have implemented several modest schemes. Scant information is available on a 3,350-MW facility in the Bay of Tokyo and another site in the Inland Sea of Japan.

Rangoon, one of the most important commercial cities of South Asia in pre-World War II days, particularly for rice trading, is located on the Martaban Gulf, about 34 km up the Rangoon River, the eastern outlet of the Irrawaddy River in the Irrawaddy delta. Tidal ranges reach 5.5 m at Mergui and 5.8 m at Rangoon.

While major plants could be constructed along the coast near Amoy (Szeming), where tides have an amplitude of 4.72 m, and Shanghai, the Peoples' Republic of China instead constructed 40 small tidal power plants in 1958 whose total capacity reaches 583 kW. Apparently, use was made of existing dams and dikes. The potential tidal energy has been estimated to top 90 billion kW-hr. Scant information is available; the Chientang Kiang estuary has a 7-m tidal range and could produce 7,000 MW. Three other sites, each with a capacity of 1,000 MW, are under study for the gulfs of Fuchin Wan, Shinhwang Wan, and Sanmen Wan.

According to recent information secured from personal communications at the Fifth International Ocean Development Conference (Tokyo,

September 1978), China contemplates completing between 80 and 90 small plants with an aggregate capacity of 7,000–7,200 kW. The current largest facility is the Taliang plant (Shunte River) equipped with eight turbines: three of these placed between the upper and lower basins have a 144-kW capacity, and the other five are placed between the low pool and the Shunte River and account for the remaining 160 kW of the combined scheme. There are probably now 128 small plants functioning with a combined capacity of 7,638 kW.

According to a 1976 survey of alternate energy sources conducted by Canadian researchers, the government of India is also reexamining tidal power (see Table 6-2). Considered for several decades are sites on the Gulf of Cambay, with tides of 12 m, north of Bombay, and near Kandla on the Rann of Kutch. The government included the Rann's use study in its five-year plan. The Gulf of Cambay, however, has a silting problem which has diverted trade from the seaport located at its northern end; rather favorable tidal and economic conditions prevail near the estuary of the Narbada River on the gulf. Bhavnagar (tidal range 7 m) has over 100,000 inhabitants and Ahmadabad, on the River Sabarmati, at less than 100 miles from the head of the gulf with a population exceeding one million inhabitants, is the center of an industrialized region. Power could definitely be used here. A preliminary surveying has been made. Somewhat less favored is a site on the Gulf (Rann) of Kutch, an inlet of the Arabian Sea, which adjoins at its head the Little Rann of Kutch, part of a large salt marsh. All these sites are on India's west coast. To the east, possibilities are said to exist for Dublat, but tidal ranges approximate only 3.40 m on the Hooghly River; Dublat is located on the commercially most important channel of the Ganges River, navigable to Calcutta. (Fig. 6.14)

The first study conducted in the Gulf of Cambay aimed at installing a 30-MW facility that would provide some 75 GW-hr annually. Both the Gulf of Cambay and the Rann of Kutch were rated as suitable for large schemes, while small plants could be built in the Sunderbans area of the Ganges delta (West Bengal), in a 1975 analysis by Wilson (Subrahmanyam, 1978). However, the Gulf of Cambay scheme would require an extremely high capital investment and gigantic engineering works; e.g., 40 m-high structures stretching over 26–32 km. The 15 TW-hr produced represent 17% of current generation in India and would equal 60% of anticipated annual energy requirements of Gujarat for the early 1990's. The Rann of Kutch scheme, though less expensive,

Table 6-2. Characteristics of Suggested Tidal Power Schemes in India. *Source: Subrahmanyam, K. S., Indian Journal of Power & River Development, May 1977.*

Characteristics	C_1	C_2	K_1	K_2	K_3	S_1	S_2	S_3	S_4	S_5
Mean tidal range (m)	6.8	6.8	5.3	5.3	5.3	3.53	3.54	3.54	3.54	2.97
Length of structure (km)	26	32.1	26	31	34	*	*	*	*	*
Deepest water level above low tide (m)	29	27	13	13	13	*	*	*	*	*
Area of the basin (km^2)	1972	1751	639	538	278	0.76	0.31	0.83	0.45	13.57
Installed capacity (MW)	7364	5510	1187	1182	586	1.73	2.29	1.85	2.25	15.4
Annual energy generation (GW-hr)	15394	11583	3037	2984	1266	3.105	1.544	3.466	1.744	31.193
Estimated cost (R $\times$ 10^7)**	1925.1	*	593.5	*	468	2.8	*	2.2	2.5	15.7
Cost of energy generation (p/kW-hr) with interest rate of:										
10%	14	*	21	*	40	0.96	*	0.67	1.51	0.54
5%	8	*	12	*	22	0.55	*	0.38	0.85	0.30
	Gulf of Cambay		Gulf of Kutch			Sunderbans Area Durgaduani Creek		(West Bengal Sites) Belladona Creek		Pitts Creek
	Single Basin	Single Basin	Single Basin	Single Basin	Two-Basin System Without Pumping					

*Data not available

**In 1980, 1 Rupee = $0.1254.

174 TIDAL ENERGY

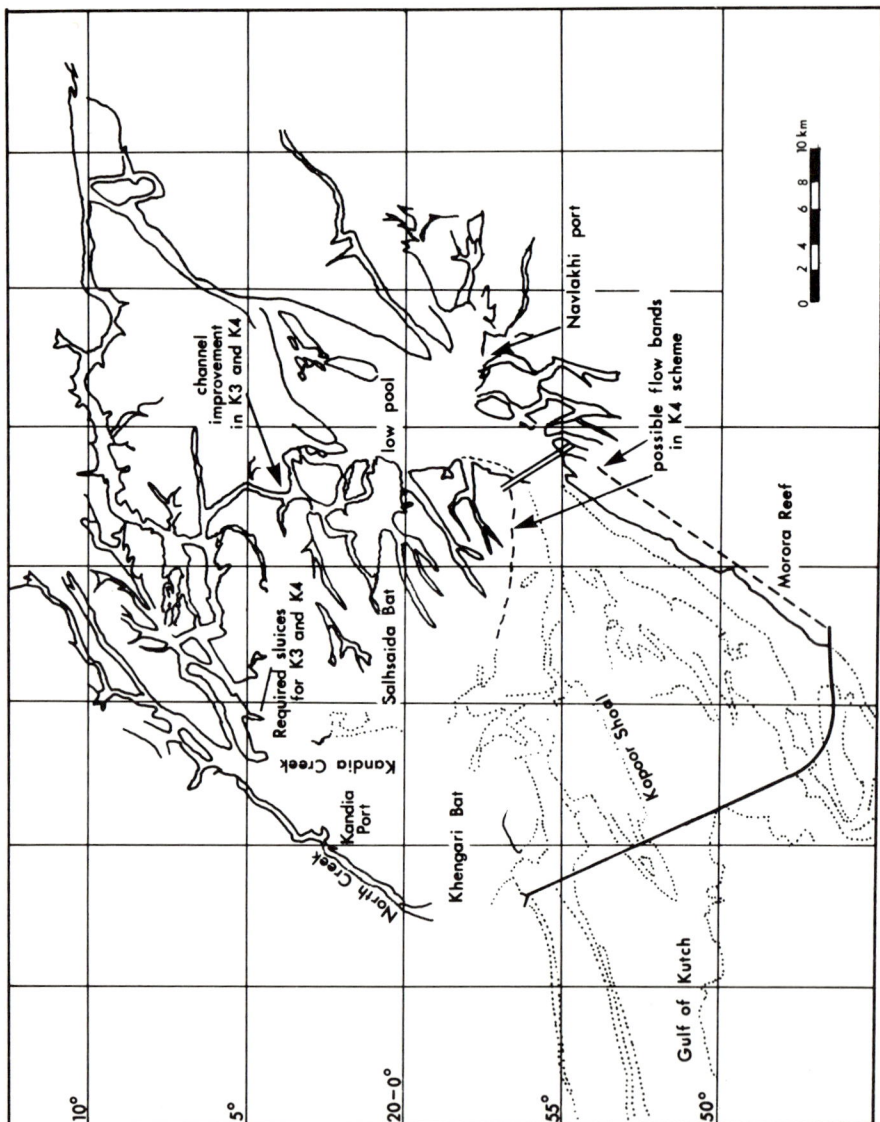

Fig. 6.14. India. Tidal power plant sites in Western India. (Adapted from *Water Power and Dam Construction*, June 1978)

involves also huge engineering undertakings, but a medium-sized plant near the port of Navlakhi would require only 5 km of barrage, provide a 260-km² basin and produce 1,300 GW-hr/yr, with the possibility of changing the proposed single-basin scheme into a dual basin operation at a later date. As for the Ganges River sites, a small plant could well serve as a pilot scheme, and if several small plants could be integrated in a grid used to develop the rather backward Sunderbans area, then what perhaps appears now as a marginal operation, may well become a worthwhile undertaking. (Fig. 6.15)

While the firming up of tidal energy poses problems, the saving on imported fuels is substantial.

AFRICA

The Arabian Sea has some high tidal amplitudes, but no proposals to tap this tidal energy are on record. Traveling westwards looms the African continent in need of energy and generally short of it. Although thalassothermal possibilities are numerous here, favorable sites for tidal power harnessing seem lacking. If modest-sized plants of local significance were to be considered, then areas south of Dakar might be suitable, as well as locations on the east coast, on Madagascar and on the Comoro Islands, where ranges reach 3.30 m at Manjunga (Comoros). However, such amplitudes fall well below the generally recommended head. The site of Porto Gole, in Guinea-Bissau, with tides of 5.5 m is occasionally mentioned but no plans for construction have ever been put forth.

EUROPE

Tidal energy tapping originated in Europe, a continent particularly well endowed with favorable sites. Obviously, it is also the area that has been most often examined. Besides locations in Spain, Germany, France, England, and Wales, and the Soviet Union, there are also sites in the Netherlands, Norway, and Scotland. A combination current and tidal energy plant was built and operated for a while on the north coast of Iceland. A Canadian study reported that the Norwegian Government, though maximum tides here are only 3.8 m, was reexamining possibilities of tidal power harnessing. The Dutch, on the other hand, decided as far back as 1957 not to include a tidal power scheme in their Delta

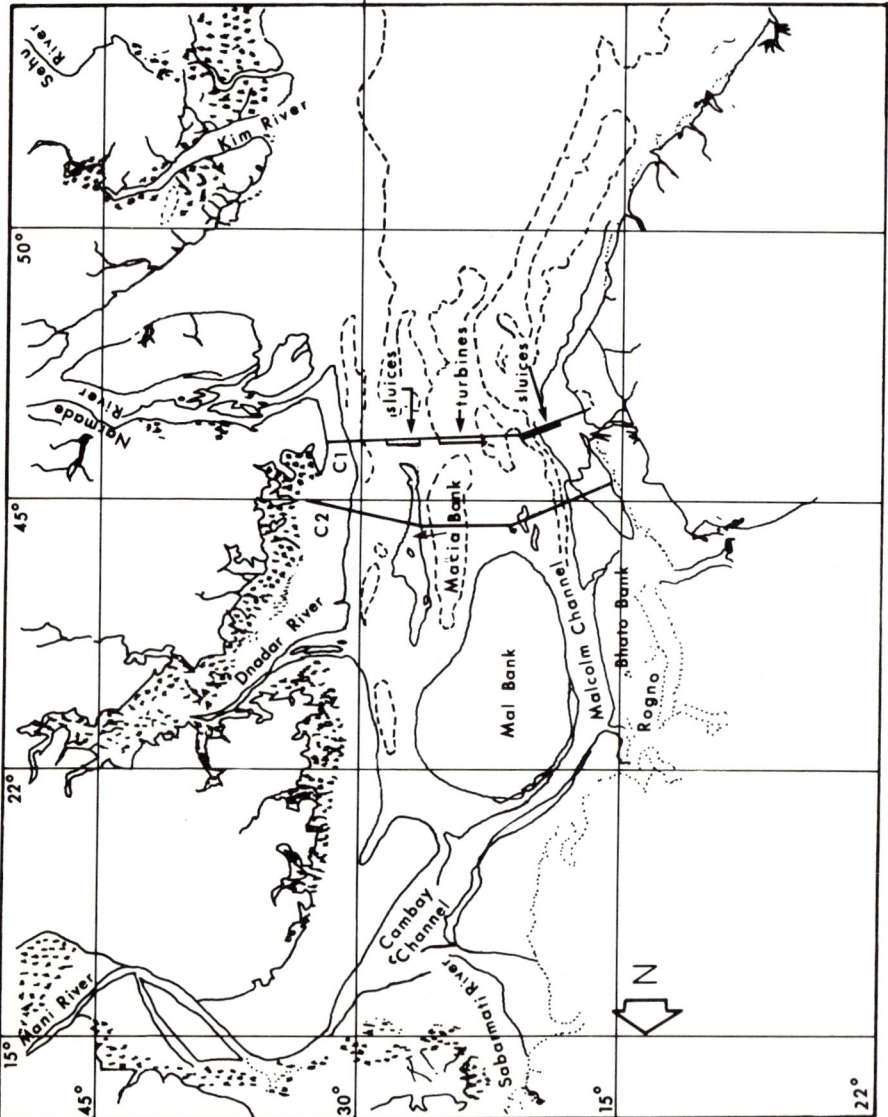

Fig. 6.15. India. Tidal power plant sites in the Gulf of Cambay. (Adapted from *Water Power and Dam Construction*, June 1978)

Plan, which is changing the complete Scheldt delta hydrological configuration. Their decision was based on the impact upon the shellfish industry. Such considerations shelved as well all plans to use the strong tidal current of the Bay of Arcachon. Arcachon, on the French Atlantic coast, lies some 60 km south of Bordeaux, and of the Pointe du Verdon at the Garonne River's mouth; an important oyster production center, it received much attention in around 1902 as a potential location for a tidal power station. (Figs. 6.16, 6.17, 6.18)

Great Britain

In Great Britain, at least two sites, the Severn River estuary and the Solway Firth, are under active government consideration. The Severn has Europe's highest tidal amplitude—14.5 m—and its potential energy output could average as much as 20,000 GW-hr. The Bristol

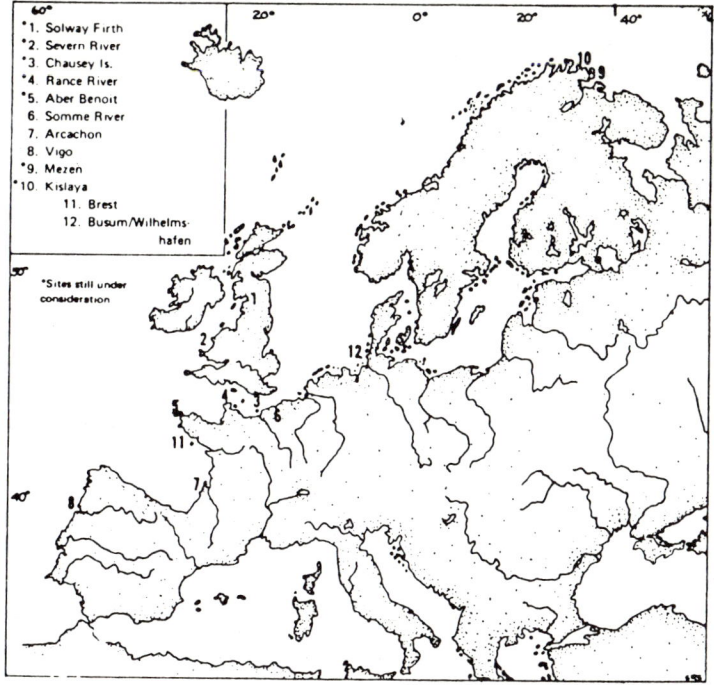

Fig. 6.16. Europe. Sites of proposed tidal power plants.

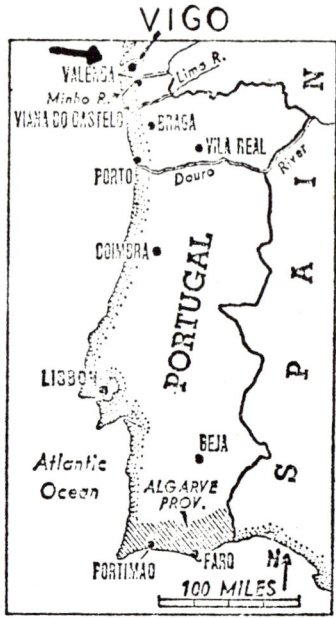

Fig. 6.17. Spain. Tidal power site.

Channel could even produce 50,200 GW-hr, according to Bernshtein (1965). The Solway Firth, with mean tides of 5 m on the average, could provide perhaps as much as 3,200 GW-hr of energy. Two other sites, Norecambe Bay and Carmarthan, with mean tidal ranges of 6 and 5.5 m, would theoretically have outputs of, respectively, 10,000 and 7,000 GW-hr. (The Severn project is examined in detail in a later chapter.)

Netherlands

The tidal energy in the Meuse-Scheldt-Rhine delta, in the Netherlands, was actively tapped in the past. Historical records show a tide mill functioning in the early thirteenth century; most of these mills were used for grinding wheat. As elsewhere, in the twentieth century, the idea to convert tidal into electrical energy came to the foreground, but tidal ranges of less than 4 m and the lack of near-site customers were discouraging factors. All projects dealt with the Western Scheldt: a single-effect scheme generating electricity as the basin emptied, at Veeregat-Zandkreek, and two-basin projects at Veeregat-Zandkreek-Western Scheldt,

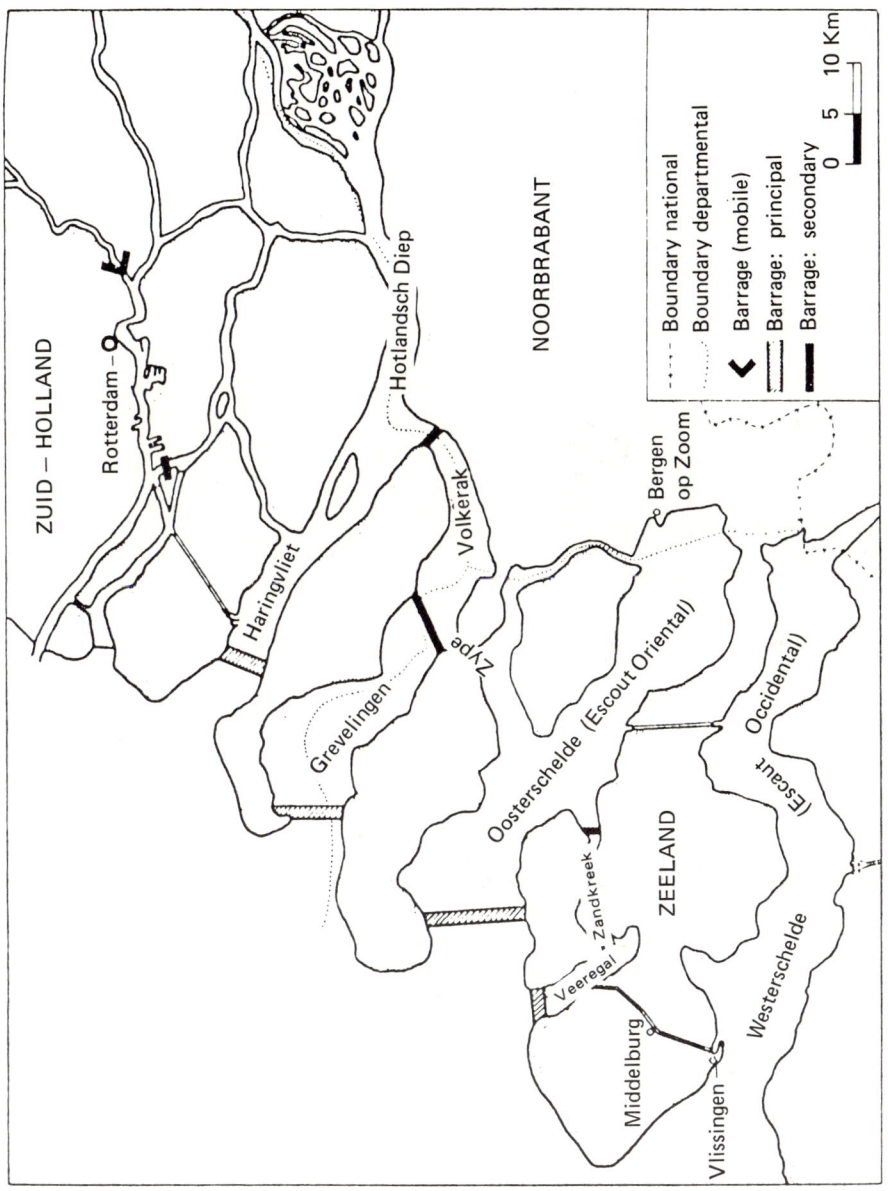

Fig. 6.18. Delta barrages. (F. J. De Vos, "Energie de la mer," *Comptes-Rendus des IV* Journées Hydrauliques*, 1975, p. 466)

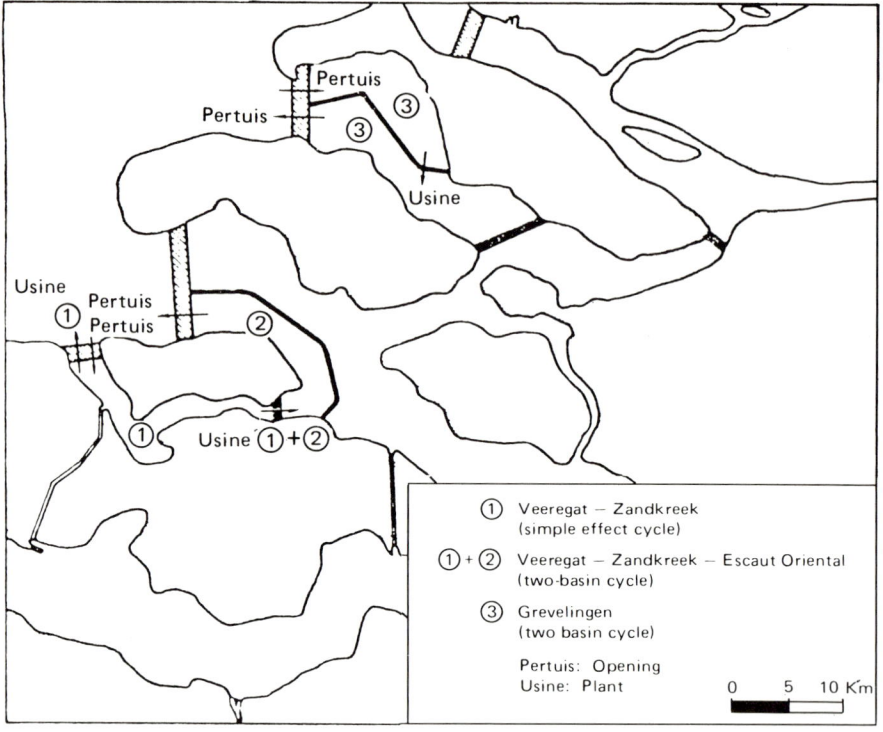

Fig. 6.19. Dutch tidal power plant proposals. (F. J. De Vos, "Energie de la mer," *Comptes-Rendus des IVes Journées Hydrauliques*, 1975, p. 466)

and at Grevelingen. Installed power (in kW) and annual production (in 10^6 kW-hr) were, respectively, for the three sites: Veeregat-Zandkreek, 16,000–60; for the two-basin scheme there, 16,000–70; and, at Grevelingen, 12,000–70. (Fig. 6.19)

The study considered possible advantages for oyster and mussel production. But even if they would benefit from the creation of retaining basins, the economic results would probably not warrant construction of a tidal power plant; furthermore, agricultural planning militates in favor of as large a fresh water reservoir as possible, which conflicts with a tidal power scheme.

France

The Chausey Islands project, presently shelved, would have cost (in terms of 1975 prices) about $8 billion, and the operational costs are estimated at $0.9 billion or 11% of this amount. With dams of at least 25 km and possibly 40 km in length, turbines of 40 MW each, and installed capacity of 12 GW, annual production could reach 27 billion kW-hr. The Rance River plant is the topic of the next chapter.

The Soviet Union

Several locations have been under study both in Europe and in Asia. An experimental tidal power plant, based upon the Rance prototype, but with a novel construction approach, was built in the Kislaya Bay. That Arctic Ocean station is discussed in a later chapter.

REFERENCES

Argentina

Fentzloff, H. E., 1972. The tidal power plant "San Jose," in Gray, T. J. and Gashus, O. K. (Eds.), *Tidal Power*. New York: Plenum.
Sogreah, 1959. *Usina Maremotriz del Golfo San Jose. Anteproyecto*. Grenoble (France): Société d'Etudes et des Applications Hydrauliques.
Sogreah, 1959. *Construccion del Canal San Jose*. Grenoble (France): Société d'Etudes et des Applications Hydrauliques.
Sogreah, 1959. *Anexo. Calculado Aproximado de la Influencia de la Rugosidad de la Inercia del Agua Sobre la Caida Turbinalle*. Grenoble (France): Société d'Etudes et des Applications Hydrauliques.
1958. Tidal power; Argentina signs contract for study, *Engineering News Record* **160**, *No. 1* (October 19): 56.

Australia

Easton, W. R., 1921. *Report on North Kimberley District*. Perth (Western Australia): North West Department Government.
Gordon, F. R., 1964. *Secure Bay-Walcott Inlet tidal power scheme. Preliminary Geological Report* (May 4th). Perth (Western Australia): Government Western Australia. [Record #196416].
Lewis, J. G., 1963. The tidal power resources of the Kimberleys, *J. Inst. Engineers, Australia* **35**, *No. 12*:333–345.
Maunsell *et al.*, 1976. *Kimberley Tidal Power*. Perth (Western Australia): The State Energy Commission of Western Australia.
Raynor, C. J., 1950. *Power from the Tides with Special Reference to Western Australia*. Thesis: Perth, University of Western Australia.
Public Works Department, 1961. *Tidal Predictions, North West Coast*. Perth (Western Australia): Harbors and Rivers Branch.
Saunders, D. W., 1974. Kimberley tidal power revisited, *Tech. Conf., The Inst. Eng., Australia*. Melbourne (Vict.): 47–55.
Saunders, D. W., 1976. Kimberley tidal power, *Proceedings ANZAAS Congress*.
Scott, W. E., 1976. "Australia takes a new look at tidal energy," *Energy International* **13**, *No. 9*:41–43.

Sogreah, 1965. *Collier Bay Tidal Power Development. Secure Bay: First Stage Study Conclusions.* Perth (Western Australia): Ministry for Industrial Development.
Sogreah, 1965. *Collier Bay Tidal Power Development. Secure Bay: First Stage Study Conclusions.* Perth (Western Australia): Ministry for Industrial Development.
Sogreah, 1965. *Collier Bay Tidal Power Development. Report to Public Works Department of Western Australia.* Grenoble (France): Société d'Etudes et des Applications Hydrauliques.

Brazil

Ferreira Monteiro de Castro, J., 1956. Centrais hidroelectricas utilisando o potencial das mares [Report to the First Congress of Engineers, Rio de Janeiro, on the possibility of utiliaztion of tidal power in Brazil], *Rev. Escola de Minas* **20**, 2:23-27.

China

Ch'iu Hou-Ts'ung, 1958. The building of the Shamen tidal power station, *Tien Chi-Ju Tung-Hsin* **9**:52-56.
Hiao Ying, 1957. "Tidal power driven turbopumps," *Chunkuo-Shuili* **9**:27.
Mao, W. J. and Deng, B. L., 1980. *An Introduction to the Development of Small Hydro-Power Generation in China.* New York: United Nations Industrial Development Organization (May).
Tseng, A. A., et al., 1979. *The Role of Small Hydro-Power Generation in the Energy Mix Development for the People's Republic of China.* Palo Alto, CA: Oriental Engineering and Supply Co. (Oct. 31).
1980. *Electric Power for China's Modernization: The Hydroelectric Option.* Washington, D.C.: Central Intelligence Agency (May).
1959. *Proceedings of the All-Chinese Conference on Tidal Power Utilization.* Shanghai: Pamphlets.

Germany

Seifert, A., 1948. "Gezeitenkraftwerk in Wilhelmshafen," *Arch. Energiewirtschaft* **IV**:209.

India

Dillon, G. S., 1975. Scope of tidal power from Indian estuaries, *Indian J. Power and River Develop.* (Aug.):245-252.
Subrahmanyam, K. S., 1978. Tidal power in India, *Water Power and Dam Construction* **6**:42-44.

Korea

[Korean] Ministry of Commerce and Industry, 1978. *Prefeasibility Study of Tidal Power Development (Incheon Bay).* Seoul (Korea): Research Institute of Shipping and Ocean.
Korea Electric Company, 1978. *Tidal Power Study 1978: Phase 1.* Seoul: Korean Ocean Research & Development Institute and Shawinigan Engineering Co. (August).
Song, W., 1979. Korea tidal power project—Phase 1 study, *Proc. Int. Symp. Korea Tid. Power* (Nov. 14-15). Seoul.
Won, T. J., 1975. The tidal power resources and their power generation projects of the western coast of Korea, *Pacific Sci. Congr.* [Vancouver, B.C.] **XIII**, *No. 1*:162.
Won, T. J., 1977. Tidal power projects on the west coast of the Republic of Korea, *World Energy Conf.* [Istanbul] (Sept. 19-23).

Netherlands

De Vos, F. J., 1957. Raisons pour lesquelles aucune usine marémotrice ne sera insérée dans le nouveau projet d'endiguement dit "Deltaplan" en Hollande, *La Houille Blanche (IVes J. de l'Hydraul.)* **11**:465–471.

Lingsma, J. S., 1963. *Holland and the Delta Plan.* Rotterdam: Van Nigh & Ditmar.

1967. Power from the Rance and the Rhine, *The Engineer* **223**, *No. 5790,* (Jan. 19):74–75.

1959. The Dutch "Delta Plan," *The Engineer* **208**, *No. 5410*:337–352.

New Zealand

Keough, Mc., 1959. Tidal power, *New Zealand Elect. J.* **32**, *No. 3*:82–83.

Spain

Vallarino, E. and Castillo, C., 1960. Evaluacion del potencial mareomotriz de las costas espagnolas, *World Power Conf.* [Madrid], Report II, C-16.

Ulster

Braikevitch, M. B., 1966. Contribution to the discussion on "Feasibility study of tidal power from Loughs Strangford and Carlingford, with pumped storage at Rostrevor," by Wilson, E. M., *Proc. Inst. Civil Eng.* **34**, *No. 1*:83–100.

Wilson, E. M., 1965. A feasibility study of tidal power from Loughs Strangford and Carlingford with pumped storage at Rostrevor, *Proc. Inst. Civil Eng.* **32**, *No. 9*:1–29.

7
The Tidal Power Plant of the Rance

On November 26, 1966, near St. Malo, the first major hydroelectric plant to use the energy of the tides went into operation on the Rance River. About a year later, the plant was ready to furnish some 500,000 kW to France, a modest amount in view of plans to produce 1 million kW by harnessing tidal power. (Fig. 7.1)

The Rance River plant is equipped with turbines that have reversible blades and let water flow as it fills and empties the basin. When fed electricity, they become powerful motors that operate the turbines as a pump and also act as a flow-regulating gate in both directions.

In the world's first actually operating system, the difference of 9–14 m between high and low tide produces over 544,000 kW. Because of its reversible operation, power is tapped from the waters as they rush upstream at high tide and as they recede toward the sea. Turbines thus generate power as the reservoirs empty and as they fill. The flow amounts to about 18,000 m^3/sec. The station is linked with France's national electric grid. The reservoir's level can be raised by pumping. The storage ability of the station is a most valuable asset. The volume of water displaced reaches 718 million m^3; when in full operation, 24 generating units are put to work.

SITE SELECTION

Designed in 1959, started in 1961, and costing close to FF 570 million (then $100,000,000), the Rance River project required the removal of 1,500,000 m^3 of water and the drying up of about 75 hectares of the estuary. The cost overrun was 14%. Equipment accounted for 55%, cofferdams 13%, and civil works 28%.

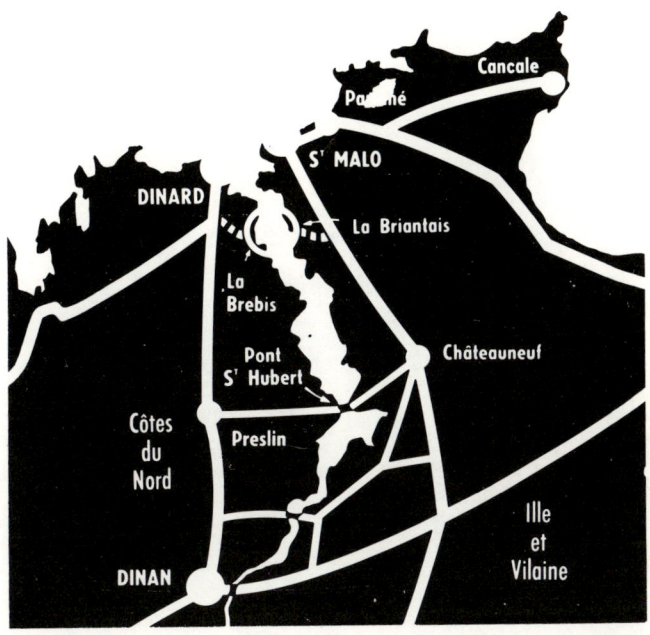

Fig. 7.1. Location map of the Rance River tidal power plant. *(Source: Electricité de France)*

The decision to select the Rance River for the plant's location was not a sudden one. Up to the very last moment, many supported a larger plant near Mont St. Michel. According to Robert Gibrat (1966), not less than 21 sites, besides the Rance, were considered at one time or another for a tidal power plant construction. Still under study are projects involving the Chausey Islands in the Bay of Mont St. Michel and the Minquiers Islands. Both would exceed in importance the Rance River plant. All locations were on the English Channel, the Dover Strait, and the Atlantic. However, the more recent schemes all deal with sites on the Normandy and Brittany coasts. (Fig. 7.2)

The Breton littoral of the Channel is one of the world's regions where the greatest tidal amplitudes occur. The phenomenon is due, in part, to the obstacle constituted by the Cotentin Peninsula which protrudes in the path of the tidal current coming from the Atlantic Ocean. In the Rance River, the height difference between high and low tide can reach 13.5 m. The tidal wave rolls in at a speed of 90 km/hr between Brest and Saint Malo.

Fig. 7.2. Rance River prior to plant construction. *(Source: Electricité de France)*

The tides are of the semi-diurnal type, with two high and two low tides within a 24-hour, 50-minute time span. At equinoctial tides, some 180 billion cm^3 are discharged per second, both at ebb and flood tide. The combination of the so-called "double-effect" and the pumping process makes it possible to regulate a "tailored" production, in function of the time of day and of tidal amplitude. Hence, exploitation tends to be freed from the tidal lunar cycle in favor of the solar cycle, the cycle of man's activities.

The principles followed in the Rance River plant construction are relatively simple: by cutting the estuary with a dam, an upstream basin was created. Water rushes through the opened sluices into the basin at incoming tide, and the flow is stopped when the high tide is reached. Thereafter, the basin is permitted to empty into the sea and energy is created. The reversible blades permit also the production of electricity as the tide flows in. Thus, round-the-clock production is obtained. (Fig. 7.3)

Fig. 7.3. Model of dam at research laboratories in Grenoble. *(Source: Electricité de France)*

Final Site Selection

Three sites were considered in the Rance estuary for the location of the barrage. A first one was discarded for many reasons: scenery, social (difficulties involving new road links), and too-close proximity to the entrance to the port of St. Malo (unfavorable effects of fast currents). For a second one, the length of the dam would have been shorter, but the storage volume of the pool was considered to be too small. The site finally selected avoided the objections to the two others. Located about 4 km from the open sea, it is well protected from wave action, but did not include Chalibert Island because of the number of sets (40) planned in the early stages, at the time of the Suez crisis.

Some years later, in the new atmosphere created by the discovery of Saharan and Lacq gas deposits, the saving of fuel did not remain a major concern and, in this new prospect, the optimization of the installed capacity led to planning 32 sets instead of 40, and finally 24. The site that was ultimately selected could then take advantage of Chalibert Island, since the length between the lock and this island is sufficient for a power station housing 32 sets.

The Rance River tidal power plant bridges the estuary between the resort and fishing towns of St. Malo and Dinard. The river has a width here of 750 m; the dam crosses between the points of La Briantais and La Brebis and passes through Chalibert islet. This location is about the narrowest of the 21-km-long estuary. The actual total length is about 910 m. The river bed is granite overlaid locally by sand and gravel, and depths do not, at the lowest tide, exceed 12 m (at most), though tides are occasionally "capricious," as occurred again on March 27, 1967, at the vernal equinox, when waters receded between 13 and 20 km twice during the day. (Fig. 7.4)

The retaining basin extends over 20 km up to the Chatellier lock near the town of Dinan. The basin covers 22 km² and a usable volume of 184 million m³.

The Cotentin presqu'isle, whose presence in the path of the incoming waters causes the high tidal amplitude here, is now directly linked with the Finistère peninsula by the road constructed on top of the dam.

The lock is imbedded in the point of La Brebis. The sas is 65 m long

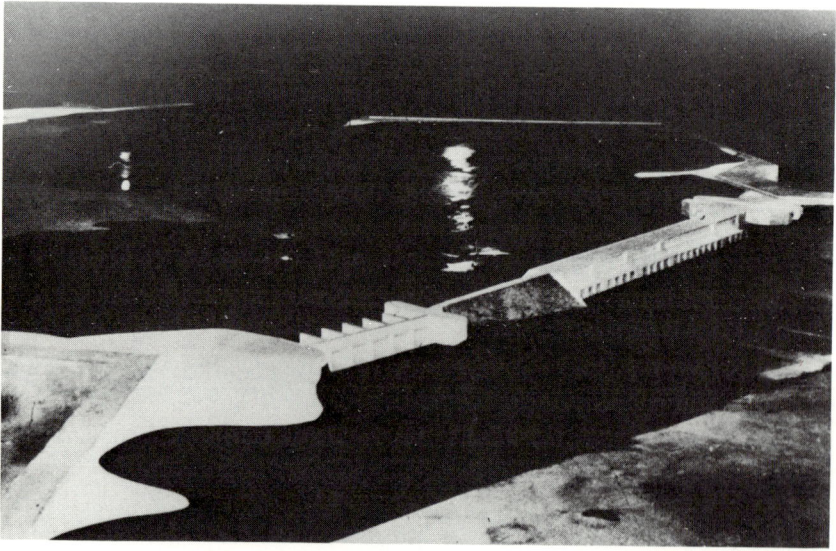

Fig. 7.4. Scaled outlay of Rance River plant in Grenoble Sogreah laboratories. *(Source: Electricité de France)*

and 13 m wide. The lock gates are vertical shaft sector gates. The plant itself is located in the deepest part of the river; it is a hollow concrete dyke, a tunnel with upstream and downstream linings reinforced by buttresses spaced 13.30 m apart. Its roof is a vault creating thrusts in the direction opposite that of the water. The plant's total width is 53 m; it is about 390 m long. Foundations are at 10 m below sea level, and the top part is at 15 m above sea level. Inside the plant are 24 bulb groups of 10,000 kW, three transformers of 80,000 VA (at 225 kV) and 4 mobile bridges of 82 metric tons each. A set of twice four groups debits on each transformer, located in a group bay especially enlarged seawards. The transformers are linked by 225-kV oil-filled cables to a substation located on the left bank (La Brebis) head, at about 300 m from the plant's western end; three lines transmit the power, respectively, to Aube and Paris, Rennes (the capital of Brittany), Landerneau, and Brest. Access to the plant is gained through a shaft on the left bank and a gallery running beneath the lock. (Fig. 7.5)

Fig. 7.5. Seaward view of the Rance River plant; to the left, the lock. *(Source: Electricité de France)*

The dead dyke links the eastern end of the plant to the Chalibert island and is approximately 175 m long. The mobile barrage or lock gates are situated between Chalibert island and the right river bank. It is 115 m long. It includes six wagon valves, or passes, enclosed by flat gates, 15 m wide and 10 m high, which enable filling and emptying of the reservoir.

CONSTRUCTION PHASES

The green light for construction resulted to a great extent from two elements: the development of reversible turbines and the capacity of the plant to "store" energy and, on the other hand, the definitely favorable conclusions reached from the operation of the reduced scale experimental St. Malo plant. The importance of the scale model plant should be emphasized. The trial runs conducted at St. Malo made it possible to foresee with utmost precision the effects of the tide's regime in the Channel and to determine the level differences to be anticipated between the sea and the estuary, for each construction stage and at different times of the day. These tryouts also provided information on current distribution and velocity, on filler rock size for the cofferdams and for the installation of these cofferdams when cutting off permanently the estuary from the sea. The actual plant construction phases showed the correctness of the indications gathered on the scale model.

Since then, in Grenoble (France), other reduced scale models have been built and studied. They showed that some tidal energy is lost by the construction of a tidal power plant. This loss, due to backwater effects, can be estimated at 15% for the projected Chausey Islands plant. (Fig. 7.4)

The entire Rance River tidal power scheme was built in the dry, inside three protective sets of weirs which were later dismantled. The first phase consisted in the building of a protective wall on the left river bank to construct the lock, and next a wall on the right bank reaching to Chalibert Island for the construction of a mobile barrage. Caissons were sunk on the site of the northern weir. The major cofferdam was then completed and the second phase started.

The lock and the valve sluices were put into service and the cut across the river was then progressively closed. Thereafter, the dam gates were shut and the water level kept at 8.5 m in the estuary, thereby permitting normal navigation to the Le Chatellier lock. The last step consisted in the completion of the southern weir. In all 400,000 m³ of earth were removed and 350,000 m³ of concrete used in addition to 16,000-ton steel and 275,000 m² of shuttering. Eleven thousand tons represent the weight of the 24 installed units, out of some 14,000 tons for all the equipment. Some 900 workers were on the site at the peak of construction. Net production is 537 million kW-hr in the basin to sea direction and 71.5 MW-hr in the sea to basin direction. Of this total of 608.5 MW-hr, the energy used up through pumping takes away 64.5 MW-hr, leaving 544 MW-hr. The total installed power is 240,000 kW, with 24 10,000-kW sets.

PLANT OPERATION

When using the turbines, the plant has—in the basin to sea direction—a capacity of 10 MW as long as the head remains 7 m. The same capacity is maintained under those conditions for the sea to basin operation. With heads of 5 and 3 m, the capacity is much lower. Pumping can be done in basin to sea and sea to basin directions; it is of great value because the difference in amplitudes is considerable. (Table 7-1) The unit performance table shows an efficiency of more than 85% for heads exceeding 7 m, in direct generation. As for pumping, the maximum input required is 6 MW per unit, or a total of 144 MW for all 24 units. (Table 7-2)

For small tides, the plant uses repeated single-effect operation with pumping, while, for medium height tides, repeated double-effect operation with pumping is utilized. With strong tides, a sesqui-cycle operation with pumping is called upon; this is a 3/2 operation that provides three periods of generation during two consecutive tide cycles. For exceptional high tides, double-effect operation without pumping is used.

The purpose of pumping is to store water and thus to increase the head near high tide; when pumping near low tide, the basin level is brought below sea level. When the head is too small for generation, the units can be used as sluices to supplement discharge. Without these technological refinements, the plant would have an annual mean output of 45 MW, versus the 65 MW that are actually obtained. Nevertheless, the output during spring tides is 2,940 MW-hr/day, for 738 MW-hr/day during neap tides.

The effect of pumping is shown in Table 7-3, and the relative proportion of operation modes in Table 7-4.

Table 7-1. Tidal Range Variations and Discharges at the Rance.

TIDE	TIDAL COEFFICIENT	TIDAL RANGE (m)	VOLUME OF WATER STORED (m³)	DISCHARGE AT FLOW (m³/sec)
Equinoctial spring tides	120	13.50	184×10^6	18,000
Ordinary spring tides	93	10.90	150×10^6	12,900
Mean tides	70	8.50	110×10^6	9,400
Ordinary neap tides	45	5.00	65×10^6	5,550
Minimum neap tides	35	3.50		4,000
Mean tidal range		8.50		
Live storage capacity at equinoctial spring tides			184×10^6	
Maximum surface area				22 km²
Dam length				750 m
Power output				500 GW-hr

Table 7-2. Unit Performance Data.

POWER GENERATION		HEAD (m)			
Direct (basin to sea)	11	9	7	5	3
Capacity (MW)	10	10	10	8	3
Discharge (m³/sec)	110	130	170	220	200
Efficiency (%)	85	87	87	72	55
Reverse (sea to basin)					
Capacity (MW)	10	10	9.5	5.5	2
Discharge (m³/sec)	13.5	160	230	200	115
Efficiency (%)	68	72	60	58	60
PUMPING (6 MW INPUT)		AGAINST HEAD (m)			
Direct (basin to sea)	1		2		3
Discharge (m³/sec)	175		168		100
Efficiency (%)	30		48		66
Reverse (sea to basin)					
Discharge (m³/sec)	200		160		108
Efficiency (%)	25		30		58

Table 7-3. Increase in Pool Level Due to Pumping. (Note: The Maximum Figure of 5 m Was Realized in August 1974 for a 5-m Tidal Range.)

TIDAL RANGE (m)	NATURAL POOL LEVEL INCREASE (m)	MAXIMUM HEAD AT END OF PUMPING CYCLE (m)
7–10	0.50–1.25	1–2.50
6–7	1.25	2.50
5–6	1.75	3.50

Table 7-4. Relative Duration of Operation Modes at La Rance River Tidal Power Plant.

MODE	%
Direct generation	56
Reverse generation	11.7
Direct pumping	14.7
Reverse pumping	1.4
Spilling (direct and reverse)	16.2
	Total 100

EFFICIENCY AND COST

Though a construction cost of 2.000 FF/kW (in 1980 about $473, currently about $345) is high, the running, including 35% government tax, is lower than the average running cost of the company's hydroplants.

Sixteen plant operators and lock keepers, 28 routine maintenance crew members, four technical and clerical employees, and the plant manager and his assistant make up the 50-man permanent staff.

Cost

Considering the total cost per kW-hr exclusively on the basis of alternative thermal capacity cost saved, thus leaving aside economic considerations which led to the decision to build, annual costs per kW-hr (in 1973) were 4.7 centimes for run-of-river plants, 7 centimes for storage plants, 8 centimes for seasonal storage plants, 10.3 centimes for conventional thermal plants, 5.3 centimes for nuclear plants, and 9.67 centimes for the Rance. This latter amount includes 8.34 centimes for amortization and 1.33 centimes for running cost. In 1973, $FF5 \cong \$1$; hence, running cost was $\pm$ 2.65 mills/kW-hr.

Efficiency

The mean net annual energy production is a little more than 500 GW-hr, and the mean capacity throughout a year is 65 MW. The Bulb Set availability is outstanding. It can reach 95%, in percentage of running time. The 5% unavailability can be broken down into 3.5% for organized withdrawal of sets for routine overhaul on neap tide weeks (i.e., one week out of two), resulting in no energy loss, and in 1.5% of forced outages, resulting in a 3.5% energy loss, in terms of output percentage. Output increased steadily, mainly between 1971 and 1972, as more pumping energy became available.

The operation policy consists in optimizing the value of the energy generated instead of generating the maximum amount of energy. On the other hand, the output of the plant has to be coordinated with that of all the other French hydropower plants, especially daily or weekly storage plants, which are operated similarly. The power plant is run following a computer program established by the Regional Control Center in Nantes, and fed into the local computer, which automatically operates the 24 units.

Mean availability of the 24 units climbed from 77 to over 95%, and unavailability of a unit is now less than 15 days a year. The gross output of 604 GW-hr, reduced by 60 GW-hr for pumping, leaves a net output of 544 GW-hr and the ratio of actual generation to design climbed from about 83% to over 91%.

Fig. 7.6. Aerial view of the four-lane road built atop the Rance River station. *(Source: Electricité de France)*

FRINGE BENEFITS

The new source of energy for France's forgotten province not only brought new pride to Brittany, it provided a chance of lifting a lethargic and underdeveloped economy into the twentieth century. (Fig. 7.6)

Fringe benefits have accrued through the building of a road on top of the dam. This new 14-m wide two-lane motorway cuts the distance between Dinard and St. Malo by 35 km. About half a million cars use the roadway in August alone.

To the oracles of doom who heralded the ruin of tourism, one of the main income sources of Brittany, proponents of the power station now

Fig. 7.7. Interior view of the Rance River plant. *(Source: Electricité de France)*

Fig. 7.8. Control room of the Rance River tidal power plant. *(Source: Electricité de France)*

point to the fact that the dam has done nothing to spoil the surrounding countryside, has even opened new vistas of the river, and has attracted crowds of tourists. The number of tourists has steadily increased, and busloads come to view both the picturesque countryside and the scientific achievement of the dam. An average 175,000 people a year flow to visit the dam, which now includes a small museum.

These facets should not be underestimated. The decision not to use the dam as a roadway was instrumental in killing the British Severn project, and the anticipated tourist visits figure predominantly in the supporting arguments of the American Quoddy region plans.

The fishing industry has remained unaffected, and original fears in

198 TIDAL ENERGY

that domain have thus been allayed. Finally, a new community, the Cité de la Gougeonnais, has been especially built for the plant's 43-man workforce. (Figs. 7.7–7.12)

In the area of technology, the new techniques have proven usable in conventional hydropower river plants and resulted in huge savings; the

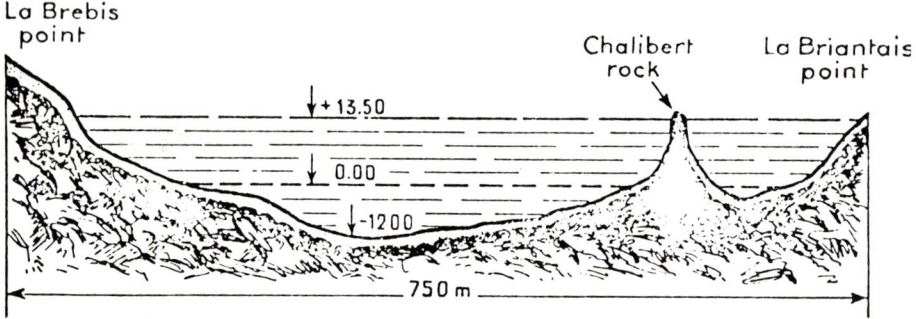

Fig. 7.9. Estuary profile at center line of structure. (Lebarbier, *Nav. Eng. J.*, April 1975)

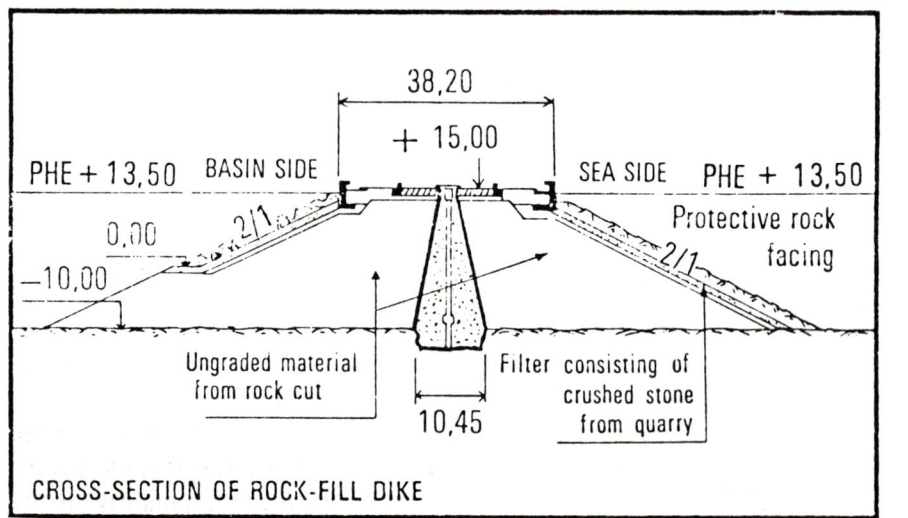

Fig. 7.10. Cross-section of rock-fill dyke. (Lebarbier, *Nav. Eng. J.*, April 1975)

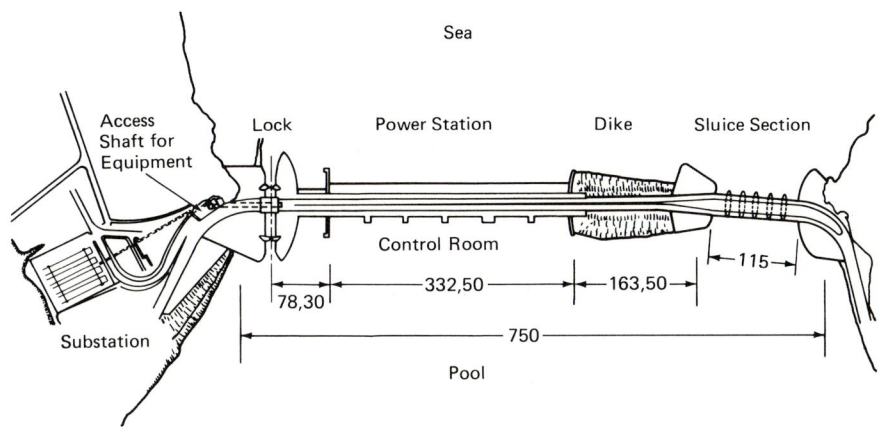

Fig. 7.11. General layout of structures. (Lebarbier, *Nav. Eng. J.*, April 1975)

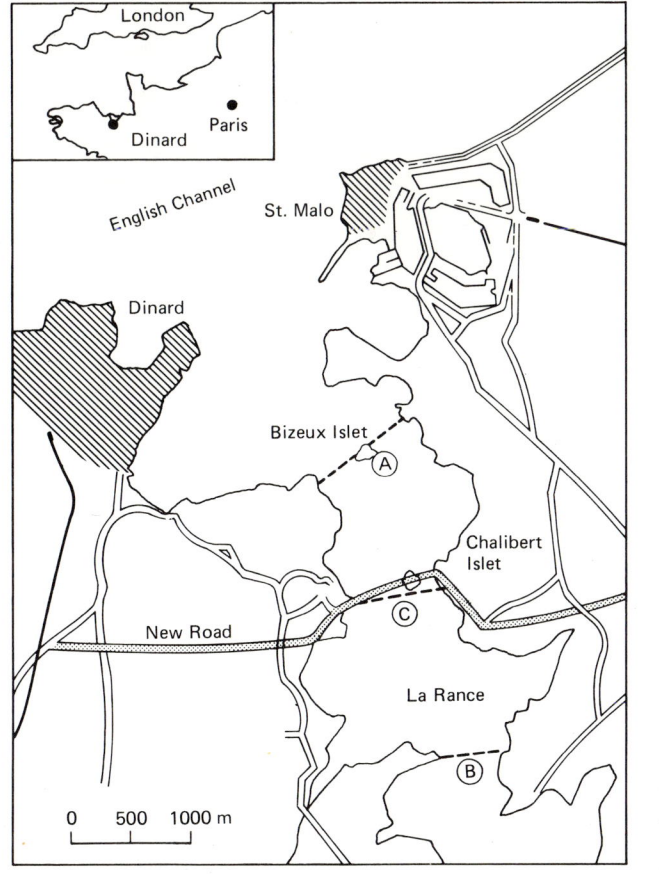

Fig. 7.12. Sites contemplated. (Lebarbier, *Nav. Eng. J.*, April 1975)

bulbs have been exported for use throughout Europe (e.g., Austria, Sweden, Switzerland, and the U.S.S.R.), as well as to Canada, and are in use in the Rock Island powerhouse on the Columbia River (U.S.). Furthermore, much knowledge has been gained in the area of construction, corrosion, and cathodic protection, the latter a matter of great concern for desalination plants.

OPTIMIZATION

Due to the high construction cost of any tidal power plant, compared with a thermal power plant taken as a reference, it is obvious that the plant cannot be justified, on economic grounds, solely on the savings in fuel oil cost. The plant must have a minimum reliable output to enable it to be regarded as an alternative to a standby thermal plant of similar output. (Figs. 7.13–7.17)

For high tidal ranges—i.e., higher than 9 or 10 m—there is no need for direct pumping, since sufficient head and water volume are available for the installed capacity of the plant; reverse generation may be used if needed, sometimes with an increased head due to reverse pumping. Reverse generation involves reduced head and storage volume for the next direct generation, but the gain associated with reverse generation largely outweighs the loss involved in direct generation. Checking many various cycles, the conclusion is that the working head is generally 5–6 m. The minimum working head depends on the moment generation occurs. This may be advanced or delayed to meet the demand of peak hours. (Fig. 7.15)

For mean and spring tides, both the ebb and flow are used for producing electricity, this generating mode being called "double-acting cycle." Electricity production is continuous, due to the reverse generation; and this makes the maximum pool level lower than the maximum sea level. In addition, pumping may be used for emptying the pool below sea level to increase the head for the purpose of reverse generation. But this operation mode is not usual. (Figs. 7.13, 7.14)

Generating sets are also used as open sluices in addition to the main sluice gates at the barrage, when spilling with a difference in head between pool and sea lower than the minimum working head (1.20 m outwards when emptying the pool or 1.60 m inwards when filling). In this case, the runner blades are feathered so as not to obstruct flow, but, for Lebarbier (1975), a tidal power plant can only be offset against the

THE TIDAL POWER PLANT OF THE RANCE 201

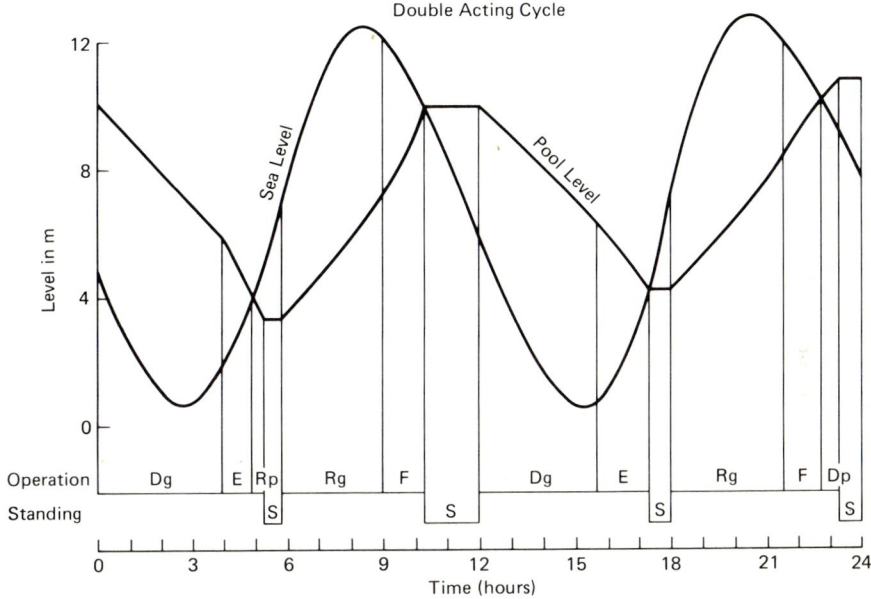

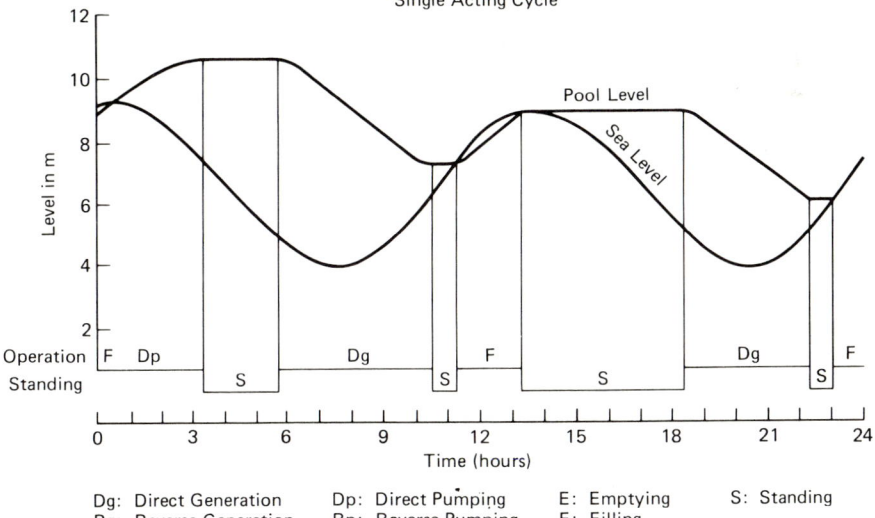

Dg: Direct Generation Dp: Direct Pumping E: Emptying S: Standing
Rg: Reverse Generation Rp: Reverse Pumping F: Filling

Fig. 7.13. Double and single acting cycles. Operation mode. (Lebarbier, *Nav. Eng. J.,* April 1975)

202 TIDAL ENERGY

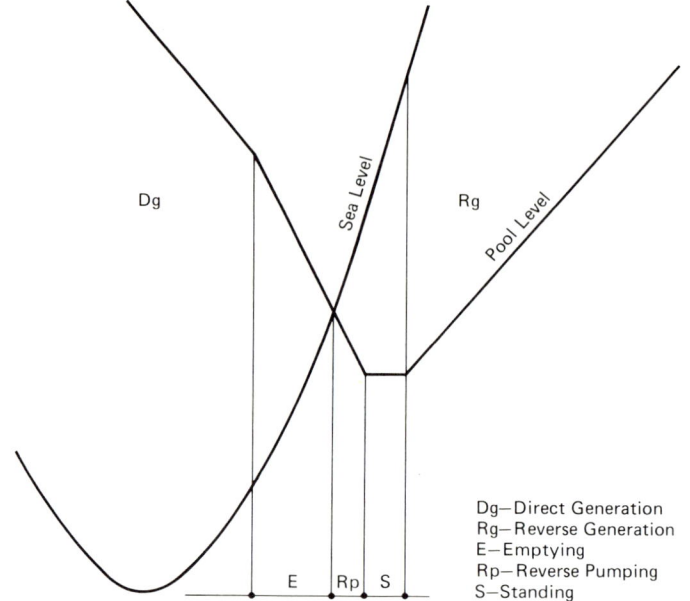

Fig. 7.14. Details on double acting cycle. (Lebarbier, *Nav. Eng. J.*, April 1975)

cost of a thermal plant in a system where there is a shortage of energy and not a shortage of power. (Fig. 7.16)

This will be the situation for another decade in France, where the electricity system comprises a large part of hydro. Not enough hydro storage plants are capable of operating at full load during the 16 weekday hours. (Fig. 7.17)

CONCLUSION

The completion of the French tidal power plant heralded man's first successful attempt at transforming ocean energy into electrical energy, and an "alternative" energy source had been put to work. Talk immediately started about more plants. Though several sites were investigated, no other French plant has been built. The technical experience (namely, the development of the bulb-type generator) was put to use in river plants, and Soviet engineers built an experimental plant in the Arctic. It is discussed in the next chapter.

The dream obviously came through, and the long way from the occa-

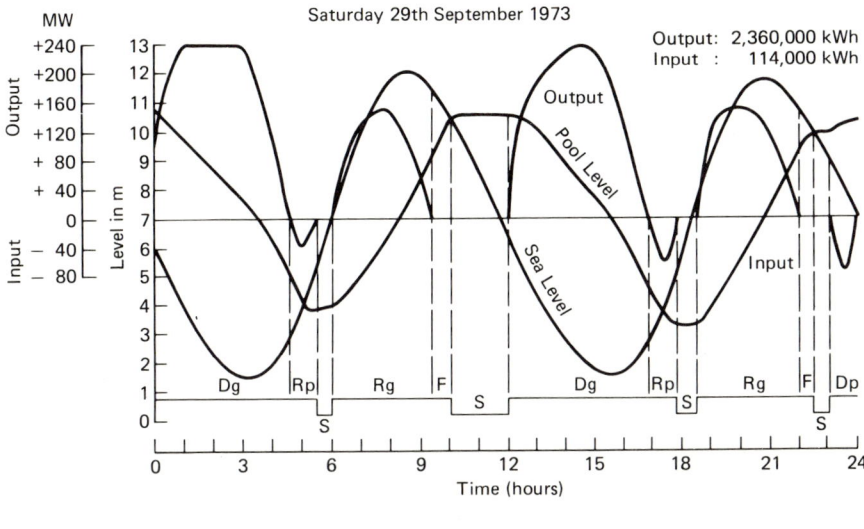

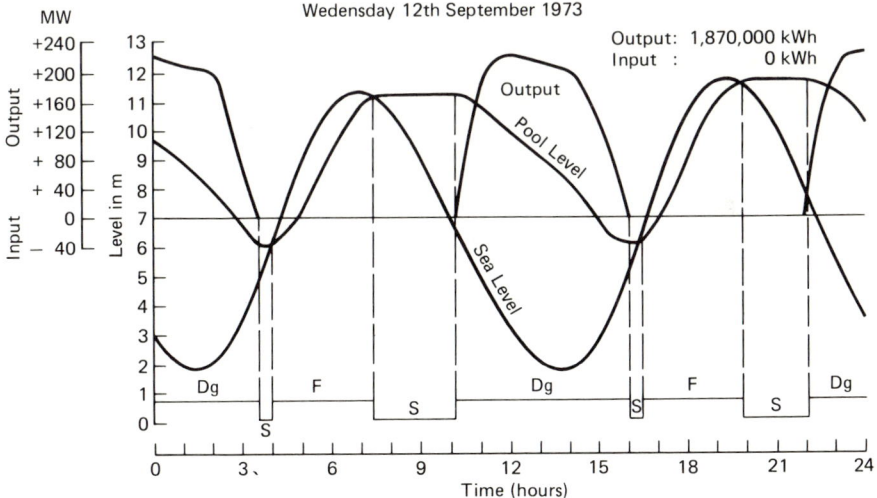

Fig. 7.15. Two different cycles for the same tidal range. (Lebarbier, *Nav. Eng. J.*, April 1975)

sionally somewhat sophisticated tide mill to the electricity-generating station has been covered. Arguments about feasibility have been settled; new ones, however, have now been brought forth—economic "rentability," in particular "staggering" construction costs. We will look at these in a subsequent chapter in some detail, but be it already said that the gap between per-kilowatt cost of thermal, and even nuclear power

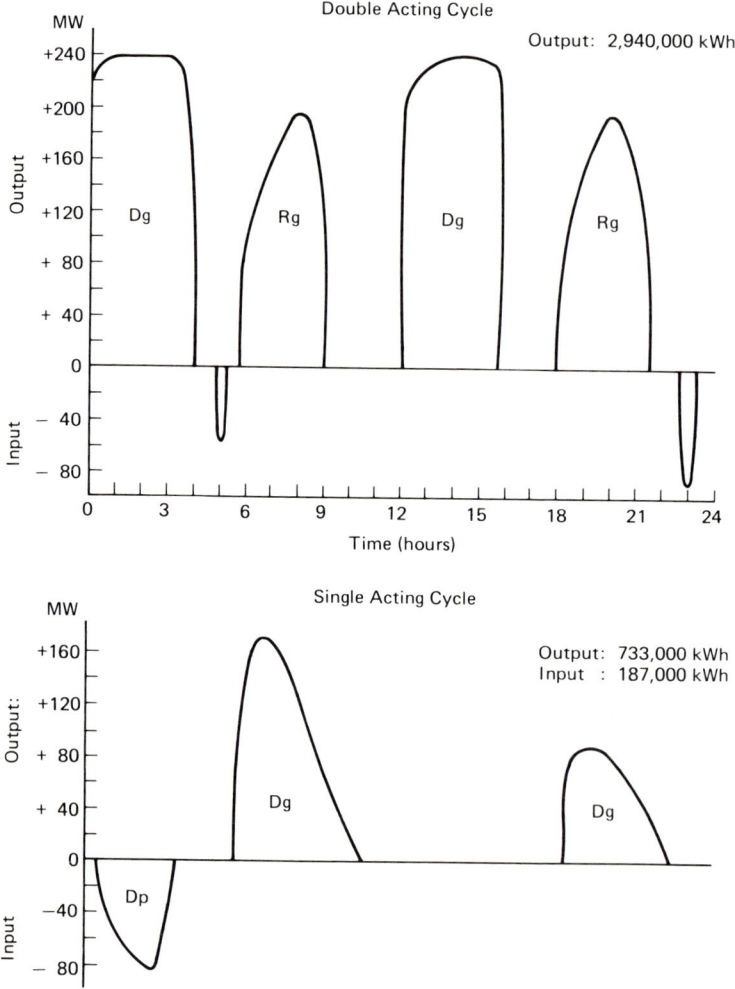

Fig. 7.16. Output. Double and single acting cycles. (Lebarbier, *Nav. Eng. J.*, April 1975)

plants, and the same kilowatt produced by tidal energy is about to (or already is, according to some Canadian estimates) be bridged.

Even since the completion of the Rance River plant technology has progressed and, apparently, some of the high construction costs of building the St. Malo station (use of cofferdams, for example) can be avoided, as the Soviets have shown and as the Canadians plan to do. Tidal power may thus well be an idea whose time has (finally) come.

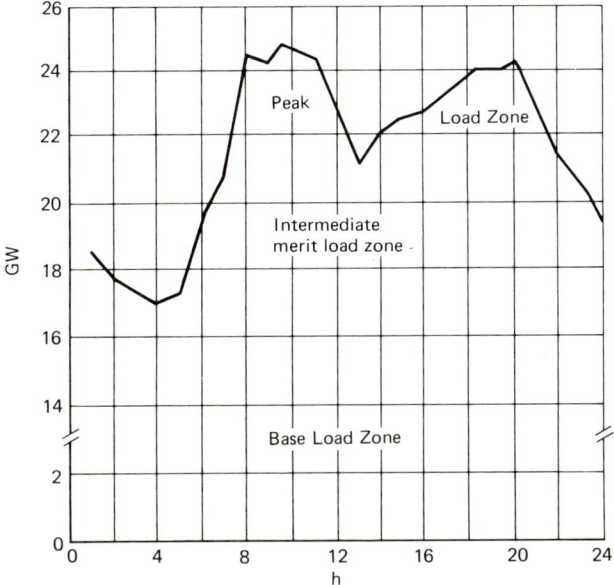

Fig. 7.17. Typical daily load curve. (Lebarbier, *Nav. Eng. J.*, April 1975)

The ever increasing demand for new sources of energy seems to be a warning not to discard the ocean as a source of energy in favor of nuclear-produced energy. To the contrary, it appears that both types of energy could be quite profitably put to use side by side.

REFERENCES

André, H., 1978. Ten years of experience at the "La Rance" tidal power plant, *Oc. Manag.* **4**, 2-4, 165-178.

Bonnefille, R., 1976. Les réalisations de l'Electricité de France concernant l'énergie marémotrice, *La Houille Blanche* **31**, 2, 87-149.

Banal, M. and Bichon, A., 1981. Tidal energy in France. The Rance tidal power station. Some results after 15 years of operation, *Int. Symp. Wave & Tidal Energy, Cambridge* **II**, *Proc.* 327-338

Gibrat, R., 1966. *L'énergie des marées.* Paris: Presses Universitaires de France, 230 p.

Gibrat, R., 1973. L'énergie marémotrice dans le monde: l'usine marémotrice de la Rance et l'environment, *La Houille Blanche* **211**, 2-3, 145-151.

Lebarbier, C. H., 1975. Power from tides: The Rance River tidal power station, *Nav. Eng. J.* **83**, 2, 57-71.

8
The Kislaya Guba Plant

The Soviet Union has probably one of the (if not the largest) reserves of tidal power resources in the world. The small gulfs alone could provide 8.2 million kW, and the White Sea by itself has a potential of 16 million kW. Tidal ranges vary between 1.2 and 11 m. Most of the tidal energy dissipation is concentrated in the Soviet Arctic, within the Kara and Laptev seas. The total dissipation for all Arctic seas reaches 2.66×10^{17} erg/sec.

A plant was built on Kislaya Bay near Murmansk, and two sites at the entrance to the White Sea and another one on the gulf itself are under study; the power possibilities reach 0.3, 1.3, and 14 million kW, respectively (Bernstein, 1965). Design work proceeds on a plant for Mezen Bay, and sites under consideration on the Okhotsk Sea, if combined, could lead to production of 130,000 GW-hr of energy. (Fig. 8.1)

KISLAYA GUBA

In the Kislaya Bay, on the Kola peninsula (U.S.S.R.), the inlet is narrow and thus water gushes in at flood tide at a speed of approximately 4 m/sec. The dam, completed by assembling prefabricated sections, takes advantage of the narrow opening between 40-m high cliffs. The experimental plant was built after the Rance plant model, but with some modifications. (Fig. 8.2)

The water surface at Kislaya (or Shalimskaya) Guba covers 1.1 km² and the depth is 35 m. The narrow communication canal with the sea is only 40 m wide and the depth merely 3–5 m. This eliminated the need to build a long dam, and the powerhouse itself constitutes the closure. The average tidal amplitude is only 2.5 m. The powerhouse is best described as a "concrete box" 36 m long by 18.3 m wide and 15.35 m high.

THE KISLAYA GUBA PLANT 207

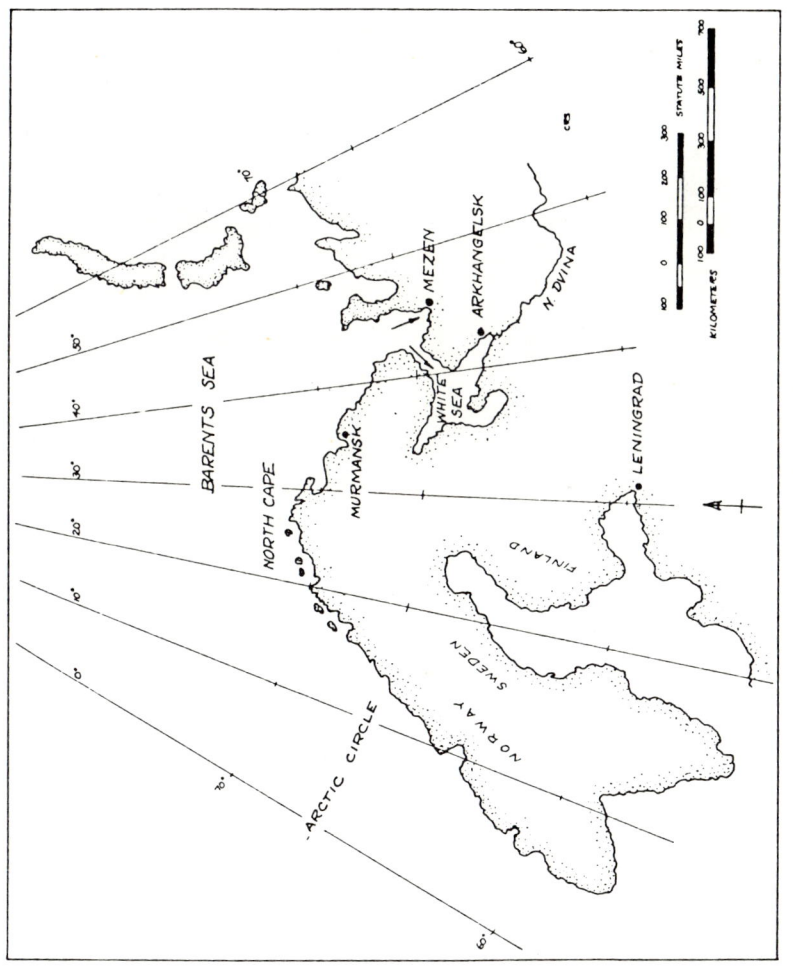

Fig. 8.1. U.S.S.R. tidal power sites in Europe.

Fig. 8.2. Display at Oceanexpo '75, Soviet pavilion, Kislaya tidal power plant.

This "box" was built of frost-resistant reinforced concrete beside the Kislogubsk River on Prityka Cape, near Murmansk (of World War II fame). Its compressive strength reached 750 kg/cm². The concrete was made with sulfate-resistant cement with 0.2 per thousand of compound and air-entraining additives, pure quartzic sand, and hard igneous rock aggregates of which 35% had a diameter of 5–10 mm and 65% had a diameter of 10–20 mm. (Figs. 8.3, 8.4)

The Soviets precast concrete cellular units, brought them to the site, and sank them into place by filling some cells with sand. The remaining empty cells were then used to house the turbogenerators. This method removes the need to build cofferdams and the mechanical equipment is not affected by differential settlement.

Their approach to dam building is also different from that used by the French in other instances. When the inlets are narrow and rocky, an atomic blast could close off a basin in a few seconds when placed in both abutments; the Soviets instead used TNT. The pilot tidal power scheme on the Kislogubskaia is using the French-type submersible bulb units. (Fig. 8.5)

Keeping the structure in place once it had been sunk was managed by filling empty cells with sand, but to waterproof the soil and to add more weight, heavy fuel oil was added in the proportion of one to four. The site itself was prepared by rock removal by underwater blasting

THE KISLAYA GUBA PLANT 209

Fig. 8.3. Artist's drawing of Kislaya plant site.

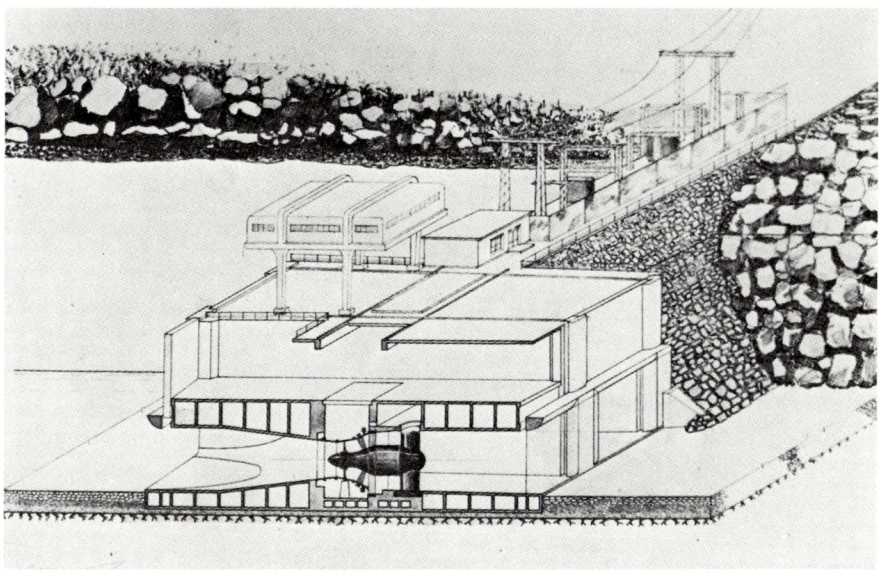

Fig. 8.4. Artist's view of Kislaya plant station.

Fig. 8.5. Bulb turbine at the Kislaya tidal plant.

and loose rock removal, using a barge-mounted clamshell crane, thus creating an excavation of 25,000 m³. A sand and gravel bed was then put in place, leveled, smoothed, and compacted. Settlement of the bed in the first three years (1969–1974) has been 20 mm. Steel blades extending a quarter of a meter below structure's bottom edges protect the bed against wave action.

Resistance of Materials

The plant being located in Arctic climate, materials must resist thermal stress besides seawater corrosion and bio-fouling. The ET climate has temperatures as low as −34°C; three months have temperatures which range between 0°C and 10°C, but may reach from 20°C to 27°C. Water temperatures, however, vary little, although, at the sea-air contact zone, problems would arise. To insulate the structure, glass fabric-reinforced foam epoxy resin panels line the concrete; these 50-mm by 700–900-mm sheets are glued to it by foam injection. Beyond the thus covered areas, a dual coat of vinyl chloride paint (exterior) and epoxy tar (interior) provides corrosion and bio-fouling protection; the steel blades anchoring the structure in place, those of the turbines, and the steel of the vertical lift gates received cathodic protection.

Ferrying Problems

The idea of transporting to the operational site a plant built elsewhere was not new. This technique had been used with underwater tunnels, lighthouses, locks, and breakwaters. The novelty consisted in its application to a hydraulic plant. It has also been used in connection with drilling platforms and in the Netherlands' Delta plan. However, the weight problem had to be resolved; hence the decision to build a relatively light structure and to weigh it down when implanted at the site. Lev Bernshtein is credited with designing a "floating" powerhouse. Thus, the structure is a set of jointly working cells. The powerhouse included floorings and walls, penstocks and foundation slabs; bottom water outlet and open spillway are accommodated inside the powerhouse. (Fig. 8.6)

Equipment

The Soviets adopted the reversible-blade turbines installed by the French in their Rance River plant; runner diameter is 3.8 m and rotational speed 69 rpm. The generator to which the turbines are coupled has a rotational speed of 600 rpm. However, work is being done to develop a variable-speed reversible power set. (Figs. 8.7 and 8.8)

OTHER LOCATIONS

Mezen Bay, on Cape Tolstik (White Sea), northeast of Archangelsk, is another very favorable site included in the Soviet grand design. The dam necessary here would be but a fraction of the length of the one built on the Rance: 62 m would suffice. Tide amplitude reaches 5.5–8.3 m. The total capacity is estimated at 8 million kW and the power per length unit of dam is about 1.3 million kW. The site has been considered for over a decade; a power station could be built here by isolation of a lagoon covering 2,000 km^2. Such a plant would provide 14 million kW-hr. The combination of two stations under consideration on the White

Fig. 8.6. Powerhouse being towed to site (August 28, 1968).

Fig. 8.7. General view of the tidal power plant in Kislaya Guba Bay.

Sea would raise this amount to approximately 40 million kW-hr. For the White Sea, the total exploitable potential is 40 billion kW-hr.

The Asiatic sites of Tugur and Penzhinsk would contribute to the industrial opening of Siberia; both Khabarovsk and Komsomlsk lie within 400–600 km of Tugur, and even Yakutsk is at only 1,000 km away from Tugur and 1,500 km away from Penzhinsk. Anadyr and Magadan are about 750 km from Penzhinsk. The author's communications with Soviet sources did not disclose additional possible implementation information; however, there has been some talk about the possible construction of a 320,000-kW plant at Lumbovskaya, where a short dam could cut off a 180-km^2 bay. Bernshtein himself wrote in 1974, however, that work had begun on a 5,000-MW power plant in Mezen Bay.

Estimated potential for Tugur is 10,000 MW and, for the Penzhinsk plant, 35,000 MW. Production at facilities in the Sea of Okhotsk could reach 170,000 GW-hr/yr, and in Lumbovskaya Bay, on the Bering Sea, a 380-MW-capacity plant could produce about 915 GW-hr/yr. Other

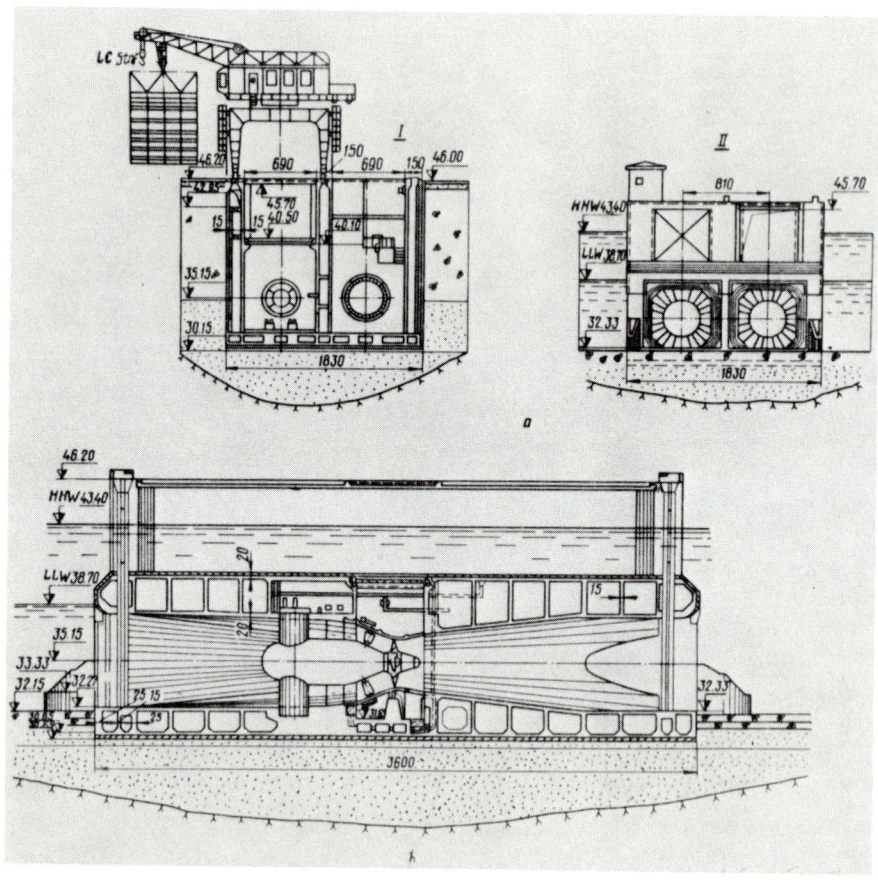

Fig. 8.8. (a) Powerhouse: I. view from basin side (cross-section); II. view from seaside; (b) powerhouse: longitudinal section.

sites at Kulsi and Mezen and in the White Sea could produce annually (respectively) 1,300, 2,600, and 36,000 MW-hr (Bernstein, 1961 and 1965). (Fig. 8.9)

CONCLUSION

A most important feature of the Soviet achievement is certainly the use of caissons that are floated in place. *In situ* construction of cofferdams and sand islands, as recently used in the Dutch Delta Project and previously at the Rance site, is a time-consuming and very costly phase of sea structure building. Concrete gravity structures built in-shore and then floated into place appear an excellent alternative.

Obviously, a large-scale tidal power plant would require larger and more numerous caissons than near Mezen; nevertheless, several generating units could be grouped in a single caisson. According to a paper by Severn and Campbell read at Canterbury in 1978, a broad multiple-unit caisson would present less stability problems while afloat and its control might be easier during the sinking process. Even navigation locks could be thus built on land as caissons or using caissons as components.

Currents would increase in strength as the embayment is being closed by the barrage so that permanent openings in powerhouse and sluice structure become insufficient to permit passage of sufficient flow; providing supplementary, albeit temporary, openings in the dam (i.e., temporary sluice caissons) would alleviate the situation. This procedure has been used in the Dutch Delta plan. The caissons could incorporate provisions for ductways, a sea wall, and a roadway, the latter an economic feature whose value has been repeatedly stressed.

It is not clear whether the Soviets will implement any of their additional projects, but interest remains very high in the U.S.S.R. A meeting on tidal power had been planned in Khabarovsk for mid-1979; however, it was postponed. Currently, the Kislaya originally-installed generator is being improved by the All Union Research Institute of Electrical Engineering. The use of a new generator will enable the development of a variable-speed reversible power set, which could encourage development of more tidal power plants and conventional-type low-head hydroelectric stations.

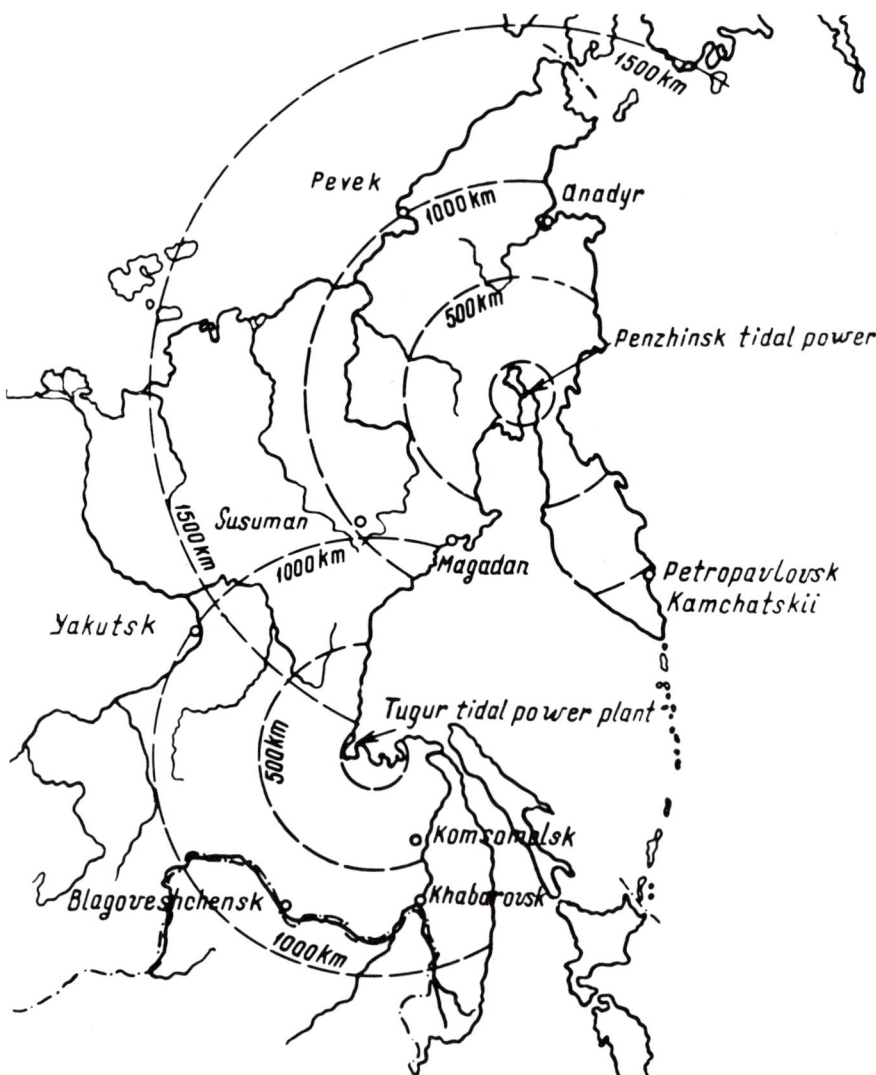

Fig. 8.9. Proposed sites for Asia plants on the shores of the Sea of Okhotsk and zones of electric power consumption.

The design of the floating powerhouse of the Kislogubskaya tidal power plant has served as a model for proposals of large tidal power plants in Canada and Great Britain.

The Soviet position on the use of tidal power was recently stated as follows (in a pamphlet distributed at the Soviet pavillion at OCEANEXPO '75 in Okinawa): "Utilization of the tidal power seems urgent from the standpoint of reducing the environment pollution as well as for countries poor in power resources since tidal power plants are the source of eternally renewed 'pure' energy, do not require flooding of cultivated lands and do not disturb the ecological balance.

"The total tidal power potential, technically feasible for utilization, amounts to about one trillion kWh per year [kW-hr/yr]. Taking into consideration the fact that the present power consumption amounts to about 10 trillions kWh per year and the power consumption in the year 2000 will make about 30 trillions kWh, one may say that the tidal power by itself is of not decisive importance. Still the distinguishing features of this kind of power (independence from the annual variations of the run-off volume) and the possibility of its utilization during the peak load hours make the tidal power plants an important component of a power system which ensures the optimum utilization of electric power plants of other types."

REFERENCES

Bernshtein, L. B., 1972. Kislaya Guba experimental tidal power plant and problems of the use of tidal energy, in Gray, T. J. and Garhus, O. K. (Eds.), *Tidal Power.* New York: Plenum, pp. 215–238.

Bernshtein, L. B., 1974. Kislogubsk: A small station generating great expectations, *Water Power* **26**, *No.* 5:172–177.

Bernstein, L. B., 1961. *Central Tidal Power Stations in Contemporary Electricity Production.* Moscow: State Publishing House.

Bernstein, L. B., 1961. *Prilivniye Elektrostantsu v Sovremyennoy Energetikye.* Moskva-Leningrad: Gosud energeticheskoye izdatyel'stvo.

Bernstein, L. B., 1965. *Tidal Energy for Electric Power Plants.* Jerusalem: Israel Programme for Scientific Translation.

Bernstein, L. B., 1974. Russian tidal power station is precast offsite, floated into place, *Civil Engineering* **44**, *No.* 4:46–49.

Kagan, B. A., 1974. Dissipation of tidal energy in the Arctic Seas, [*Akad. Nauk SSR.*] *Bull Atm. Ocean. Phys. Ser.* **9**, *No.* 6:375–376.

Logvenov, V., 1968. Prilivy stuzhat chelovyeku, *Pravda* **29**, *No. XII.*
Nekrasov, A. M. and Posse, A. V., 1959. Work done in the Soviet Union on high-voltage long-distance direct current power transmission, *Trans. Austral. Inst. Electr. Engin. Part III-A, Power Apparatus and System* **78**:515–521.
Sorokin, V. N., 1959. *Prilivnaya Elektrostantsiya v Lumbovskom Zalive* [*Tidal Power Plant in Lumbovsku Bay*]. Moscow Energy Institute, Thesis.
Zhibra, R. V., 1966. Energiya prilivov, Vknigye, *Myezhdunarodnaya Assotsiatsiya po giravlicheskim issledovaniyan, XI Congress 1965, tom 6.* Leningrad, pp. 223–242.

9
United States Projects

If only major schemes are considered, then the United States has only two very promising sites for construction of tidal power plants: one on the East Coast, perhaps best developed jointly with Canada, and another one on the Pacific Ocean in Alaska. Closest to ever having been built is the Passamaquoddy Project on the Atlantic.

THE ALASKAN SITES

Far less attention has been given to a site at 60° of latitude North, Cook Inlet, Alaska. According to the site selected, 6,000–18,600 GW-hr of energy could be generated per year. The most recent investigation appears to be one conducted about ten years ago by Wilson and Swales (1972).

Historical Background

First reported by Captain James Cook of Sandwich [Hawaii] Islands fame, the tides of Cook Inlet are repeatedly mentioned in tidal energy literature. However, an assessment of harnessing feasibility was not made until Wilson and Swales' study.
  Militating against serious consideration as a site are remoteness of Alaska, a limited market, drift ice in the winter, silt-loaded water, and the active seismic nature of the area. However, reconsideration of that location may one day occur since the technological achievements of the Soviets in the Arctic, and their announced plans for tidal power developments in Siberia as well. Yet, Alaska's recently discovered oil wealth and undeveloped hydropower will hamper tidal power development for the near future.

220 TIDAL ENERGY

Site Characteristics

The area has a Dcf climate, which means that precipitations occur year-round, and only three months a year do temperatures exceed 10°C. In such a subarctic climate, drift ice has to be reckoned with for up to six months a year and it travels in large packs into and out of inlets. Silt occurs in large quantities and deposits in the Arms; any site selected for a plant must be reasonably free of sediment accumulation.

The inlet is 370 km long narrowing from 96.5 km at its outlet into the Gulf of Alaska to about 21 km at the entrance of Turnagain Arm. The semi-diurnal tidal range at Port Graham, near the entrance, is 4.39 m, far less than at Anchorage (7.65 m). Because of diurnal inequality, level differences reach as much as 1.80 m in succeeding high and low waters (Fig. 9-1).

Six barrage sites were considered. The largest possibility would be near Kustatan-Nikishka, where tides have a mean range of 9.15 m. This

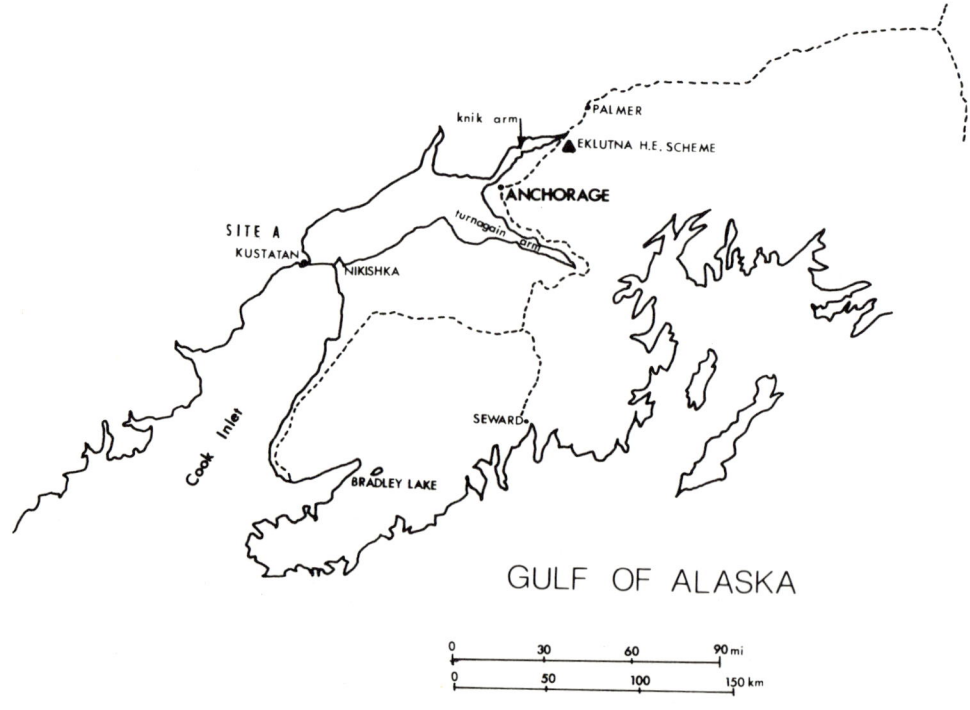

Fig. 9.1. Map of Cook Inlet, Alaska.

scheme would entail the impounding of an area of 3,100 km² and a barrage of 15.77 km. Annual output would be of the order of 75,000 GW. Though this is technically feasible, an impact study is necessary to ascertain the consequences of damming on tidal amplitude.

Using Turnagain Arm as a reservoir would require dams at Point Possession, linking it with Fire Island, from Fire Island to a point east of the Little Sustina River's mouth, or from Fire Island to Point Campbell. While the powerhouse would have to be placed inside the first dam (Point Possession-Fire Island), the third alternative (Point Campbell-Fire Island) is necessary to use Turnagain Arm. Using both the first and second barrages would permit combining the reservoirs of Turnagain and Knik Arms (Fig. 9-2).

The two last sites are in Knik Arm: one location up the Arm from Anchorage is unfavorable because of great depths, but the other linking Point Mackenzie and Point Woponzof would be built in waters only 21.3 m deep.

A 750-MW project with a 4.35-km-long dam would span 6.44 km of Knik Arm at about 12 km north of Anchorage. This single-pool plant would be built in waters averaging (at mean low water) between 3 and 12 m. The basin area ranges from 5,500 hectares (55 km²) at low tide to 23,760 hectares (237.6 km²) at high tide.

About 11 km west of Anchorage, the Turnagain Arm project is a 2,600-MW single-basin scheme requiring an 8-km-long main dam and a subsidiary 4.8-km dam between Fire Island and Point Campbell. Basin area would vary from 46,540 (low tide) to 85,800 hectares (high tide). For a two-basin, 2,600-MW Fire Island plant, a 12.9-km dam would close Knik Arm, and a 14.5-km dam would close Turnagain Arm. In addition, a navigation lock for access to Knik Arm would be needed. Basin area varies with tides from 21,000–42,000 hectares.

Wilson and Swales (1972) retained for examination three schemes, all with mean tidal amplitudes of 7.47 m. The Knik Arm proposal would provide 6,000 GW/annum. Combinations of the Point Possession barrage with the Little Sustina area dam, or with the Point Campbell dam, would provide, respectively, 12,500 and 18,600 GW annually.

Still another site has been under study: about half-way up Cook Inlet, some 65 km west of Point Possession. A dam could be built that would link the West and East Forelands; it would be 15.7 km long and would separate Redoubt Bay to the west from Trading Bay to the east. Power generated could reach 75 billion kW-hr annually.

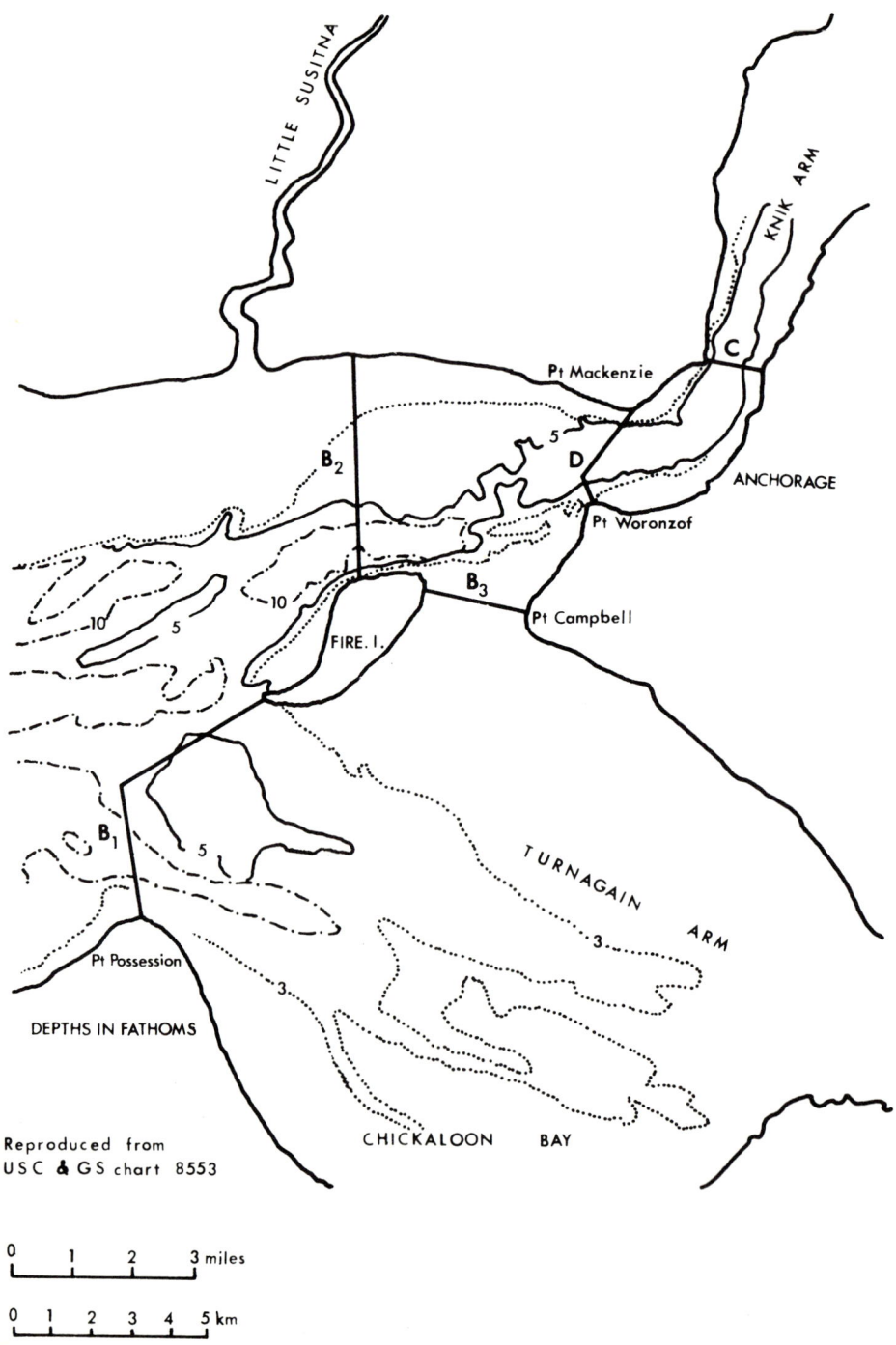

Fig. 9.2. Map of Turnagain Arm, Alaska.

Finally, a tidal power barrage could be inserted between the village of Angoon and Turn Point, on Admiralty Island, off Chatham Strait, and separating Favorite Bay from Kootznahoo Inlet. The powerhouse would be located at Turn Point and the dam would be only 460 m long. The basin area would cover 30 km^2. With mean tidal amplitudes of about 3.3 m, a one-pool scheme and ten units, this modest scheme would have a gross energy potential of 315 GW-hr/yr and an annual power production of 80 GW-hr. This last project is a native American proposal.

Technological Characteristics

Though no decision as to the type of plant to be built was made, assessments were based on ebb-flow generation coupled with pumped storage and, on the use of fixed-blade, straight-flow turbines with rim generators of 9.15-m turbine-wheel diameter. The Knik Arm scheme was considered most attentively, keeping in mind current power needs in Alaska.

The actual structure of the barrage would consist of a few major components and many identical units. Because of icing problems, submerged sluice gates would be preferable; submerged venturi sluices would provide refilling, and so would secondary crest-gate sluices above the turbines.

Conclusions

Though the potential of tapping tidal energy in Cook Inlet is considerable and apparently no unsurmountable engineering problems exist, construction of a plant is, at this time, very doubtful. Economics militate against it. However, were new tidal plants to be constructed elsewhere in the world, particularly in the Soviet Arctic, and were Alaska to "open up" (as are the northernmost reaches of Soviet Europe right now), one might expect renewed interest for Cook Inlet's tides. With Alaska's yearly requirements estimated at one billion kW-hr and Cook Inlet being capable of providing that much, it may be worth rethinking the idea. Additional benefits include deep water for Anchorage, reduced currents, perhaps less ice in winter, and certainly a permanent road across Knik Arm.

THE QUODDY PROJECT

Construction of a tidal power plant on the United States-Canadian border has been envisioned as a one-pool and two-pool undertaking, as a one-nation and two-nation scheme. Interest in such a plant has risen and waned over the last 50 years. At present, it is at best considered a low-priority project in the United States, but interest in tidal power is much keener in Canada.

The most recent assessment of the potential of the 5.5-m tides in the Passamaquoddy Bay in eastern Maine is a capacity of 1,800 MW with an annual generation potential of 15,800 GW-hr.

Historical Background

Dexter Cooper's plans consisted of building dams and sluiceways in openings of the Bay of Fundy and powerhouses in openings between Passamaquoddy and Cobscook Bays. Passamaquoddy Bay would fill at high tide and be kept as the high-level pool of a two-pool system, with Cobscook Bay the low-level pool to be emptied at low tide. Water moving from the higher to the lower pool would pass through a power plant, while the upper pool level would drop and the lower pool level would rise between filling and draining periods. (Fig. 9.3)

Passamaquoddy Bay lies, for the greater part of its 260 km^2, in Canada, but Cobscook Bay, covering about 106 km^2, inclusive of connecting waters, is entirely within the State of Maine. Both bays are part of the Bay of Fundy, known for having the largest range of tides in the world. Most of the water circulates in and out through channels north and south of Moose Island, site of Eastport, the United States' easternmost city.

Although actual work was never undertaken on a project until Franklin D. Roosevelt became president, according to *The Public Papers and Addresses of Franklin D. Roosevelt* (1938), he had discussed the matter with Owen D. Young of General Electric, and an application was filed with the U.S. Federal Power Commission on January 2, 1924, for a preliminary permit for an international project, all this independently of Dexter Cooper's further efforts. In fact, Maine granted Cooper an incorporation act in April 1925, approved by referendum a year and a half later, and Canada followed up with a charter in May 1926. No restrictions were imposed by the Maine Act, but Canadian government

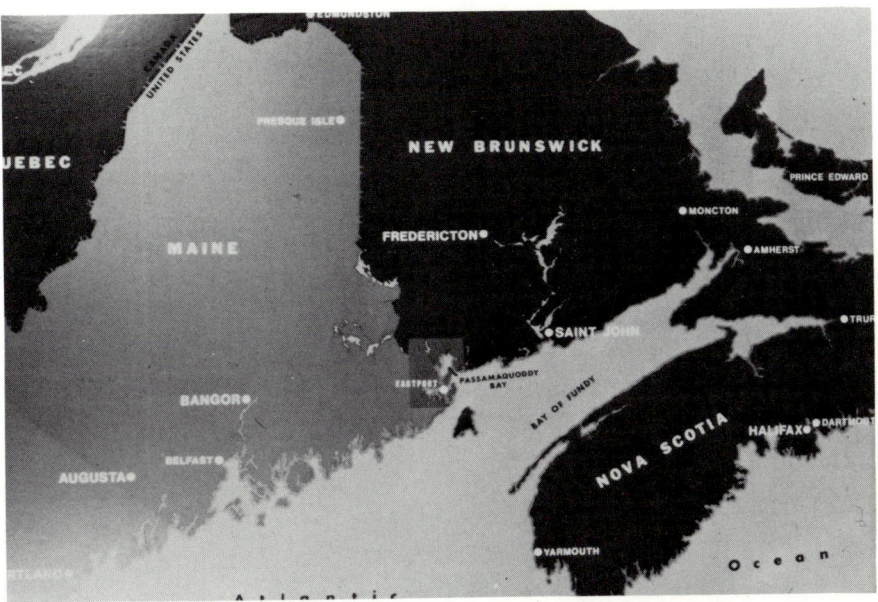

Fig. 9.3. Passamaquoddy region in relation to New England and the Canadian Maritime provinces. *(Source: U.S. Corps of Engineers)*

approval for dam construction was required as a precaution against fishery damage. This latter concern has remained quite alive, is included in recent studies and will be discussed later. (Fig. 9.4)

Cooper was eventually joined in his efforts by Westinghouse, Boston Electric, General Electric, and the Middle West Utilities. The Dexter P. Cooper Company asked the U.S. Federal Power Commission, in 1928, for permission to build a facility that would provide 464,000 hp to be eventually increased to 1,087,000 hp. However, because no dams were built within three years and, obviously, because of continued worries about environmental effects, Canada did not renew the charter and Cooper applied in September 1929 for a permit to construct a two-pool project located entirely within the State of Maine; 80,000–240,000 hp would be generated. However, no action was taken by the U.S. Federal Power Commission. (Fig. 9.5)

In 1934, Secretary of the Interior Ickes created a Passamaquoddy Bay Tidal Project Commission. It was to consider the project as a work relief undertaking. The Commission, of which Cooper was a member,

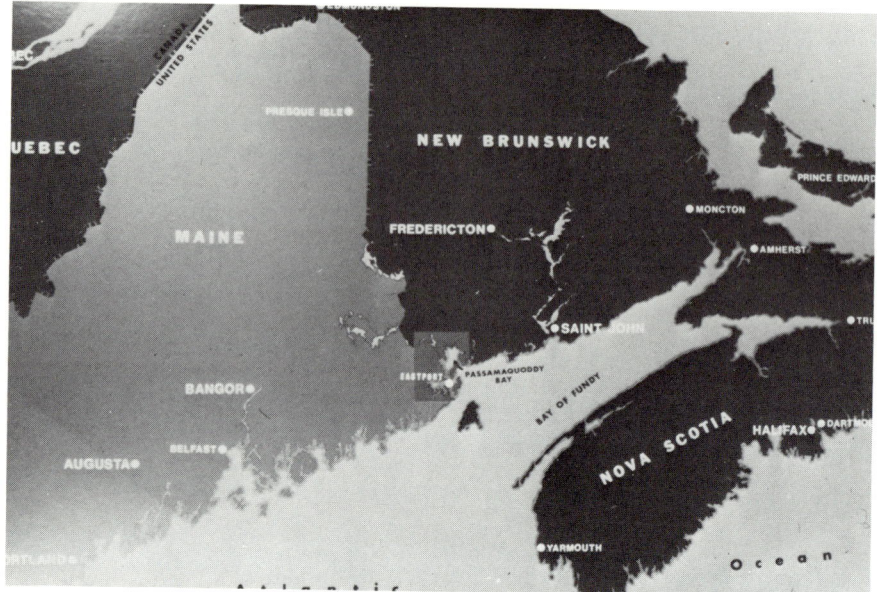

Fig. 9.4. Passamaquoddy tidal project location map. *(Source: U.S. Corps of Engineers)*

recommended in 1935 a Cobscook Bay project utilizing, in addition to the tidal power plant, a pumped storage facility; but the same month, the U.S. Federal Power Commission, on purely economic grounds, rejected the proposal. Yet, four months later, under the Emergency Relief Appropriation Act of 1935, an allocation of $10,000,000, reduced afterwards to $7 million, was approved.

The 1935 Project

Actual work started on July 5, 1935, under the direction of the U.S. Corps of Engineers. The 1935 project concerned only United States territory and dealt with a single-pool scheme using Cobscook Bay, a pumped storage facility near Haycock Harbor, and a 16-mile (26-km) electric transmission line. The ultimate project, however, aimed at a two-nation, two-pool plan. Five dams were to be built where Cobscook Bay connected with the ocean and Passamaquoddy Bay. The power plant, originally involving 10 turbines (later to be 22), was to be located in the Eastport dam; power would be generated as water entered Cobscook Bay, and sluices in the dam closest to the ocean would permit

UNITED STATES PROJECTS 227

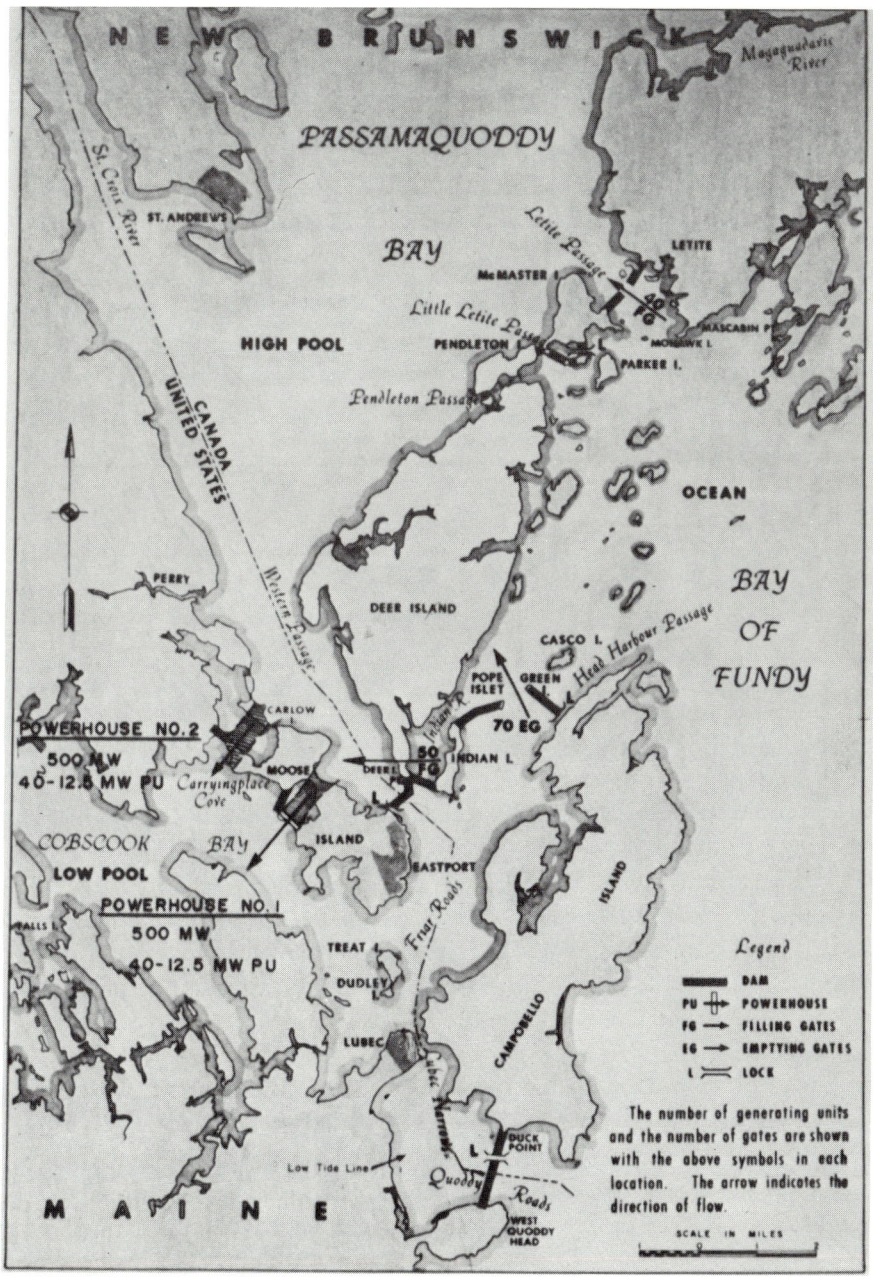

Fig. 9.5. International Passamaquoddy tidal power project.

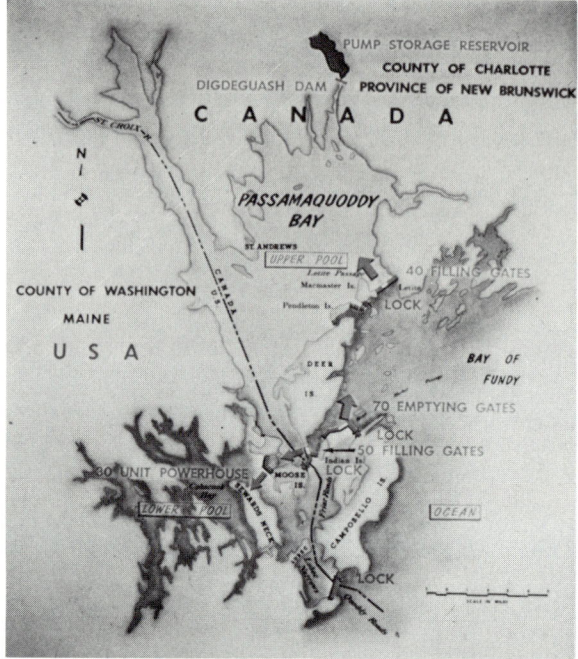

Fig. 9.6. Passamaquoddy tidal power survey map. *(Source: U.S. Corps of Engineers)*

water return at low tide. A navigation lock in the Lubec dam would permit ocean to Cobscook Bay access. A head of 5.5 m between ocean and bay was thus utilized. (Fig. 9.6)

To utilize more fully peak production, and to ensure continuous power generation, some of the peak power would be used for pumps to bring water from the ocean into Haycock Reservoir, about 40 m above sea level, and to the south and west of Cobscook Bay. At Haycock Harbor, situated on the Bay of Fundy, another power plant would take advantage of that 40-m head, at low tide, as the water was released.

However, this site was eventually abandoned when it appeared that some dykes built on clay were shifting and others on moraines were porous, letting water work itself into the areas' fresh water supplies.

On June 30, 1936, all work came to a halt: a bill which included a call for the appointment of a board to study and report on the Passamaquoddy project was never acted upon by the Senate, and no funds or authority for further construction ever came from Congress. Five thousand workers, who had been brought to the area and accommodated in the specially created town of Quoddy Village, headed homewards. Two storm-damaged dams, already completed, were kept under repairs until April 1937. (Fig. 9. 7)

From an economic point of view, the Corps of Engineers estimated the project to cost $37 million and bought the Cooper Company rights for $60,000, although the company had spent close to $407,000. Cooper himself remained as a consultant. Additional costs were occasioned by housing to be constructed for the workers, administration, warehouses, heating and utility plants, and even a school. A well constructed village of 239 buildings was erected and streets laid out at a cost of a million and a half dollars.

Of the actual work completed during the 1935-1937 period, by the U.S. Army Corps of Engineers, little remains of the dam linking Dudley and Treat Islands. However, two dams carrying a single-track railroad and a state highway are still in good shape: they connect Corlow and Moose Islands, and Corlow Island and the mainland at Pleasant Point. Of Quoddy Village, built for the construction teams, nothing remains except a segment of land still owned by the federal government. (Figs. 9.8, 9.9)

President Roosevelt made another attempt, in January 1939, to relaunch the project. He expressed the view that it would be justified to fund test borings and determination of the advisability to construct a small experimental plant in Maine. With the economic situation in Eastern Maine at its worst, forests depleted and fisheries all but exhausted, "such project might be beneficial, and even Canada might join a two-nation undertaking." (Fig. 9.10)

Fig. 9.7. Artist's concept of the Passamaquoddy tidal power plant. *(Source: U.S. Corps of Engineers)*

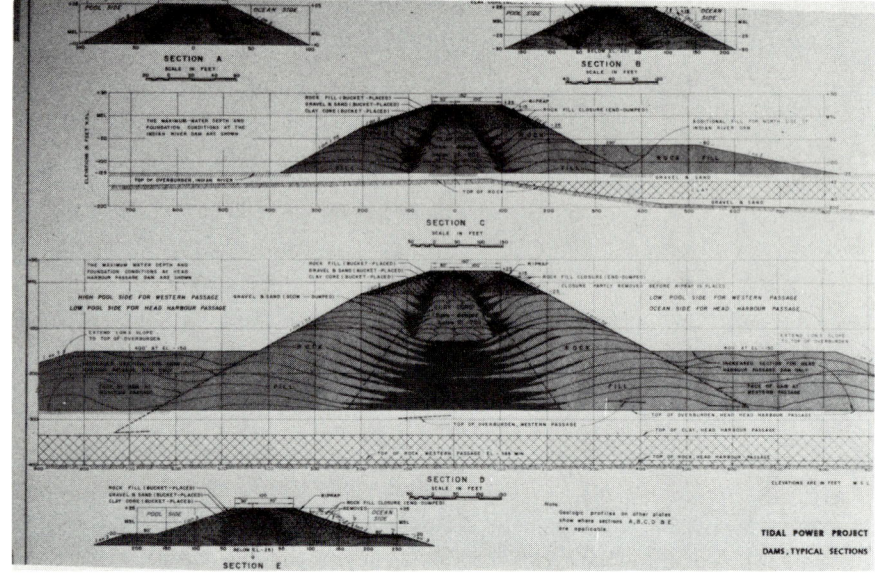

Fig. 9.8. Cross-section of proposed dam. *(Source: U.S. Corps of Engineers)*

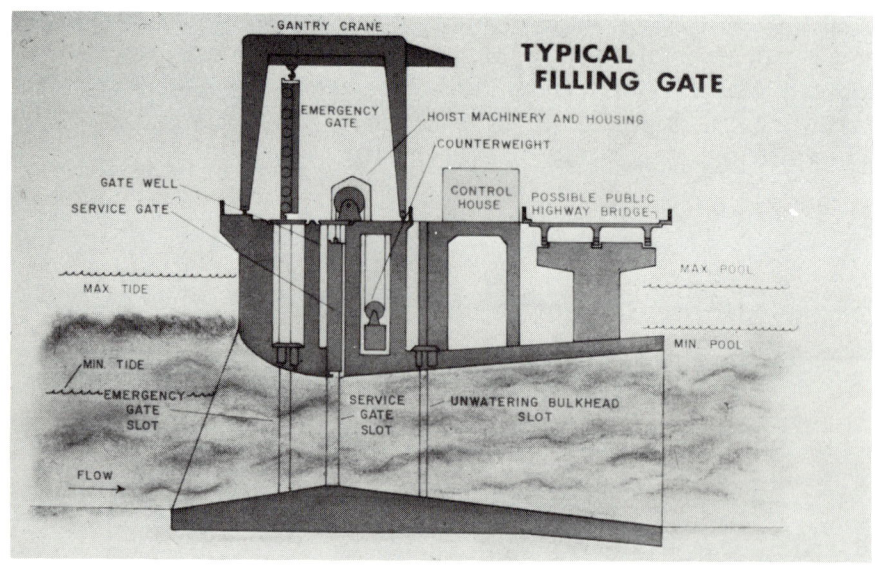

Fig. 9.9. Filling gate, cross-section. *(Source: U.S. Corps of Engineers)*

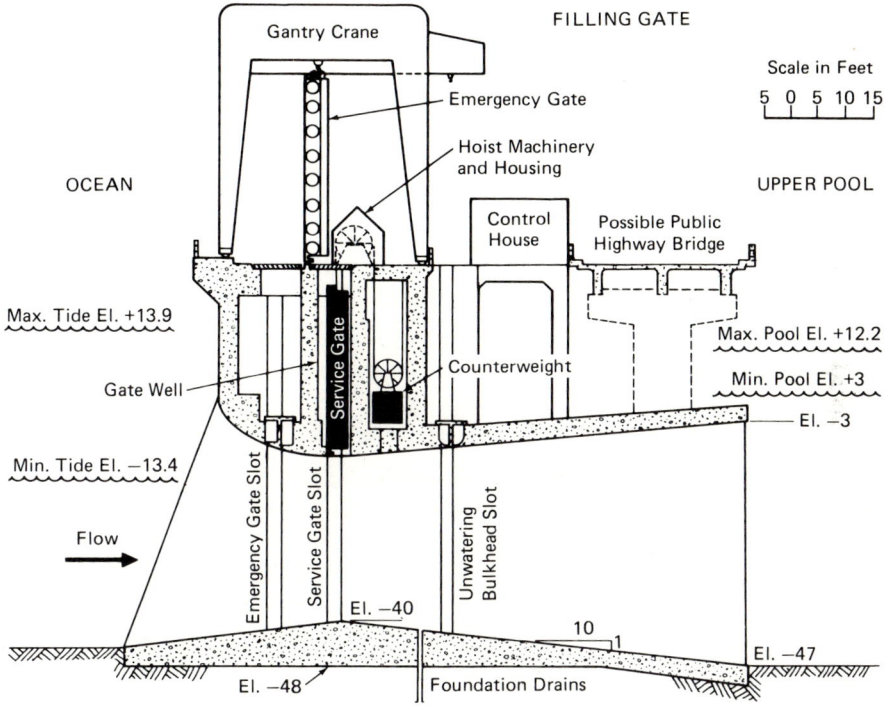

Fig. 9.9. *Continued*

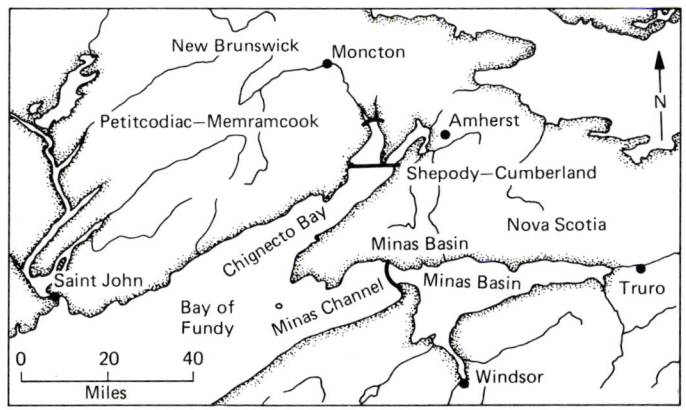

Fig. 9.10. Location of possible tidal power schemes in the Bay of Fundy (1935 proposal).

Finally, in 1941, the Federal Power Commission spoke out against an exclusively United States plant, finding it economically non-competitive but it favored further exploratory work for a bigger international Passamaquoddy scheme.

It admitted that tidal power had advantages over other sources of power, the water supply being dependable and accurately predictable, its head precisely known, and be unaffected by floods, droughts, or ice jams. Unfortunately, the potential 578 million kW-hr had no local or export market, not even in Canadian New Brunswick.

Comparing construction costs of a tidal power plant and the hydroelectric development of 17 plants on the Penobscot and Kennebec Rivers, the report found an outlay required of $87,854,000 for a two-pool project, $40,884,000 for a one-pool, ten-unit project, versus $48,715,000 for river projects. Annual costs were estimated, respectively, at $3,844,000 for the two-pool scheme, $1,292,000 for the one-pool scheme, and $4,518,000 for the 17 river plants.

In 1945, the government of Canada found uneconomical a project with a powerhouse between the Petitcodiac and Memramcook estuaries, which would have provided an annual energy of 1,300 kW-hr. A more recent study recommended implementation, but a research team advised, before any further steps were to be taken, to investigate the Minas Basin in Nova Scotia, site of the world's maximal tidal ranges, as a still more promising site. (It should be mentioned in passing that this scheme had been suggested way back in 1919 by W. R. Turnbull of Saint John, New Brunswick!)

Later Proposals

The study conducted over the period 1956–1959 involved underwater mapping and exploration, coring, aerial and hydrographic surveys, and tidal observations. Ultimately, the best scheme appeared to be the one including the 260 km^2 of Passamaquoddy Bay (high pool) and 106 km^2 of Cobscook Bay (low pool), with a 30-unit powerhouse at Carryingplace Cove. Each unit would be rated at 10,000 kW, yielding 95,000–345,000 kW in total, or an average energy production of 1,843 million kW-hr/yr.

The project envisioned construction of about 11 km of rock-filled dams in an area of widely varying bottom conditions: in some locations, bedrock is exposed, while elsewhere, clay strata reach a thickness of

30.5 m. The dams themselves were to be constructed of a clay core flanked by dumped-rock fills moved (as much as possible) from material removed and accumulated at locks, gates, and navigation sites. The clay for the core would come from the 13 million m^3 of clay excavated at Carryingplace Cove. Eight percent of the 10,880 m of dam would be in water of depths ranging from 38–91 m, where a granular core would be used. Construction would require building of cofferdams of exceptional magnitude; such cofferdams were successfully made in the Rance estuary in 1964, but are costly and should be dispensed with.

The plan included 90 filling gates and 70 emptying gates. The venturi throat was selected because, among other advantages, it permitted maximum discharge for a given gate area.

The powerhouse would have contained 30 turbine generator units, each placed in a concrete monolith 23.77 m wide and 53.34 m long in the direction of flow, with the bottom of the draft tube 33.52 m below the powerhouse deck. Metal parts of the powerhouse would be exposed to salt water and protected by coatings, special alloys, and cathodic protection. Turbines were to be fixed-blade propeller Kaplan type with a throat diameter of 97.5 m and a 40-rpm speed. The Rance River plant is equipped instead with horizontal-axis turbine generator units that "triple" as turbine, pump, and sluiceway, and flow in both directions. The choice of Kaplan units for the American project was based on unresolved maintenance problems, the need to compensate for low rotative inertia, and lower cost.

Consideration was given to auxiliary power sources because, at times, the minimum output from the tidal power plant may coincide with peak demand for power. Among these sources were river hydroelectric plants, pumped storage plants, and steam-electric auxiliaries. Least favorable was the latter. The best river plant could be located at Rankin Rapids, on the Upper St. John River in Maine at some 282 km as the crow flies from the tidal project. A powerhouse of eight units would provide a capacity of 460,000 kW and generate 1,220 kW-hr/yr. Combined with the tidal power plant, capacity would be 555,000 kW and generate 3,063 kW-hr annually. The energy borrowed from Rankin Rapids would be "repaid" when the tidal output would be greater than the load.

The best pump storage site was found east of St. Andrews, New Brunswick, near the outlet of the Digdewash River into Passamaquoddy Bay. Using 400 million kW-hr of energy from the tidal plant for pumping, the combined scheme would generate 1,759 million kW-hr annually.

Production and Building Costs

The tidal power generation would vary from 65,000–345,000 kW, with a dependable capacity of 95,000 kW. The annual energy, in kW-hr, would range between 1,738 and 1,923 million. The power would be very reliable, because actual tides cling closely to predictions, and equipment failure would be quite rare. While tidal power would follow the lunar cycle, power use would follow the solar (or man's) cycle.

By 1975, all of the power produced could have been used up by Maine and New Brunswick.

The estimate, 20 years ago, of construction of the tidal power plant ran up to $484,000,000. With auxiliary developments, costs would climb to $696,000,000. By today's standards, an estimate of $900,-000,000 is probably conservative. The value of the power, in 1959, would have been close to $11,000,000 (of which $4,700,000 for Canada alone). Fringe benefits, as in the Rance River plant, would also accrue: besides some downstream benefits on the St. John River, below Rankin Rapids, the tidal power plant would save 1,280,000 tons of coal (5,700,000 of oil) annually, and recreation and tourism would be boosted; some disadvantages would result, but they would be either minimal or compensable, or could be remedied. (Table 9-1)

The Passamaquoddy Site

The area under consideration for a half-century seems to be very favorable as a tidal power plant site.

The Passamaquoddy tidal project area is a region that has been the locale of crustal and tectonic movements and that has undergone alterations due to intrusions and volcanic flows. The passages around the islands involved in the project are partially submerged old stream valleys; the islands are parts of the pre-glaciation landmass that remained above sea level.

The overburden in the area is made up from unconsolidated glacial surface deposits, weathered material, and peat. The bedrock consists of sedimentary and igneous Paleozoic rocks. The region is considered stable and earthquake danger is very remote. Sand, gravel, and crushed rock are used for road building and could thus, as in the Rance project, be used as fill for the tidal power project.

Geologically, the situation is thus good. From a hydrological viewpoint, conditions are equally satisfactory.

Table 9-1. Construction Costs and Economic Analysis of United States Tidal Power Projects (in 10^6). Projects A_3, M_1, M_2: Area for Upper Pool Only. Half-Moon Cove: For a 4-MW Single-Pool Scheme.

SITE	TOTAL INSTALLED CAPACITY (MW)	SLUICE GATES	UNITS (POWERHOUSE)	CONSTRUCTION COST AND INTEREST	ANNUAL COSTS	BENEFIT/COST RATIO (POWER)	POWER COST (¢/kW-hr)	ANNUAL POWER PRODUCTION (GW-hr)	GROSS ENERGY POTENTIAL (GW-hr/yr)	MEAN TIDAL RANGE (m)	MEAN AREA (km^2)
MAINE:											
International (M_1)	500	160	40	1,790	138.7	0.50	7.19	1,930	8,030	5.52	256
International (M_2)	1000	160	80	2,888	223.4	0.32	10.64	2,100	8,030	5.52	256
Treat Island (M_3)	180	25	18	492	38.4	0.37	5.65	680	2,710	5.52	88.6
Cooper Island (M_4)	180	25	18	441	34.6	0.36	5.85	590	2,350	5.52	77
M_3 + 180 MW Pumped Storage at Dickey-Lincoln	360	25	18	557	44.1	0.51	7.57			5.52	
Half-Moon Cove	5.2		6	10	0.9	0.40	5.52	30	84	5.52	2.7
ALASKA:											
Knik Arm (A_1)	750	75	50	1,957	153	0.15	5.34	2,870	12,200	7.92	207
Turnagain Arm (A_2)	2,600	250	170	5,960	466	0.15	5.18	9,000	42,300	7.47	754
Two-Pool Scheme (A_3)	2,600	600	170	8,195	637	0.31	5.82	10,950	58,440	7.47	287
Angoon (Indian)	30		10					80	315	3.23	30

Ocean tides are strongly influenced by the resonance of the Gulf of Maine Bay of the Fundy system, though the tidal system in the Bay of Fundy itself is not highly resonant. Construction of a tidal power complex would modify the flow in and out of Passamaquoddy and Cobscook bays and possibly influence tides in the Bay of Fundy, hence at the project's location as well, and consequently influence the generated amount of power. An M.I.T. study, however, predicted that Bay of Fundy tidal ranges would only increase slightly or not at all. Tidal velocities would increase during and after project completion but could not be estimated without a model.

Significant slopes would occur in the tidal project pools only in one location (Falls Island, Cobscook Bay) in the lower pool but power production would remain satisfactory.

When pitting the advantages of greater tidal ranges at the mouth of the Bay of Fundy against the smaller ranges, but topographical advantages, at the Bay's head, the latter is a better site. According to the experts' survey, this is one of the few sites in the entire world where a two-pool plan can be easily and profitably built. (Figs. 9.11–9.13)

Single-pool projects, with the low head available, were abandoned in favor of the two-pool schemes because the cost of power would be higher, peak generation rates would increase, and power would be gen-

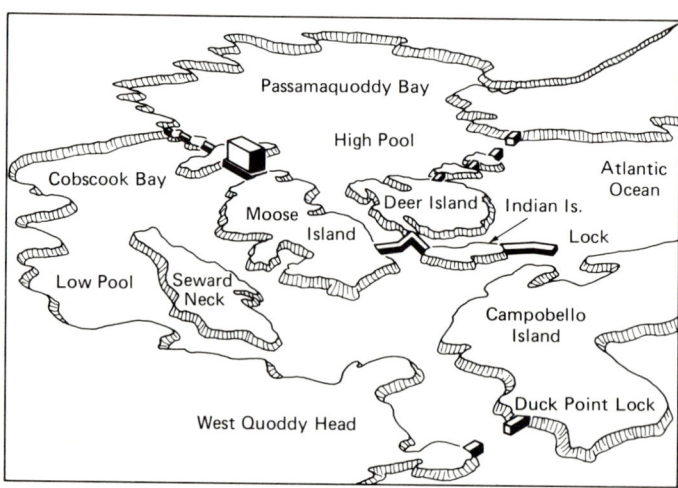

Fig. 9.11. Schematic map showing upper and lower basins at Passamaquoddy without water. *(Source: U.S. Department of the Interior)*

UNITED STATES PROJECTS 237

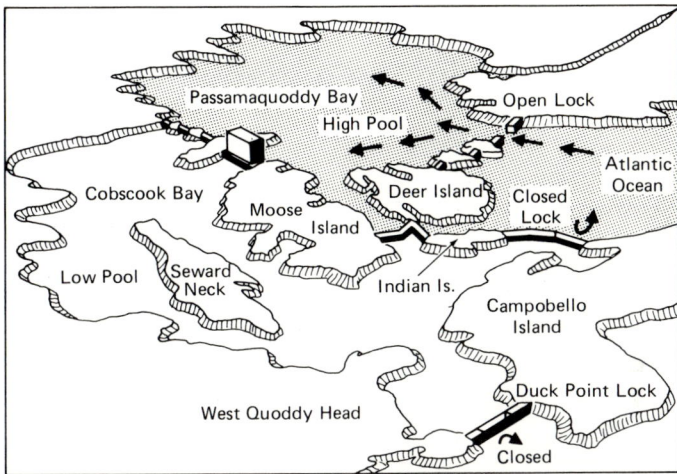

Fig. 9.12. Schematic map showing the flow of incoming tides to fill the upper basin. Lower basin locks are closed to keep out flood tide. *(Source: U.S. Department of the Interior)*

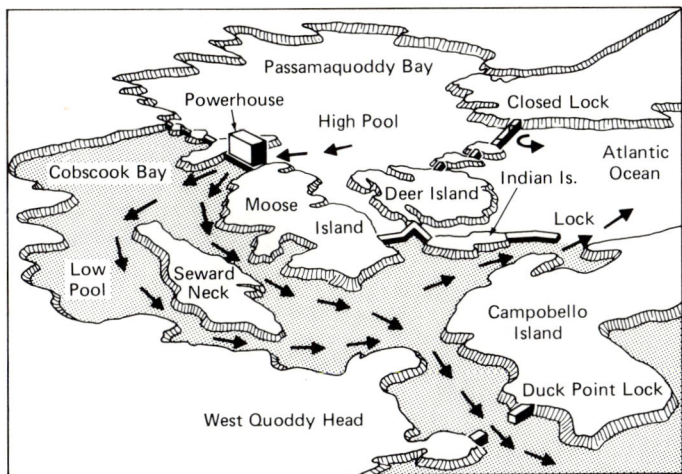

Fig. 9.13. Schematic map release of store flood tide waters through powerhouse and draining into lower basin and back to the sea. Locks to upper basin are closed until the next incoming tide.

Fig. 9.14. High and low tide near Eastport, Maine, between Passamaquoddy and Cobscook bays.

erated intermittently. Bulb-type turbines were discarded in the design principally because of cost and lack of experience. Now that the Rance River plant is a reality, a second look at the bulb turbines might be advisable. (Figs. 9.14–9.16)

Production could still be enhanced by coupling a pumping facility with the proposed tidal power plant. The pumped-storage auxiliary most favorable location would be at Digdewash, where a fresh water inflow from the drainage area exists. Auxiliary river hydro developments have also been considered.

Passamaquoddy Bay would be the high pool which would be permitted to fill up at high tide. The low pool would be Cobscook Bay: water in this pool, retained by gates, would be allowed to flow to the ocean during low tide.

This arrangement would provide true flexibility for the Passamaquoddy tidal power plant, because water could be "saved" (retained) and put to work to produce power at the most advantageous times. Such

Fig. 9.15. High tide near Grand Manor Island, Maine. *(Source: U.S. Corps of Engineers)*

Fig. 9.16. Low tide near Grand Manor Island, Maine. *(Source: U.S. Corps of Engineers)*

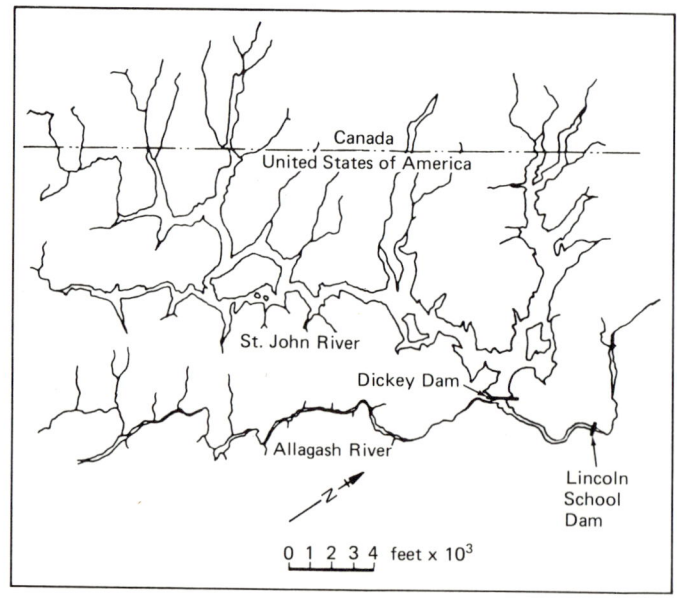

Fig. 9.17. Dickey-Lincoln school lakes project.

tidal power is thus remarkably suited to generate power, at peak and at off-peak times.

Because computers permit to "program" power output, it can be integrated into any electrical grid so that maximal use of the tidal plant can be ascertained.

The 1963 Recommendation

The detailed investigation conducted from 1956–1961 by the International Joint Commission led to the conclusion that a tidal power project was feasible. A two-pool scheme was favored, and the integration of a hydroelectric site on the St. John River was selected. But ultimately the project was turned down for economic reasons: unfavorable interest rates and continuous operation as a baseload plant. Senator Edmund S. Muskie of Maine and President John F. Kennedy recommended a careful analysis of the Commission's report. The study committee discovered that the power produced could be used up and that the Quoddy project is no longer a local but rather a regional scheme. New extra-high voltage transmission lines were now in use and a closer look at the bulb type, axial-flow turbines showed them to be less expensive. Finally, thermal plants were operating at peak-capacity. In 1963, Senator Muskie commented as follows.

> The conclusions and recommendations of the International Joint Commission on the proposed Passamaquoddy Power Project will require the most careful and rigorous analysis. In my opinion, the joint engineering board report and the Commission's recommendations should be treated as the basis for discussion in planning a program of sound utilization of the power resources of the Maine-New Brunswick area.
>
> I am recommending to the President that the Administration explore all unanswered questions in the report, and that serious attention be given to those suggestions made by the Commission which would affect a final determination on the feasibility of the project.
>
> Imagination, patience, and sustained effort will be required to achieve effective utilization of the power resources of the area covered in the Quoddy report. Although there may be differences of opinion as to the validity of its assumptions, the Commission's report does provide a check point from which we can move in planning for the future.

Subsequently, the late President John F. Kennedy asked Secretary of the Interior Stewart L. Udall to evaluate the IJC (International Joint Commission, 1961) report. The President asked Secretary Udall to investigate changes which "might result in making the project economically feasible." President Kennedy also noted that the Passamaquoddy Project "has challenged engineers and students of water resources for many years, and I am hopeful that the current [International Joint Commission] report . . . can be as useful as possible in formulating a sound policy for the development of resources in the area covered by the report."

One of the apparent errors in assessing the Passamaquoddy project value and feasibility is that its tides' utilization was steadily considered under the optic of conventional around-the-clock generation baseload of electricity, thereby limiting Quoddy's installed capacity to about 300,000 kW, of which about 95,000 kW were dependable power. However, the same amount of water can yield 1,100,000 kW of peaking power, approximately 200,000 kW of baseload power, and off-peak energy, when combined with the Dickey Dam area on the Upper St. John River.

The so-called "Passamaquoddy project" actually embraces several schemes: Passamaquoddy itself, Dickey, and the Lincoln School—the last two authorized in 1965. While aiming at power production, several fringe benefits would accrue of the project's completion: flood control, fish and wildlife benefits, area redevelopment, recreation, and tourism. The Passamaquoddy-St. John River Study Committee (1963) examined alternatives to tidal power—federal steam, nuclear, and pumped storage plants—and concluded that none offered equivalent new employment and sustained economic opportunity. Pointing to the certainty that more power would be needed and that there is no other way to conserve the ever-wasting energy of the tides, the authors stated that the project was "a first and necessary step toward vitally needed economic rejuvenation for people whose ancestors did so much to help frame our national heritage." (Fig. 9.17)

If ever completed, this project would dwarf the Rance River plant, as would several other plants in various locations throughout the world.

The last report on the Passamaquoddy tidal power project dates from August 1964, and nothing, except for a 1979 reassessment, has been done since that date. On the other hand, the United States Congress authorized, in October 1965, the Dickey-Lincoln School project in Maine, on the St. John River. Design was started but is now again at a standstill, because, as in 1936, money failed to be appropriated for this project since 1966. The 1964-study had supported a twin-basin plant and two centrals, each producing 500 million kW. This tidal power plant was to be linked with hydroelectrical and thermic plants and the total power production would thus have reached 1,250,000 kW.

The 1964 Report resulted, in fact, from a four-year study whose recommendations can be summarized as follows.

1. Immediate authorization, funding, and construction of the Dickey and Lincoln School projects on the St. John River and their associated transmission system. Construction would be contingent upon completion of necessary arrangements with the Canadian government. This would also have the immediate and major by-product of preserving the famed Allagash River in Maine, one of the few remaining wild rivers east of the Mississippi River.

2. Authorization of continued study, reexamination, and possible redesign of the Passamaquoddy project, taking full advantage of the latest technological advances with possible reductions in capital costs. Further consideration should be given to the economic benefits of Passamaquoddy associated with recreation, economic development, and elimination of poverty in the region—as well as its uniqueness and contribution to technology and creation of a new source of energy for the United States. This study should be related to an extra-high-voltage (EHV) transmission grid in New England. Any study of an EHV grid for New England should consider participation of federal, non-federal public, and privately owned utilities in a joint venture and the benefits that could accrue from further coordination of the transmission network with systems of neighboring provinces in Canada.

3. Preservation and development of the famed Allagash River for recreational use by the federal government, the state of Maine, or cooperatively by the state and federal government on a matched-funds basis. Preservation of the Allagash should include acquisition of sufficient land in public ownership surrounding the river to insure its preservation in perpetuity, and recreational usage and attendant development should be consistent with the wild rivers concept. Additional efforts should also be instituted to increase the recreational potential of Dickey Dam, to develop other recreational resources in Maine, particularly in Washington County, to review recreational potential at existing federal flood control projects in New England, and to plan for inclusion of recreational resources at future federal projects where feasible.

4. The Roosevelt International Park Commission has under consideration recommendations to the two governments for a more extensive improvement program, which would affect the jointly-owned property and other parts of Campobello Island. In accordance with your objective of a cooperative review and planning of the region's recreational and other resources, full and sympathetic consideration should be given to any recommendations by the Commission in which the United States would share responsibility for action with the Canadian government under the terms of the agreement establishing the park.

5. The federal government, in full participation with state and regional planning groups, should continue and intensify a comprehensive program already planned and initiated for the multiple-use of the area's natural resources including river reregulation, outdoor recreation, hydroelectric development, fish and wildlife conservation, particularly by restoration of the Atlantic salmon fisheries, municipal and industrial water supply, and other constructive uses. This review should also consider the impact of natural resource development in the overall economic improvement of New England.

Current Projects

A two-basin project within Cobscook Bay, investigated by the U.S. Corps of Engineers, would include a dam from Seaward Neck to Shackford Head. Its annual power production would be 300 million kW-hr. The Cooper Island project would be a single-pool (also in Cobscook Bay), one-way scheme; the dam would extend from Cooper Island to Shackford Head. (Figs. 9.14, 9.15, 9.16)

Two dams would be required for Treat Island and Carryingplace Cove Project—one from Treat Island to Estes Head at Eastport, and the other from Dudley Island to Lubec.

The international project proposes either a one-powerhouse scheme (at Carryingplace Cove) or a two-powerhouse scheme (with second powerhouse on Moose Island). This is a two-basin project, with Cobscook Bay as the lower pool, entirely in the United States, and Passamaquoddy Bay, mostly in Canada. The lower pool includes also Friar Roads and Quoddy Roads, partly in both countries. Some 11.270 km of dams are needed: for the high pool across Western, Letite, Little Letite, and Pendleton passages, and for the low pool across Head Harbor Passage and Quoddy Roads.

Conclusions

The Passamaquoddy project was found, in 1959, to be economical for the United States but uneconomical for Canada. The Petitcodiac-Memramcook scheme, as said before, had been ruled uneconomic in 1945. The Shepody Bay-Cumberland Basin project was reported, in 1959, to be the best physical location and to offer the most favorable economic conditions at half the cost of the Passamaquoddy scheme. The size of the flow, exceeding a hundredfold that of the Lawrence River, and the marketing of the power produced, tagged the Minas Basin project, as uneconomical and wasteful in 1956. However, the scheme has been under renewed study since 1965, in view of new technological advances. A final report is expected shortly and is exclusively concerned with wholly Canadian possibilites. The Canadians are concentrating on Chignecto and Annapolis-St. Mary's bays, Minas Basin, and the Bay of Fundy.

Today's tidal power projects are not any longer those envisioned by Dexter Cooper and Franklin D. Roosevelt. The new "Passamaquoddy project" would provide large benefits to the United States and Canada. Low-cost power would finally become available to New England and support development of New Brunswick. Flood damage caused by the St. John River would be checked in Maine and New Brunswick, while navigation would be improved in both Passamaquoddy and Cobscook Bays. Tourism will provide an increment of revenue, the dams would make highway link possible between Maine and New Brunswick, and the experience gained would be invaluable to develop power plants on the Bay of Fundy.

Finally, expendable resources should be conserved, while renewable resources should be put to use. Energy from the tides is an eternally renewable resource; it should have been placed, a long time ago, at the service of Americans and Canadians.

Meanwhile, back in the "old" world, the Rance River plant already has successfully provided energy for more than a decade.

HALF-MOON COVE TIDAL PROJECT

One could argue whether this project, in which the Passamaquoddy Tribe at Pleasant Point Reservation (Perry, Maine) has been involved since 1976, should be discussed as part of the Quoddy scheme. This research effort has been funded by the U.S. Department of Energy and is assisted by the Bangor Hydroelectric Company in order to investigate potential tidal power production integration with public utilities. The project itself could serve several purposes.

1. Generate electricity without damage to the environment for local and regional use.
2. Research and develop tidal power at an operating demonstration installation, as a companion to already existing plants.
3. Establish an economic base upon aquaculture, tourism, and research laboratories.
4. Develop tidal power within the United States.

Characteristics of Site

The project involves construction of a small tidal power plant by building a 348-m dam across Half-Moon Cove, adjacent to Cobscook Bay and Pleasant Point Reservation, where the average tidal range reaches 5.5 m. The cove would serve as retaining basin. Neap tides have 4-m ranges and spring tides have an amplitude of 7.92 m.

Half-Moon Cove has a surface area of about 3.24 km^2 at mean high tide, with an intertidal zone covering nearly 2 km^2. The basin has a shallow entrance; the approximate depth is 8.23 m from mean low water elevation to bedrock; the entrance is narrow, cutting down on dam construction costs. Entrance width varies from 341 m for extremely high tide levels to less than 137 m at neap tide. Other favorable site characteristics include a bedrock foundation with small sediment deposition, construction accessibility by a roadway still in existence, and an electrical service area demand appropriate for the proposed power plant's output. Furthermore, the emptying and filling operations will be facilitated by a regular geometric configuration of the cove.

Plant Characteristics

Four proposals for single-pool units with reversible-blade turbines were made. They would generate, respectively, per year 18, 22, 24, and 38 million kW-hr. (Table 9-2)

However, a two-basin system has also been considered, using Half-Moon Cove as one reservoir and Straight Bay as the second pool, located at a distance of about 10 km on Cobscook Bay. A computer analysis provided the data summarized in Table 9-3. Dependable power production appears not economically feasible for greater capacity plants than 5 MW$_e$, except, perhaps, under a long time duration mode. Power production could also be increased by adding more basins to the scheme. Costs including capital investment, contingencies, and interest charges during construction result in a kW-hr price still higher than for conventional plants: 5–6¢ for a single-pool system and 6.75¢ for a paired-basin system, versus about 2.7¢ for public utilities. However, comparisons are

Table 9-2. Half-Moon Cove Proposed Plants. Generation and Operational Times Per Tide Cycle for 5.5-m Tidal Amplitude.

CAPACITY (MWe)	MILLION kW-hr	% TIME AT FULL CAPACITY	% TIME IN OPERATION
4	18	38.6	69.8
5	22	35.8	69.2
10	24	22.5	56.7
12	38	22.4	56.7

Table 9-3a. Power Level for 4- and 5-MW$_e$-Capacity Plants.

TIDE RANGE (m)	LOWEST DEPENDABLE POWER LEVEL (MW)		% TIME AT FULL CAPACITY	
	4 MW$_e$	5 MW$_e$	4 MW$_e$	5 MW$_e$
4	0.85	1.00	49.2	26
4.75	1.50	1.20	70	65.3
5.76	2.80	2.40	91.9	81.8
6.16	3.20	3.10	95.8	90.7
6.83	4.00	5.00	100	100

Table 9-3b. Total Capital Cost for Half-Moon Cove Proposed Plants.

CAPACITY (MW)	TOTAL COST ($10^6)	NORMALIZED CAPACITY COST ($10^6)
4	9.5	2,374
5	11.09	2,217
10	19.09	1,908
12.5	22.46	1,797

inaccurate because 1974 prices were used; using a 6% escalation rate (for a 5-MW$_e$ installed capacity), the cost of a kW-hr of tidal-generated electricity would climb between 1984 and 1998 from 6.57 to 7.72¢ while conventionally produced power would go from 4.54 to 10.27¢.

Project Impact

Completion of a project would have a favorable socioeconomic impact on a region of endemic high unemployment and for a reservation with a current 58% unemployment rate. Decreased turbidity and tidal fluctuation coupled with increased summer temperatures would be favorable for aquaculture projects; touristic draw power has been proven by the Rance experience and would exert a ripple effect; and, finally, some enterprises able to use an intermittent energy supply could be attracted, as could desalination plants and cryogenic gas producers. The savings in oil is estimated at 43,000 barrels annually for the smallest-capacity plant.

A comparison may be attempted here between the Pleasant Point Reservation and sites in developing countries. Here, the public utilities power plants are over 160 km away, and transmission losses would be drastically reduced by a regional scheme; a population of 4,000 people within a 16-km radius could be supplied by a 5-MW$_e$ tidal power plant with about 58% of its needs (21.65 million kW-hr) exporting 7.99 million kW-hr as surplus. Such schemes could permit, in some locations, development of industry hitherto problematic due to fuel price increases.

In addition, as in the Rance project and as proposed for the Severn River scheme, the local transportation system could be greatly enhanced by installing a roadway across the tidal plant between Eastport and Perry. Such a road would not only accommodate the anticipated influx of tourists but provide an updated local traffic pattern.

Economic Aspects and Current Status

An evaluation was made by Laberge (1976) in terms of life cycle cost analysis. He used a 4% annual escalation rate, for a 5-MW$_e$ plant, for annual benefits, operation, maintenance, and the value of tidal power production, over a 50-yr period. Assuming that electricity production began in 1985 based on a baseline of 2.15¢ kW-hr (at 1976 rates), he

Table 9-4. Plant-Related Benefits (50-yr Period and 4% Escalation).

YEAR	ANNUAL POWER PRODUCTION INCOME ($10^6)	RECREATION ($10^6)	AQUACULTURE ($10^6)	RESEARCH AND DEVELOPMENT ($10^6)	TOTAL ANNUAL COST
1985	0.692	0.385	0.411	0.270	1.422
2009	1.774	0.988	1.052		1.958
2014	2.159	1.202	1.280		2.149
2024	4.789	2.632	2.805		3.420

concluded that annual income exceeds annual cost after 29 years, that total benefits exceed annual cost by $336,000 during the first year of operation, and that, within 25 years, recreational and aquaculture benefits exceed total annual cost. (Table 9-4)

At this time, a feasibility study is planned, in cooperation with two Maine engineering firms—E. C. Jordan Engineering Co. and Maritec, Inc., by the Passamaquoddy Tribe Energy Office. This study envisions the construction of a demonstration project to quantify the parameters related to small tidal power schemes as forerunners to larger projects and the establishment of a paired-basin mode of development.

REFERENCES

Boardman, W. F., 1963. Quoddy and Rankin Rapids as a multipurpose project, *Public Utilities Fortnightly* **71** (April 11):19–25.

De Rouville, A., 1957. General report on the utilization of the tidal mechanical energy, *La Houille Blanche IVes J. de l' Hydraulique* **11**:435–455.

Friedlander, G. D., 1964. The Quoddy question: Time and tide, *IEEE Spectrum* **I**, No. 9:96.

Huntsman, A. G., 1928. The Passamaquoddy Bay power project and its effect on the fisheries, *Saint John Telegraph Journal*.

International Joint Commission, 1961. *Investigation of the International Passamaquoddy Tidal Power Project, Report of the International Joint Commission, Docket 72; April*. Washington, D.C.: The Commission.

Laberge, N., 1976. *Passamaquoddy Tribe Tidal Project*. Point Pleasant (Maine): Passamaquoddy Tribal Council.

McNaughton, A. G., 1960. Passamaquoddy not feasible? *Electrical World* **153** (March 28):60–61.

Passamaquoddy-St. John River Study Committee, 1964. *Supplement to the 1963 Report, The International Passamaquoddy Tidal Power Project and Upper Saint John River Hydroelectric Power Development*. Washington, D.C.: U.S. Department of the Interior.

Roosevelt, Franklin D., 1938. *The Public Papers and Addresses of Franklin D. Roosevelt*. New York: Random House, pp. 272–273.

Smith, L., 1959. The Quoddy project stirs again, *Public Utilities Fortnightly* **64** (November 5):753–765.

Smith, L., 1961. The status of power supply in Maine, *Public Utilities Fortnightly* **68** (December 7):873–882.

U.S. House Committee on Foreign Affairs, 1953. *Survey of Passamaquoddy Tidal Power Project* (83rd Cong., 1st sess.). Washington, D.C.: Government Printing Office.

U.S. House of Representatives, 1965. *Communication from the President of the United States Lyndon B. Johnson* (89th Congress, 1st Session). Washington, D.C.: Government Printing Office.

Wilson, E. M. and Swales, M. C., 1972. Tidal power from Cook Inlet, Alaska, in Gray, T. J. and Gashus, O. K. (Eds.), *Tidal Power*. New York: Plenum, pp. 239–256.

1945. *Report on Tidal Power, Petitcodiak and Memramcook Estuaries*. Frederictown: Government of the Province of New Brunswick.

1965. White House offers Maine power plan, *New York Times* (July 11):1, 51.

1966. *Report of the Hydro Resources Sub-Committee of the Alaska Advisory Committee of the Federal Power Commission. Alaska Power Survey*. Washington, D.C.: U.S. Government Printing Office.

1967. Bay of Fundy tidal power study, *Engineer* **223**, *No. 5802* (April 7):509, 513.

10
Tidal Power and Canada

Tidal mills and "moulins à marée" were brought to Canada by European settlers and, as elsewhere, gradually fell into disuse. Yet, as early as 1919, the unusual tides of New Brunswick triggered thoughts of electricity production. Canada projects that its electrical energy demands for the Maritime Provinces substantially exceed the 10,000-MW peak load forecast for 1995. Blackout could occur, particularly in January (more likely if a thermal unit goes out); when days are short, industrial and domestic demands overlap, and the river flow is low. Current primary demand exceeds 180 billion kW-hr/yr. Sites elsewhere have been studied, too, for possible implantation of a tidal power plant; some attention was given to Ungava Bay, where tides of great amplitude and a favorable river system are found. According to Godin (1973), the combined mean power output of tidal energy on Ungava Bay could reach 5,400 MW. If harnessed in conjunction with river hydraulic energy, estimated at 12,000 MW, the location constitutes a feasible scheme for the Quebec coast.

UNGAVA BAY

Ungava Bay is located north of 58° and between 70° and approximately 66° of longitude. Tides are about as great as in the Bay of Fundy. The Payne, Leaf, Kaniapiskaw, Whale, and George Rivers drain a surrounding plateau before debouching in the Bay. The largest ranges occur in the southwestern part of the bay. Though about constant over half a month's time, the importance of the range is four times as great during spring tides than during neap tides.

Considering the power potential of the rivers emptying into the Bay, a combined bay/river scheme appears the most attractive formula; tidal and hydropower total 17,600 MW. Godin (1973) calculated the power potential of the tidal basins: the mean power output would total 10,018 MW, but involves 13 sites; assuming that one project—the larger-scale one—is carried out for each basin, the total mean output is 5,621 MW.

Climatic conditions are not favorable for energy storage; hence, the dual tidal-hydro system is the most practical solution.

Although the potential is considerable, and the technology necessary exists and is relatively simple, prospects for development of the power are not promising. There is no large population center close by, distances to users are considerable, and the economic development of Quebec has slowed down over the last decade. Tapping of the Ungava Bay (and the rivers' potential) is dependent upon the needs of northeastern North America, its economical development, and the extension of both the railroad and road net.

BRITISH COLUMBIA

Though some sites have occasionally been mentioned in British Columbia, where a 1979 study came out against tidal power, overwhelming emphasis, when considering tidal power in Canada, is placed on the east coast. High amplitudes include 4.11 m at Nottingham Island, 6.20 m on Kingua Fjord, 9.15 m in Ashe Inlet, 4.45 m at St. John (N.F.), 7.28 m at St. John (N.B.), and 10.98 m at Port Burwell. Favorable tidal power sites, with 7-m tides, exist on Ungava and Frobisher Bays, near the Hudson Strait mouth; by using several Ungava Bay sites, a total output of 5,600 MW could be reached. Thirteen potential tidal power sites, with a total estimated output of 13.100 MW-hr/yr, have been identified along the coast of British Columbia. Among these, the largest output would be attained at the Portland Canal, the Jervis and Sechelt Inlets, and the Observatory Inlet; tidal ranges reach only 5, 3.5, and 4.7 m, respectively. Of these, Observatory Inlet, near Prince Rupert, and Sechelt Inlet, near Vancouver, were the subject of a recent study. The first site, for an annual output of 2.03 MW-hr would require a capital outlay of $ (Canada) 2,498,000, and a modest 167-kW-hr plant at Sechelt Inlet would cost $ (Canada) 243,000. Tidal power does not appear an economically attractive option for British Columbia, because the electricity generated would cost 7–9 times more than electricity from conventional hydroelectric projects.

But it remains the Bay of Fundy that shows most immediate promise and that has been getting the closest scrutiny. Eight sites, four with annual energy hovering around 20,000 GW-hr, are currently being investigated in New Brunswick and Nova Scotia; Cobequid Bay (N.S.), Shepody Bay, and the Cumberland Basin (N.S.) have been suggested. In Shepody Bay, St. Mary's Point would be joined to Cape Maringouin; in Cumberland Basin, Pecks Point to Bass Point; and in the Minas Basin, Economy Point to Cape Tenny. (Fig. 10-1)

The geometric shape of the Bay of Fundy amplifies tides. Ocean tides ranging from 60 cm to 1.80 m increase by resonance to 6–15 m inside the Bay. The Fundy embayment, about 265 km long, mouth to head, has a ratio of one to four for length to tide wavelength, and resonance is additionally enhanced by depth and width decrease.

Mean tidal amplitudes and average annual energy output are, respectively, 9.5 m (8.8–14.3 m) and 2,967 GW-hr, and 10 m (9.75–14.3 m) and 2,352 GW-hr for Shepody Bay and the Cumberland Basin. Often

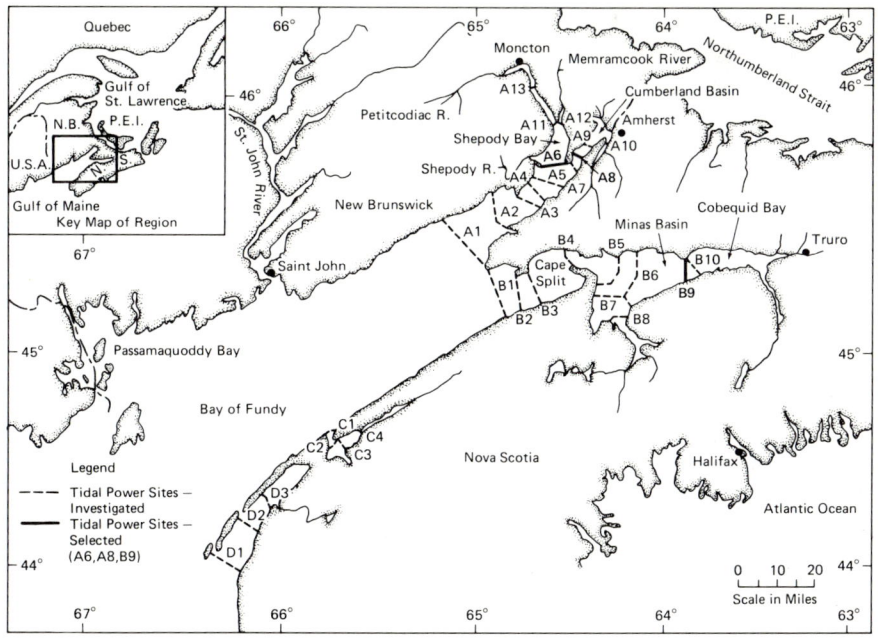

Fig. 10.1A. Bay of Fundy. Location of various sites examined. (R. H. Clark)

254   TIDAL ENERGY

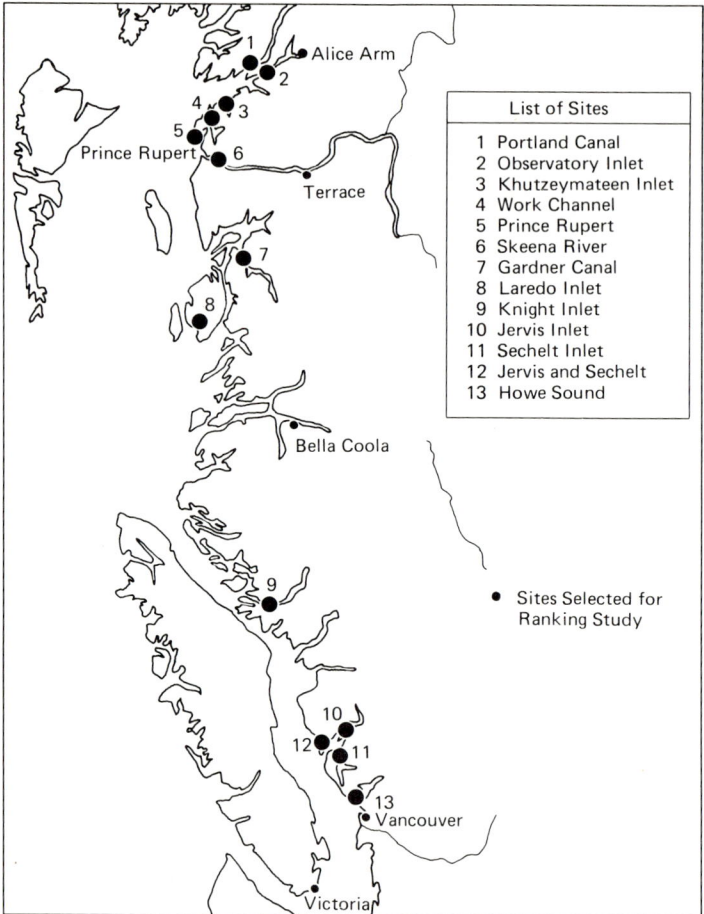

Fig. 10.1B. Tidal power in British Columbia.

mentioned, the Minas Basin (N.S.) has tides of 11.5 m, with spring tides reaching a 16.2-m amplitude and a potential of 10,374 GW-hr. The Canadian Tidal Power Review Board determined that a Cobequid Bay mouth site could furnish Maine with electricity at a cost of 32 mils (1977 U.S. $) per kW-hr; tides vary here from 7.3–16.1 m, with an average of 11 m and annual energy would be 10,374 GW-hr. (Figs. 10.2 and 10.3)

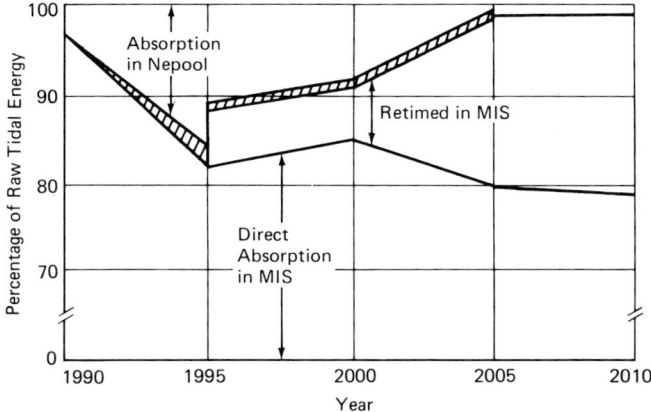

Fig. 10.2. Utilization of tidal energy from site A8. (Furst and Swales, *Proc. Wave and Tidal Energy Symp.*, Canterbury, 1978)

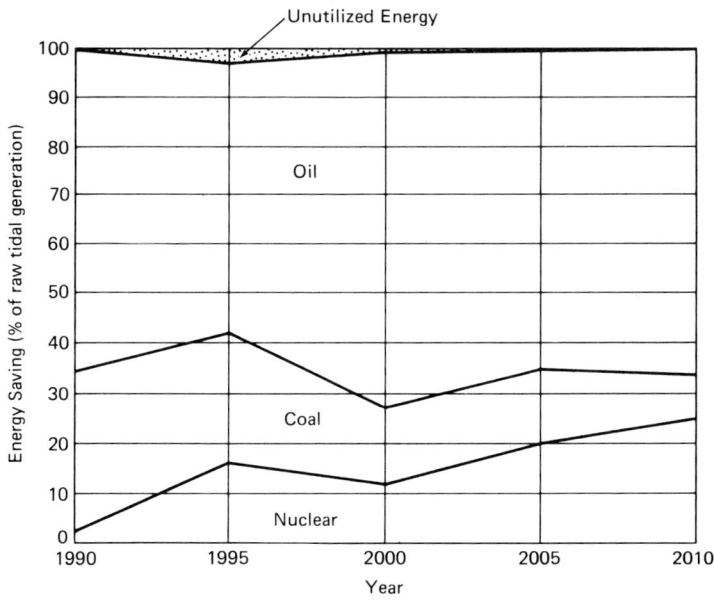

Fig. 10.3. Energy savings by energy sources with site A8. (Furst and Swales, *Proc. Wave and Tidal Energy Symp.*, Canterbury, 1978)

Table 10-1. Final Ranking of Sites with Their Hydraulic Characteristics ("A" Sites in Chignecto Bay; "B" Sites in Minas Basin). Source: Warnock and Tanner 1978. Selection of Optimum Sites for Tidal Power Development in the Bay of Fundy, Proc. Int. Symp. Wave & Tidal Energy (Canterbury) I, No. 1, E1:1–22.

SITES FINAL RANKING		COMPARATIVE INDEX (Mills/kW-hr)	GROSS ENERGY POTENTIAL (10^6kW-hr/yr)	INSTALLED CAPACITY (MW)	CONSTRUCTION COST (10^6)	ANNUAL COST (10^6)	ANNUAL ENERGY PRODUCTION (10^6kW-hr/yr)	TIDAL RANGE (NATURAL) MEAN HIGH		MAXIMUM TIDAL CURRENT (m/sec)	LENGTH OF ARTIFICIAL BARRIER AT HIGH TIDE (km)	AVERAGE BASIN AREA (km²)	GEOGRAPHIC LOCATION OF RETAINED SITES	BULB-TYPE 7½ m-DIAMETER TURBINE CHARACTERISTICS				
ORDER	NAME							(m)	(m)					OUTPUT	RATED HEAD (m)	RATED DISCHARGE (m³/sec)	SPEED (rpm)	UNIT POWER
1	B9	26.7	57,600	2,460	1,766	136.0	7,350	11.8	16.0	2.1	17.8	175	Cobequid Bay	38	7.5	738	72	32.9
2	B6	30.5	111,000	5,531	5,171	398.2	14,400	11.4	14.9	2.1	26.8	557						
3	B5	37.9	148,600	5,030	4,564	351.4	21,000	11.0	14.5	2.1	15.8	864						
4	A8	39.0	15,700	784	601	46.3	2,042	10.5	14.4	2.1	2.6	73	Shepody Bay	31	6.5	694	67.7	33.3
5	A6	45.3	22,600	1,510	1,422	109.4	3,900	10.2	14.2	1.3	7.1	128	Cumberland Basin mouth	31	6.5	694	67.7	33.3
6	B3	45.5	195,900	9,794	8,332	641.6	25,500	9.6	13.0	2.6	14.2	1371						
	A																	
7	A4	45.7	61,500	3,073	2,831	217.9	8,000	9.8	13.6	1.4	10.1	03						
8	B4	79.2	159,200	7,951	8,517	655.8	20,700	10.2	13.9	4.1	8.4	1031						

THE BAY OF FUNDY PROJECT

Studies of the Bay of Fundy received major attention from 1944 on, when a proposal was made to link the Petitcodiac and Memramcook Rivers by a canal in which a powerhouse would be installed, a two-basin concept. Stamped as uneconomic, other proposals were made for Chignecto Bay and the Minas Basin in 1950–1960. The Atlantic Tidal Power Programming Board examined 23 sites, in the sixties, in Chignecto Bay, the Minas Basin, and the St. Mary's-Annapolis Basin. They retained three sites, one each in Shepody Bay, the Cumberland Basin, and Cobequid Bay, characterized above. An economic unfavorable report ensued, but with recommendations of continued monitoring which has been done since 1972 by the Bay of Fundy Tidal Power Review Board, an instrumentality of the Governments of Canada, New Brunswick, and Nova Scotia. (Table 10-1; Fig. 10.4)

Financed in December 1975, a first phase of the latest study, launched in 1976, encompasses the following topics.

1. *Tidal power plant design.* A review of all potential sites and power schemes to update cost estimates, consider new construction methods, and reevaluate available equipment—especially turbines.

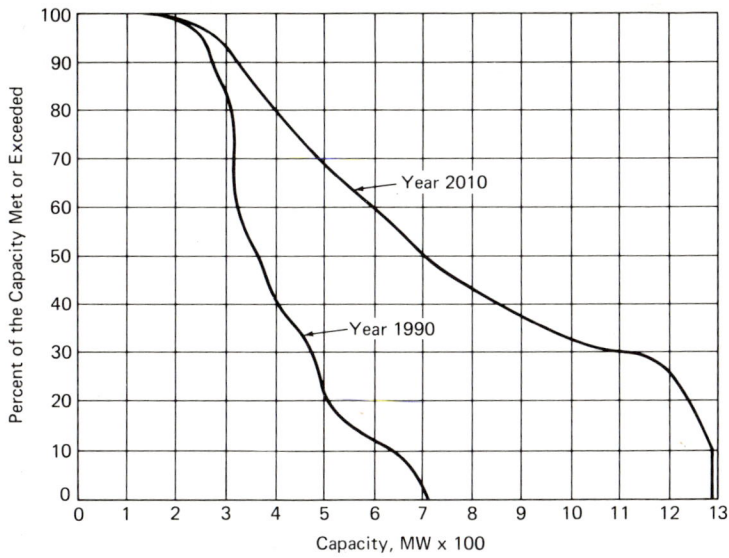

Fig. 10.4. Tidal emergency capacity duration curve for site A8. Single basin double effect scheme. (Furst and Swales, *Proc. Wave and Tidal Energy Symp.*, Canterbury, 1978)

2. *Tidal power generation.* The development of new mathematical models of the tidal regime and tidal generation potential at selected sites to assess capabilities under alternative basin effect operating schedules.
3. *Markets and utility systems.* An examination of alternative growth scenarios of the Maritime, Quebec and New England utility systems, with and without tidal power, to answer the following questions: What is the real cost of each alternative? Is tidal power competitive? If so, how will it fit into the system? What transmission facilities will be required?
4. *Socioeconomic aspects.* What will be the real costs of alternative fuels, including oil, coal, and nuclear, in the long range? What will be the real cost of capital and of labor? What effects will a very large—and lengthy—construction program have on the lives of local people?
5. *Environmental implications.* What effects could a tidal power development have on the environment and the creatures and people that depend on it? Will these effects, in general, be harmful or beneficial? What can be done to minimize the damages and to maximize the advantages? If, upon completion of the first phase of the study, the question, "Is tidal power competitive?" is answered, "Maybe," then assumptions, calculations, and judgments will have to be carefully reexamined and refined. Particular emphasis will be placed on regional, social, and economic environmental effects in coming to a conclusive answer. (See Figs. 10.5-10.7).

In the second phase, the details of all aspects of tidal power development will be worked out, and a plan and schedule presented for bringing to reality the long-held dream of harnessing the tides.

SELECTION OF SCHEME AND SITE

Based upon the numerous studies completed, the only scheme that would be acceptable here is a single-basin arrangement utilizing the created tidal heads and low-head electric turbines. Although the Rance plant is a double-effect operation, generating at ebb and flood tides, the Canadian report opted for single-effect, ebb-flow generation because it apparently offers the greatest net system savings. Because absorption of energy from an entirely predictable intermittent generating source now

Fig. 10.5. High tide at "The Rocks," Hopewell Cape, New Brunswick. *(Source: New Brunswick Department of Tourism)*

poses no problem, no attempt has to be made to create an equivalence between a tidal power plant and conventional ones. The studies thus focused on tidal output integration into the total system. (Table 10-2) and Figs. 10.2 and 10.9.

Drawing upon the experience gathered in the implementation of the Dutch "Delta plan" (see Chapter 6), construction of North Sea offshore installations, and perhaps the construction of the Soviet Kislaya plant, cofferdam construction used for the Rance plant will be dispensed with in favor of construction of to-be-floated-in caissons of powerhouse and sluiceway elements. Though geophysical exploration confirms usability of this method, further geological study of the load-bearing competency of the underlying strata remains to be carried out.

Fig. 10.6. Low tide at "The Rocks," Hopewell Cape, New Brunswick. *(Source: New Brunswick Department of Tourism)*

Renewed funding in June 1977 permitted pursuing investigations to their full conclusion. The Board found that tidal power, in its most economic role, would eliminate some fossil-fueled generation from the Maritime Provinces: 4 million barrels of oil and 450,000 tons of coal with a Shepody Bay plant, 3 million barrels and 380,000 tons with a Cumberland Basin scheme, and 12 million barrels plus 1.3 million tons for Cobequid Bay. However, it would neither eliminate nor reduce the need for nuclear generation. Assuming continued use of coal in current proportions, it is obvious that needs for nuclear generation would be reduced. (Figs. 10.8, 10.9)

The Board also selected the Cumberland Basin site as the best suited

Fig. 10.7. Arrival of tidal bore at Moncton, New Brunswick. *(Source: New Brunswick Department of Tourism)*

Table 10-2. Minimum at Site Costs for Sites B9, A6, and A8 (in 1976 $). Source: Warnock and Tanner, 1978.

SITE	B9	A9	A8
Number of powerhouse units	106	53	37
Number of sluices	60	30	24
Number of spare powerhouse units	6	3	2
Sluiceways cost ($10^6)	277	135	113
Powerhouse cost ($10^6)	1,438	745	527
Dyke cost ($10^6)	308	297	99
Fixed site cost ($10^6)	19	19	19
Total direct cost ($10^6)	2,042	1,196	758
Indirect cost, interest, contingency ($10^6)	1,633	964	505
Grand total capital cost ($10^6)	3,675	2,160	1,263
Annual charge ($10^6)	228	134	78
Annual energy (GW-hr)	12,653	4,533	3,423
Indicated cost of energy (¢/kW-hr)	1.80	2.95	2.28
Net plant capacity (MW)	3,800	1,550	1,085

262 TIDAL ENERGY

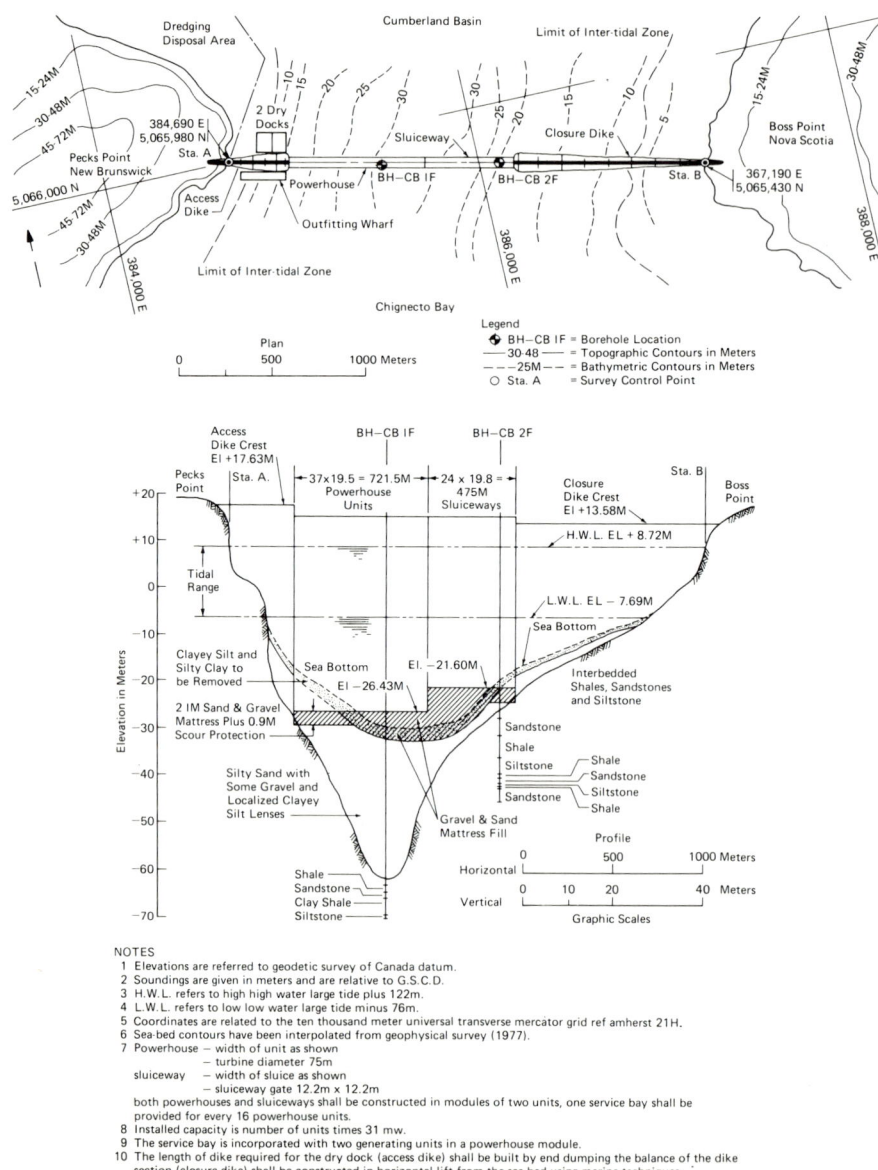

Fig. 10.8. Plan and profile of development for Cumberland Bay, site A8. *(Source: Bay of Fundy Tidal Power Review Board)*

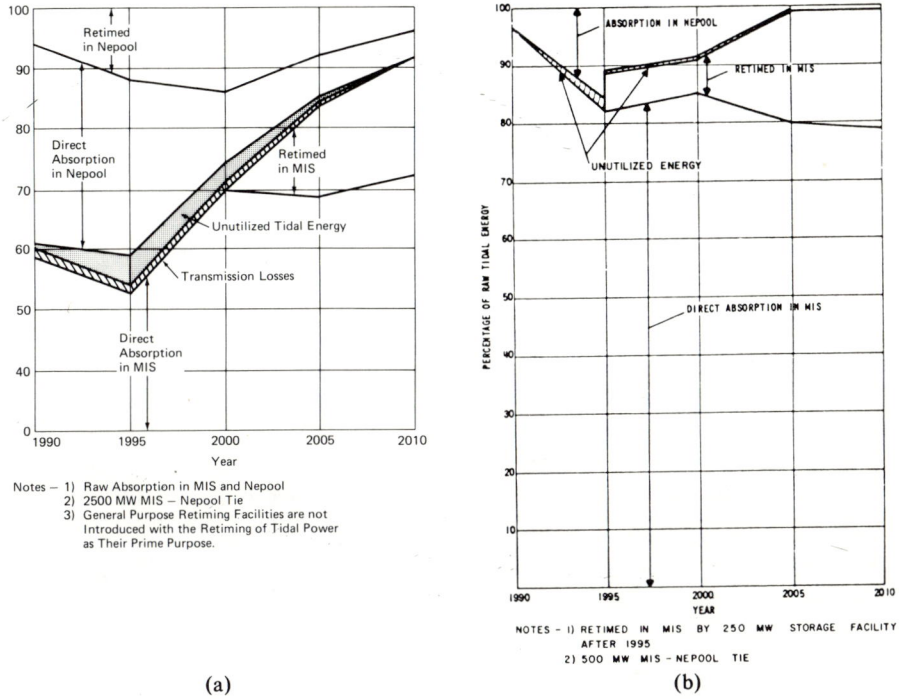

Fig. 10.9a. Absorption capability of the MIS for energy from a tidal power station. Utilization of tidal energy from site Bg (3800 MW). *(Source: Bay of Fundy Tidal Power Review Board)*

Fig. 10.9b. Utilization of tidal energy from Site A8 (1085 MW).

for an initial project. The term *initial* must be stressed because the Maritime area is capable of absorbing virtually all energy from sites with potentials of 1,000–1,500 MW. Why Cumberland? Eight reasons are given. 1) A joint site, it affords an opportunity to equalize benefits between New Brunswick and Nova Scotia. 2) It could be socioeconomically preferable. 3) Being the smallest project, technical problems would be minimized. 4) It is the largest and most economical site for which benefits are at least 90% derived from within the combined New Brunswick, Nova Scotia, and Prince Edward Island power utilities system, within which, additionally, adjustments (e.g., retiming devices) can be foreseen to eliminate dependence on an export market, for surplus energy. 5) The scheme would not cause tidal regime modifications beyond the limits of Canadian jurisdiction. 6) Environmental impacts, possibly minimized, are less than for the Shepody and Cobequid Bays. 7) Capital investment requirements would stress the least the Maritimes utilities' borrowing capacities, as it causes the least risk of involving introduction of a new technology. 8) The project is large enough to provide a substantial contribution to the energy requirements of the Maritimes region and, if developed by 1990, it would demonstrate the practicality of the expansion of tidal power use, precisely at a time when oil resources may show indications of depletion.

Characteristics of Scheme

Studies carried out particularly during 1975–1977 have shown that the single-basin type of generation is the most economically attractive, with more complex modes providing a lesser overall energy output and a higher-cost energy. At the current stage of technology, horizontal bulb-type turbine generators, similar to those of the Rance and Kislaya, appear best suited. Double-effect units result in an energy cost very close to that of a single-effect unit. The caisson construction system was recommended to avoid the cost of building cofferdams; it would be used also for the sluice gates. The most economic barrage would be made of dumped rock-fill, of adequate size to resist scour, placed in superimposed horizontal layers.

The area provides ample space for the land-based operations, but access roads would have to be built. Both sites—$A8$ (Chignecto Bay) and $B9$ (Minas Basin)—could provide energy that could be delivered at an advantageous price. Furthermore, such cost could be further reduced as construction operations are improved and as designs of turbine generators are further refined.

Auxiliary pumping would only increase energy production on an annual basis of 5%, so it is not being considered, though not definitely ruled out. That, and the discarding of reversible blades, will differentiate the Canadian from existing plants in Europe.

K. E. Sorensen, in a personal communication, expressed his belief that a tailbay spillway facing the ocean would allow both easier closure of the main barrier at the end of construction and construction of the plant in still water. It would, furthermore, provide protection for the power plant against ocean waves and limit the extreme maximum and the extreme minimum ocean levels at the power plant. These should contribute to the reduction of power plant and equipment costs.

The earliest possible date for the inauguration of a tidal power plant to serve the Maritime Integrated System for New Brunswick, Nova Scotia, and Prince Edward Island is 1990. By then, peak demand would have probably reached 7,000 MW and would climb by 2,000–13,000 MW. Hence, whatever energy a tidal power plant could furnish could be absorbed by the market; additionally, any eventual surplus energy could be sold to the New England Power Pool.

Furst and Swales reported in 1978 that the present worth cost of the expansion sequence for the period of 1985–2010, plus a steady use for 2010–2050 at 2010 level, with an introduction of tidal output, results in a saving as compared to the same sequence using conventional output only. The saving is due to displacement of fuel and capacity credit. When dividing the saving by the total present capital cost of the tidal plant, one gets for the selected site a benefit cost ratio. (Table 10-3) Estimated consumption of oil is 8.3 million barrels and 3.8 million tons of coal annually by 1990. (Figs. 10.2, 10.3, 10.4)

In their calculations, the authors assumed, in 1976 $, prices of $15.60/barrel of oil, $29.00/ton of coal, and $99.20/kg of uranium. They also considered the life of a thermal plant to span 30 years and that of a tidal plant to span 75 years. Site evaluations were made by "adding" to the existing system fossil fuel generation of 475 MW (1985), nuclear generation of 750 MW (1990), and nuclear generation of 1,250 MW (2000). Tidal plant output for site $A8$ can be simultaneously absorbed (at least 90%); for site $B9$, the estimation is 60% in 1990 and 90% by 2010. The benefit-cost ratio for site $B9$ varies from 1.09–1.45, for site $A8$ from 0.93–1.22, and for site $A6$ from 0.68–0.90, depending on the parameters used, (e.g., the importance of nuclear development, the cost of money, or oil price increases). Specifically for site $B9$, 36% of the worth benefits are contingent on a sales contract with the New England Pool System.

Table 10-3. Estimated Annual Amounts of Fuel Displaced in the Maritime Integrated and New England Pool Systems. *Source: Furst and Swales, 1978.*

SITE	INSTALLED CAPACITY (MW)	FUEL OILS (barrels $\times 10^6$)	COAL (tons $\times 10^6$)	URANIUM (kg $\times 10^3$)
$B9$	4,028	12	1.30	54.4
$A6$	1,643	4	0.45	181.4
$A8$	1,147	3	0.38	9.1

SOCIOECONOMIC IMPACT

Based on a preliminary analysis, further studies are indicated in regard to agriculture, floods, and drainage. Ecological investigations dealing with bird habitat, andromous fish, and other commercial fish are needed, followed by minerals, land erosion, and navigation matters. Recreational impact requires looking into.

To this end, a portion of the scheduled budget for further design and specification study is set aside for environmental impact research, and the Environmental Assessment Panel has been formed for that purpose.

SENSITIVITY ANALYSIS

The Bay of Fundy Tidal Power Review Board carried out a sensitivity analysis to provide as complete a picture as possible of the future consequences of including or not including tidal power in the generation expansion programs of MIS. With site $A6$ uneconomic, the effects of changes in assumptions on the study findings were detailed only for tidal developments at sites $B9$ and $A8$. The significant parameters were the extent of nuclear development (penetration) in the MIS expansion program, load forecast changes, marketing strategies for tidal energy in the secondary market, fuel costs, and interest rates.

Nuclear Penetration

The greater portion of the tidal benefits would be derived from the displacement of thermal energy; hence, the higher the nuclear penetration in the expansion program, the greater the proportion of nuclear energy displaced and, consequently, the lower the resultant tidal benefit.

The base case MIS expansion program results in an increase in nuclear penetration, as a percentage of MIS peak load, of 50–73% from 1990–2010, respectively. This *intensive* nuclear scenario foresees the extent of nuclear generation limited only by technical constraints of nuclear cycling. Incremental nuclear generation additions would be required to operate at about 60% capacity factor, considered to be the minimum technically feasible capacity factor for this type of generation. In an all-nuclear generation scenario, a range of 50–79% nuclear penetration as a percentage of MIS peak load from 1990 to the year

2010 would result. This scenario would violate the nuclear cycling constraints, since some units would be required to operate at a 35% capacity factor. Therefore, it is very unlikely to be developed. A low nuclear scenario, with nuclear penetration limited to about 50% of the peak load might be seen as a possible result of resistance to continued development of nuclear plants. However, this would be contrary to the objective function of the least-cost generation program.

The Board concurred with the Committee that the most likely magnitude of the benefits from sites $A8$ and $B9$ would be those calculated as the base case program of intense nuclear penetration followed by the low in all nuclear scenarios and order of likelihood. The benefit-cost ratio could vary by $+30$ to -20%, depending on the scenario chosen.

Load Forecast Changes

Variations in projected load growth were tested only for site $B9$ by making alternative assumptions regarding future demand in the Maritimes market.

The base case load forecast assumed an average rate of growth of the MIS peak load of 7.2%/annum after 1985. For the sensitivity evaluation of alternative load growths, rates of 4% and 8.5% were used. The minimum rate was assumed to result from both a strong conservation program accompanied by a general slowdown in the economic growth, while the high rate reflected an increase in the market share of electricity due to substitutions from fossil fuel to electric power use, as well as a sustained high economic growth rate. The effect of the alternative assumptions regarding load growth is modest. The base case provided the lowest benefit-cost ratio, principally because the near-optimum generation expansion programs developed to meet the three load growth cases resulted in different mixes in generation. The Board considered that the assumed base case load growth rate may be slightly high.

Marketing Strategies

The base case program assumed that the full value of tidal energy to the secondary markets was credited to the tidal power project. In reality, revenues generated by surplus tidal energy in the secondary energy market would, in all probability, be less than full value and would depend on the sales contract agreement between MIS and the external market.

Various marketing strategies for sales to the secondary market were developed. It was assumed that the tidal energy sales to NEPOOL would be based on an economic energy contract, with a price for energy equal to 50% and 80% of the cost of the energy displaced.

Contributions made to the total benefits by the NEPOOL secondary market are very significant (i.e., 36%) with respect to the large tidal development at site $B9$, but insignificant with respect to tidal site $A8$ since the MIS system alone is large enough to absorb virtually all of the tidal output.

For tidal site $B9$, the optimum scenario chosen was that of raw tidal absorption into the systems. As the contribution of benefits of the secondary market decreases, retiming facilities to enable more of the tidal energy to be absorbed within MIS, resulting in less surplus energy for NEPOOL, would become more attractive: a 500-MW storage facility would reduce the net contribution by NEPOOL from 30% down to 15%.

Based on the optimum scenario for each marketing strategy (i.e., full credit), the benefit-cost ratio for site $B9$ would be reduced by about 8% under the 80% value assumption and by about 12% if the 50% value assumption were used. On the other hand, the benefit-cost ratio would only be reduced by 1% for site $A8$ under the 50% value assumption case.

For the purposes of the feasibility reassessment, the Board considered that a value of 50% of the cost of energy displaced would be the most likely condition.

Fuel Costs

The value of tidal power is also very sensitive to fuel costs, since a significant portion of the benefits would be derived from the displacement of fossil-fueled thermal energy.

The base case scenario assumed that the price of fuel for oil and coal would rise at the same rate as the general inflation to the year 1990 and then would increase by 1% annually over inflation thereafter.

Projections of future long-term oil processes were subsequently reexamined to take into account the most recent studies by Canadian and United States authorities, the Organization for Economic Cooperation and Development (OECD) and several specialized consultants. The results of this reappraisal indicate that the high fuel price scenario of

2% real escalation over general inflation after 1990 is a more realistic view of the future cost of oils.

However, short-term relative changes in the price of fuel with respect to general inflation may also have an effect on the relative long-term economics of tidal power. The likely price of fuel in the period under review is speculative, but any increase over the rate of rise in fuel costs

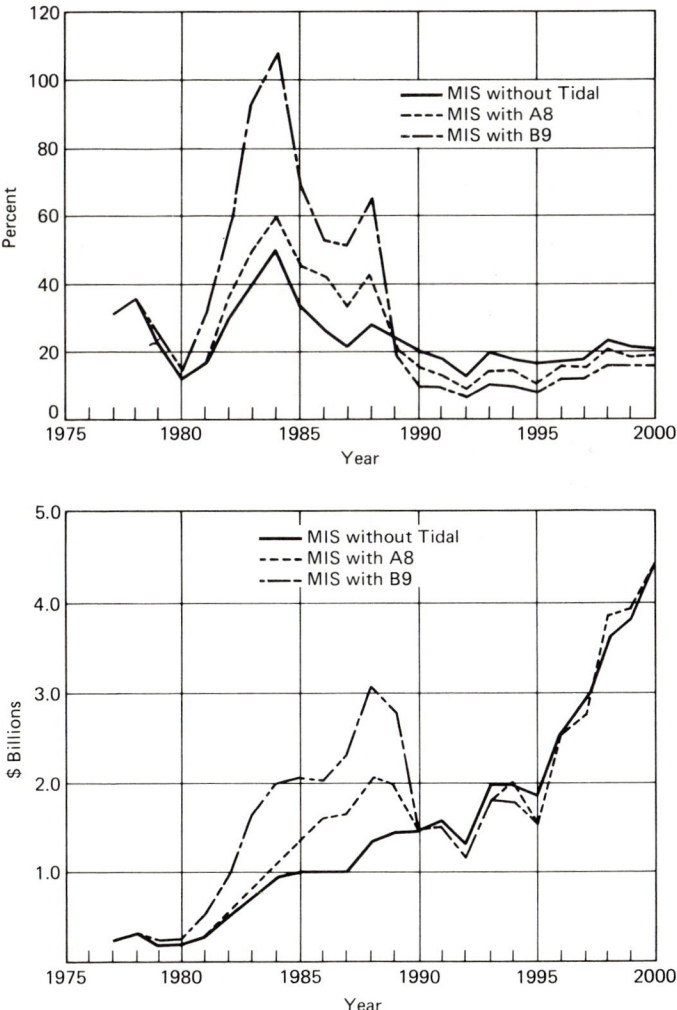

Fig. 10.10. Projected capital requirements for generation facilities. *(Source: Bay of Fundy Tidal Power Review Board)*

assumed for the base case will enhance the economic feasibility of a tidal power development. The benefit-cost ratios can vary by about +10% for the range of fuel prices assumed.

Interest Rates

Interest rate assumptions exert great influence on the benefit-cost ratio of a tidal power development because the cost of output consists almost completely of the annual capital cost, while the benefits, largely derived from displacement of thermal energy, are unaffected by interest rates.

An estimated actual interest rate was applied and interest computed in current dollars, taking into account an appropriate rate of inflation. A different approach was used in the economic analysis. Inflation was removed from both costs and revenues through the use of a *real* interest rate, and constant dollars derived by eliminating from the actual rate of interest that part which is considered to represent the effect of general inflation.

Real rates of interest on the government of Canada borrowings over the past quarter-century have ranged between 5% and -1.5%, with an average of just over 2%. Risk factors result in higher real interest rates for borrowers of lower financial stability. The spread between federal and Maritime provinces' borrowings over the last quarter-century has been about 1%. (See Fig. 10.10.)

From the point of view of the opportunity cost of capital, somewhat higher interest rates than those indicated above might be appropriate for an economic analysis. A rate of 7% or more would be appropriate for more work undertaken in a period when national productive capacity was almost fully utilized. However, this view of the cost of capital relates more directly to the timing of a commitment decision than to the underlying economics of tidal power.

The Committee assumed 5.5% interest for the base case and performed sensitivity analyses for 4% and 7%, finding that the benefit-cost ratios for tidal developments would vary by +20% for each 1% change in interest rates.

A real interest rate was adopted in the economic analyses solely for the purpose of reducing the two parameters of actual or current interest rates and inflation to one parameter. In view of this, together with the fact that financial feasibility will depend upon some measure of federal participation, the Board considered a real interest rate of 4.75% sufficiently conservative for the purpose of making a firm estimate of benefits and costs.

Adjustments

The benefit-cost ratios using the base case program appear to be understated rather than overstated. (Table 10-4)

Using an intense nuclear scenario, 2% escalation of fuel prices after 1990, a real interest rate of 4.75%, a load growth of slightly less than 7.2% and export proceeds of 50% of the value of the power that is exported, and the results of the sensitivity analysis, an overall improve-

Table 10-4. Summary of Sensitivity Analyses. *Source: Reassessment of Bay of Fundy Tidal Power (Committee Report).*

PARAMETER AND SENSITIVITY RANGE	PROBABILITY RANKING OF PARAMETER*	PERCENTAGE CHANGE TO THE UNADJUSTED BENEFIT-COST RATIOS	
		SITE $B9$	SITE $A8$
Nuclear penetration			
Low-nuclear scenario	II	24	36
Base case nuclear scenario	I	0	0
All-nuclear scenario	III	−8	−22
Load growth			
4.0%	II	3	
7.2%	I	0	
8.5%	III	7	
Marketing strategy			
Surplus secondary energy valued at:			
100%	III	0	0
80%	II	−8	−1
50%	I	−12	−1
Fuel cost			
Oil/Coal 2% escalation 1990/2010	I	12	10
1% escalation 1990/2010	II	0	0
0% escalation 1990/2010	III	−12	−11
Real interest rate			
4.0%	II	36	33
5.5%	I	0	0
7.0%	III	−23	−22

*I = probable; II = less probable; III = least probable.

ment in the economic feasibility of about 10% for site $B9$ and 25% for site $A8$ would be more representative of probable future conditions. The different adjustments used for sites $B9$ and $A8$ are due principally to the different impacts in the results from changes in the marketing strategies (Table 10-4). On this basis, a final benefit-cost ratio for site $B9$ of about 1.2 and, for site $A8$, of about 1.2 would, in the Board's judgment, be realistic.

Using the benefit-cost ratios of Table 10-5, the economic feasibility of site $B9$ is improved, and site $A8$ becomes economic to the same extent as site $B9$. The long-term economic benefits exceed the costs of a tidal development integrated with the MIS by a margin of 20% for both sites.

CONCLUSION

The Canadians have formulated plans that provide for a Canada-alone scheme and their thinking has drifted away from a joint Canada-United States undertaking. They have concluded that a Cumberland Basin scheme presents the maximal amount of safeguards from economic, technical, financial, and environmental viewpoints and feel that such a plant would settle more tidal power development.

Their study recommends that funding be provided forthwith to complete investigations, designs, and specifications for a single-effect, single-basin plant. However, the study should also assess the appropriateness of future construction, specifically in Shepody and Cobequid bays, and foresee the earliest possible placing into service of the Cumberland Basin plant. The Board recommended furthermore that 1) "institutional arrangements be established for execution of the detailed investigations and definitive designs and which could also provide the appropriate basis for the development phase"; and that 2) "immediate

Table 10-5. Final Benefit-Cost Ratios for Selected Sites.

SITE	BENEFIT-COST RATIO	BREAKEVEN PERIOD
$B9$	1.2	30/35 yr
$A8$	1.2	30/35 yr
$A6$	0.9	None

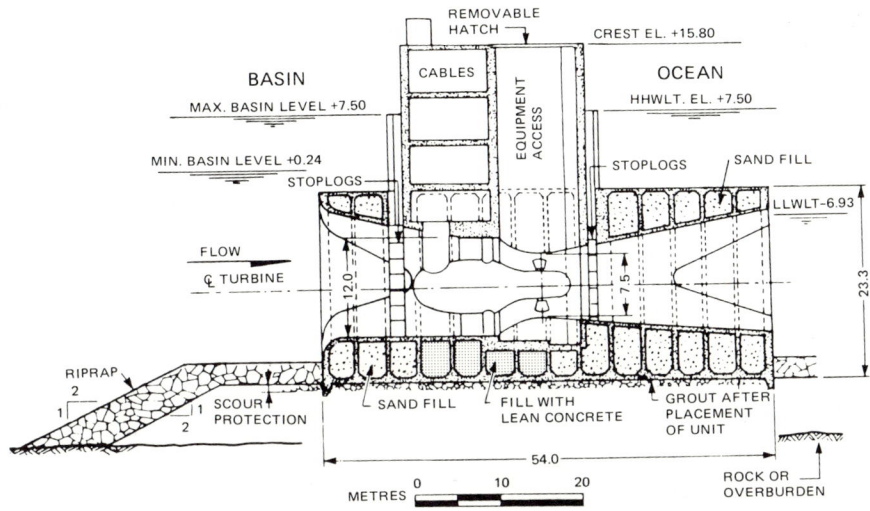

Fig. 10.11. Powerhouse cross-section site A8. *(Source: Bay of Fundy Tidal Power Review Board)*

consideration be given to the resolution of the financial constraints to developing tidal power."

In other words, the cost to the customer should not exceed what he could pay annually, if, no tidal power being utilized, the optimal expansion of generating facilities was carried out. This would insure that an unbearable burden, mainly through capital raising, not be placed on users in the Maritime provinces.

Prospects for implementation of a tidal power scheme in Canada, and even its ulterior expansion, appear to be good.

In early 1980, Tidal Power Corporation awarded a contract worth $15 million (Canadian currency) to Dominion Bridge-Sulzer, Inc., for construction of a 20-MW turbine. The turbine is to be housed in a dam near the Annapolis River's mouth on the Nova Scotia western shore. The Annapolis Basin varies in depth from 4.57–8.84 m, depending on the tide. The Swiss-Canadian company projects an 80,000-barrels/yr saving by implementing an Annapolis Basin tidal power plant.

The Straflo turbine has a 7-m diameter; after construction, in Montreal, it will be towed to its site in 1982. The rim-type turbine will have a horizontal shaft and incorporate straight water flow passage.

REFERENCES

Clark, R. H., 1978. The economics of Fundy tidal power, *Proc. Int. Symp. Wave & Tidal Energy* **I**, *No. E3*:41-54.

Furst, G. B. and Swales, M. C., 1978. Review of optimization and economic evaluation of potential tidal power developments in the Bay of Fundy, *Proc. Int. Symp. Wave & Tidal Energy* **I**, *No. E2*:23-40.

Gibson, R. A. and Wilson, E. M., 1978. Studies in retiming tidal energy, *Proc. Int. Symp. Wave & Tidal Energy* **I**, *No. H1*:1-10.

Godin, G., 1973. *The Tidal Power Potential of Ungava Bay and Its Possible Exploitation in Conjunction with the Local Hydroelectric Resources.* Ottawa (Canada): Marine Sciences Directorate (Ms. Rep. Ser. 30).

Godin, G., 1974. The energetic resources of Ungava Bay and its hinterland, *Engineering in the Ocean Environment—IEEE Int. Conf., Halifax (N.S.)* **I**:378-383.

Karas, A. N., 1977. System planning for Bay of Fundy tidal power developments, *IEEE Power Engineering Society* (Summer, Mexico City).

Laba, T. J., 1964. Potentials of tidal power in the North Atlantic Coast in Canada and the United States, *Conf. on Coastal Engineering, Proc.* **IX**:832-857.

Lawton, F. L., 1972. Tidal power in the Bay of Fundy, in Gray, T. J. and Gashus, O. K. (Eds.), *Tidal Power.* New York: Plenum, pp. 1-104.

MacCellan, H. J., 1952. Energy considerations in the Bay of Fundy system, *J. Fisheries Board of Canada* **XV**, *No. 2*:1935.

Sibley, H. K. and McNeice, W. H., 1960. Harnessing the tides, *Military Engineer* **52** (Jan.-Feb.):1-6.

Udall, S. L., 1963. *The International Passamaquoddy Tidal Power Project and Upper St. John River Hydroelectric Power Development.* Washington, D.C.: U.S. Department of the Interior.

Warnock, J. C. and Tanner, R. G., 1978. Selection of optimum sites for tidal power development in the Bay of Fundy, *Proc. Int. Symp. Wave & Tidal Energy* **I**, *No. E1*:1-22.

1969. *Feasibility of Tidal Power Development in the Bay of Fundy.* Ottawa (Canada): Atlantic Tidal Power Programming Board.

1977. *Reassessment of Fundy Tidal Power.* Ottawa (Canada): Bay of Fundy Tidal Power Review Board.

11
The Severn Project

Though several tidal power plant sites have been considered in England and Wales, the most publicized scheme concerns the estuary of the Severn River, where 14-m tides would furnish considerable energy. The original location of the dam, however, has been abandoned in favor of a new one. (Fig. 11.1)

HISTORICAL BACKGROUND

The first project for a tidal energy plant on the Severn dates from 1918. New schemes were put forth in 1933 and again in 1945. The British Parliament appointed a reexamination commission in June 1977 to study a 2–4-GW project in the Severn River estuary.

At the end of World War II, the coal shortage and the thirst for energy brought also a revival of interest for the construction of such a plant on the Severn River. A new idea it was not: the matter had been examined for several decades, even if the original proposals of Norman Davey had been ridiculed. However, the initial costs and the financing of the construction had always awed officials. Yet, in 1949, the need for new sources of energy became so imperious that the first project dating from the last throes of the First World War (1918) was pulled out of stuffy drawers. If, in 1918, few listened, more interest was shown in 1935 and again in 1945. (Fig. 11.2)

Engineers figured that a tidal power plant built in the Severn River near the Bristol Channel, where Europe's highest tides occur, would furnish considerable energy. They were thinking in terms of an 800,000-kW plant. Although a negative decision was reached each time (Headland, 1949), it is now readily admitted that had the plant been built in 1933 or 1945, it would have paid for itself within ten years of completion.

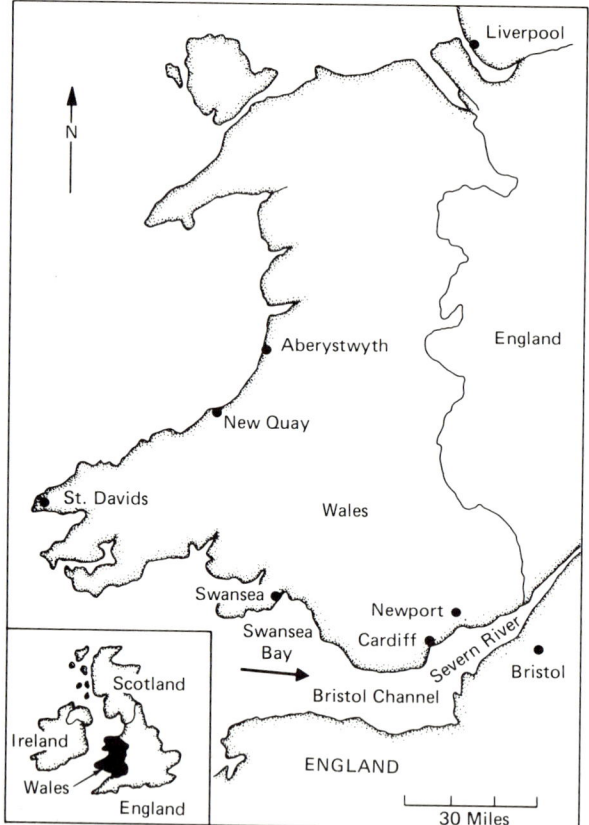

Fig. 11.1. Estuary of the Severn River.

The earlier projects supported a single-basin scheme for the Severn barrage with ebb tide utilization. A capacity of 800 MW had been proposed. Technical and economic surveys indicated, in 1949, that the tidal power plant would be of negligible value as a substitute for thermal power plants and that the tidal power station would furnish variable and intermittent output. "Consideration of direct and indirect methods of output regulation lead to the conclusion that 75% of the energy should be transmitted at 220 kV to Birmingham and London, where the tidal plant output would be small relative to the thermal plant connected to the system. Parallel operation of the tidal power station with the [National British] Grid is considered," and an appropriate transmission system was proposed by Headland.

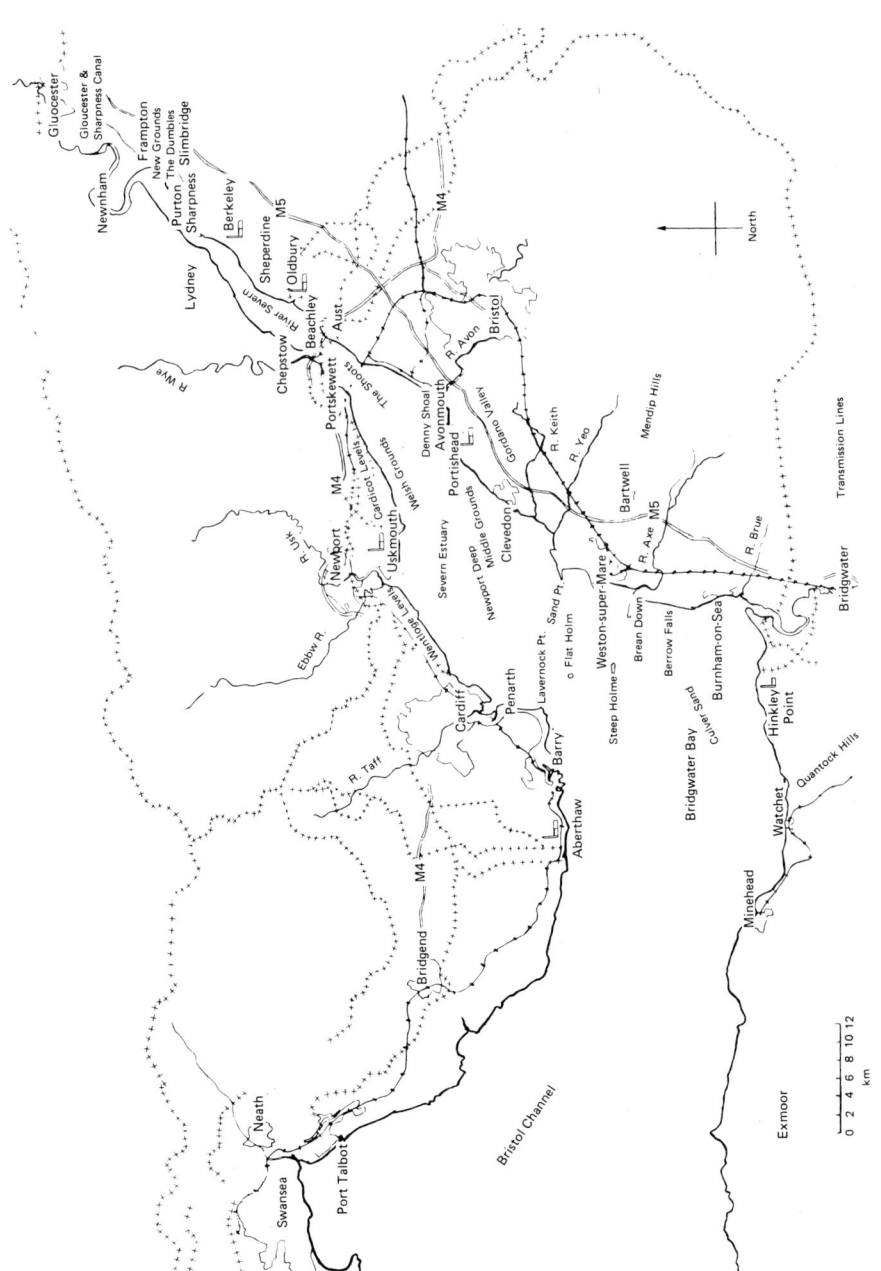

Fig. 11.2. Location map of the Bristol Channel.

ECONOMIC CONSIDERATIONS

An examination of the economics of the project showed that the estimated capital cost of the barrage and transmission system exceeded $250 million. However, this amount could have been reduced by simplifying the plant design, reducing the transmission costs, and through improvements in engineering and construction methods. Still between 2,107 and 2,207 kW-hr would be delivered per year to the load centers.

The operating costs of power stations and transmission systems would run between 5% and 14% of the estimated capital costs.

The post-World War II period was one of scarcity of coal, and thus "cheap power from the sea" seemed extremely attractive. The temptation to tap this heretofore virtually unused source of energy was further enhanced through the declining coal resources and the need to import expensive liquid fuels.

In France, a comparison of the price of tidal-produced electricity and traditional hydroelectric energy in grids wherein tidal power plants would be integrated, proved that "tidal energy" was by no means unfavorable. In Great Britain, on the other hand, tidal power-produced energy must be compared to the cost of a kW furnished by thermal and even nuclear plants (Kennedy and Headland, 1957). Hence, the problem facing the British after the 1945 armistice, in regard to the Severn River plant, was to examine, first, whether tidal energy would replace base or peak centrals and, furthermore, whether energy production by thermal plants would or would not increase if a tidal power plant were put into service.

Nevertheless, the construction of a plant on the Severn River would only be economically favorable if the cost of tidal energy remained below that of the quantity of coal necessary to produce an equal quantity of electricity in a thermal plant. Twenty years ago, the answer to that question seemed negative. Today, experts are not so sure anymore, particularly if the source of heat is oil. Yet, the considerable construction costs and those of annual production are still the main impediments to the project's realization.

If proposals were shelved again, this is also particularly due to the report of the commission of enquiry, which opposed the idea of including either road or railroad crossing in the barrage, and additionally rejected the plan to provide pumped storage facilities for the energy.

The abandoning of the plans is the more regrettable since patterns of tides differ on both sides of the English Channel, and a Severn River plant would precisely complement the Rance plant: maximum production and maximum consumption would occur at opposite times.

Further deterrents to the construction of the Severn plant were the rate of sedimentation and the production of atomic energy in Great Britain, though atomic energy would not seemingly render tidal energy useless or even uneconomical. Public opposition to nuclear power may now prove a boost to tidal power proponents' efforts.

The successful completion of the Rance River project has had favorable consequences in Britain, and research conducted by the English Electric Company shows the possibility of using the lessons from the Rance and Kislaya experiences, and those of the IJsselmeer in the construction of a tidal power station on the Solway-Firth. The British also shifted locations in the Bristol Channel. Early plans favored a barrage near Avonmouth; now a dam enclosing the entire bay from Bristol to Cardiff, is under consideration. It would improve portuary facilities of these two harbors and of Avonmouth, be the bed for a highway link between Wales and Southwest England, and provide a basin of 495 km^2.

Annual power production would reach about 13,100 million kW, or a twelfth of the total power consumption, ten years ago, of the United Kingdom. Great Britain used, in 1977, 300 million tons of coal equivalent; net energy saving of a Severn barrage would be about 8 million tons of coal equivalent.

There is an unusual conflict of positions in Great Britain. As pointed out by a proposal of David Mappin (Offshore) Management, there is apparently a large surplus of thermal generating plants currently in service in Great Britain compared with the maximum demand for power. Next, numerous new plants in excess of 12,000 MW are on order and under construction. The near-future needs in power have been overestimated, though T. L. Shaw (1977) forecast that independence from imports will end in 1990. Any tidal power station will only be fully operative after that date, and hence will help in reducing imports at the right time. Assuming that the plant would produce 4,000 or 5,000 MW, the tides would "generate" in 30 years as much oil as is regarded as proven and recoverable from the two largest oil fields in the North Sea and, in a 100-year span, as much power as all the oil the British North Sea sector can yield. If a retail value must be placed on these calculations, then the average value of the energy from a Severn River 4,000-MW plant is $2,200,000/day. (Fig. 11.3)

280 TIDAL ENERGY

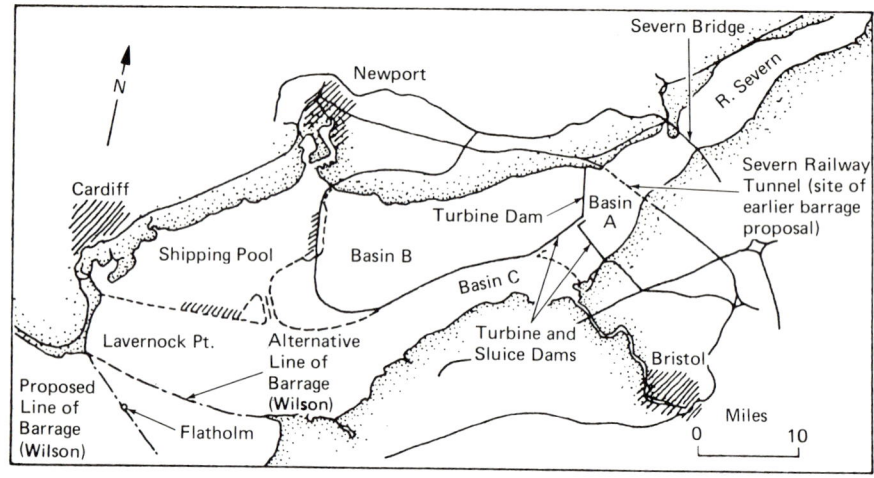

Fig. 11.3. Severn barrage: proposed sites (1945–1969).

Table 11-1. Tidal Ranges, Capacity, and Output of Four British Tidal Power Sites. *Source: The Select Committee on Science and Technology (Minutes of Evidence), May 19, 1976. London: The House of Commons.*

SITE	TIDAL RANGE (m)			CAPACITY (MW)	ENERGY PRODUCTION ESTIMATES (GW-hr)	
	SPRING TIDE	AVERAGE TIDE	NEAP TIDE		BERNSHTEIN	COLLYNS
Severn River Estuary	12.5	8.80	6.7	4,000	20,000	5,280
Solway Firth	7.3	5.11	4.8	5,000	12,600	8,800
Morecambe Bay	8.8	6.15	5.1	4,000	9,650	5,100
Carmarthen Bay	7.9	5.54	3.9	2,000	6,450	3,200

The potential of the sites under most serious consideration is quite large. (Table 11-1) A hydroelectric scheme has currently a useful life of 100 years. Hence, Shaw estimates, the investment in a Severn plant would be the equivalent of only two years of North Sea oil revenues at current prices.

The latest proposal involving the Severn River estuary recommends construction of a dam from Lavernock Point, south of Cardiff, in South Wales, to Breandown near Weston-Super-Mare, on the Somerset coast. The South Wales terminal lies near large coal mines, which supply the existing thermal power plants. This might prove to be a socioeconomic deterrent to construction of a tidal power plant because of competition with coal. Fig. 11.4.

SITE CHARACTERISTICS

The Severn River estuary places third in tidal amplitude on a world scale after Leaf Inlet in Ungava Bay on the Atlantic Coast of North America and the Bay of Fundy. The site actually encompasses a seaward segment extending 100 km landwards from the Celtic Sea and St. George's Channel to Barry and Winehead, where depth is 20 m and width 22 km. Beyond that point, the estuary narrows and becomes an elongated channel 13 m deep and 13 km wide. At Barry, the average tidal range is 8 m and the maximum amplitude occurs near Avonmouth-Aust, an area proposed for a barrage in the thirties. (Fig. 11.1)

Shaw conducted a study of the site's hydrodynamics, as some concern has been voiced that a barrage may accentuate an already inadequate

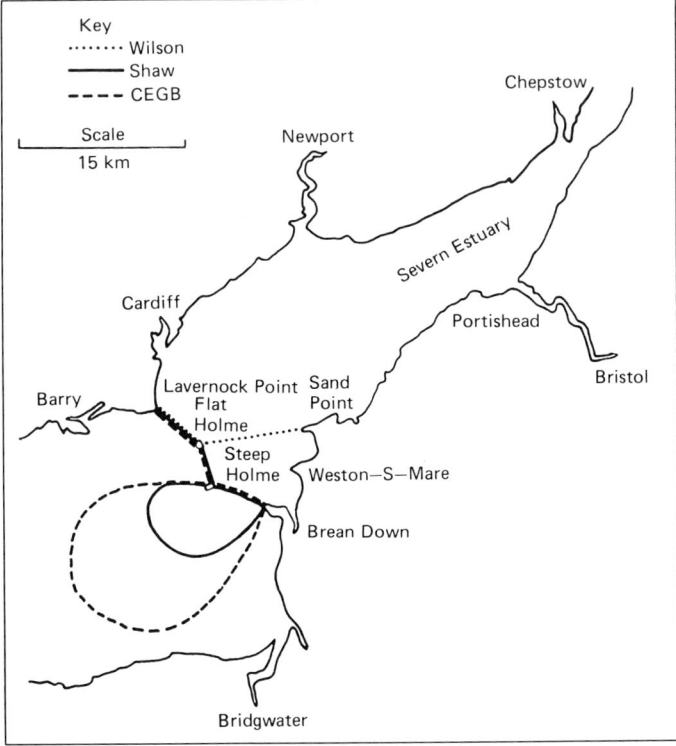

Fig. 11.4. Location map of the barrage schemes, 1975–1977. (Shaw, *An environmental assessment of a Severn barrage*)

dispersion of industrial and agricultural discharges, to which must be added piped outfalls, power station cooling water discharges, natural watercourses, suspension matter, possibly atmospheric fallout, and other effluent release. He concluded to a limited possibility only for water to leave the Severn estuary area once it is locked by barrages. Fresh water inflows from the rivers will only be significantly affected near and in river mouths; the net seaward flow near the Welsh coast is small in terms of rate of exchange, and circulation due to the Coriolis force will have insignificant influence; storm conditions through winds and waves could cause strong mixing; evaporation is a negligible factor here; and lateral exchanges between shallow and deep areas could still be effected without problems. Discharges of pollutants should not be increased, keeping in mind that the seawards ebb flow is not a flushing process, but one half of a cyclic movement only.

ECONOMIC ASPECTS IN 1978

Using caissons in construction and an inexpensive rock-fill dam, the Severn tidal power plant would cost as much as double as the lowest cost alternative, a nuclear power plant using pressurized water reactors, and still higher as a fossil fluid plant using 1977-priced oil. However, in the plus column of the tidal power plant must be placed the benefits to the harbors of Bristol and Cardiff, the use of a motorway on top of the barrage, tourism income, and other fringe benefits. By raising the minimum water level by 7.6 m, 100,000-ton bulk-carrying ships would have constant access.

The Central Electricity Generating Board and the U.K. Department of Energy estimate that a national wave energy program should be encouraged, since wave power appears to have a greater potential for energy production; however, wave power harnessing needs technical feasibility study and costing is totally unprecise.

On the other hand, the siting of the Severn plant close to existing and proposed thermal stations and transmission lines militates in its favor. Its construction would probably improve the region's prosperity, particularly in South Wales, through job creation and subsequent induced employment, and make use of 50 million tons of colliery shale waste currently stored in surface tips at 40 km from the estuary and 100 million tons of china clay tailings from Devon and Cornwall. New industries could be attracted to the region, salt water fish farming could be launched, and water recreation possibilities would attract large num-

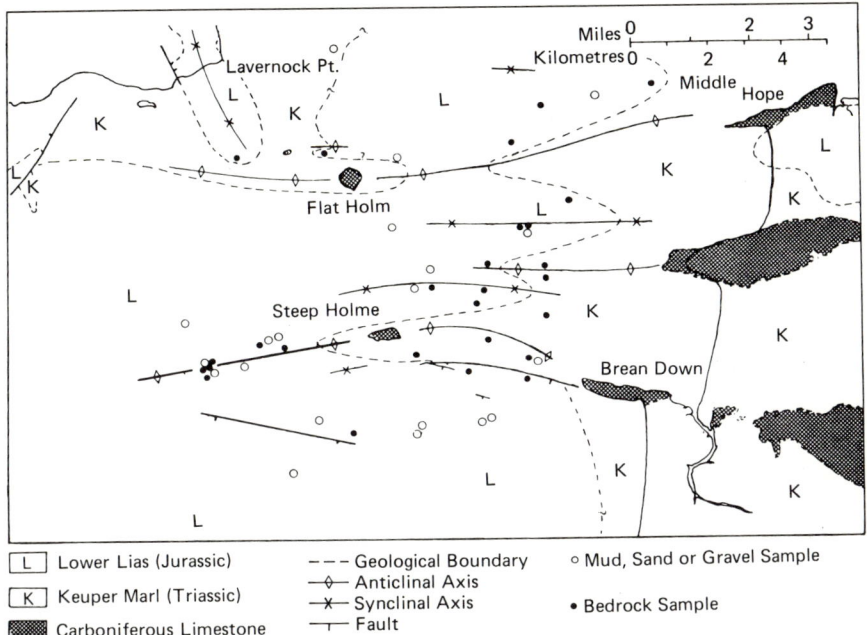

Fig. 11.5. Geological map of part of the Severn estuary. (Shaw, *An environmental assessment of a Severn barrage*)

bers of tourists. Land drainage would help agriculture and land could be reclaimed as slag heaps and china clay tips are used up. (Fig. 11.5)

The studies were conducted for the United Kingdom Department of Energy Advisory Council on Research and Development for Fuel and Power, one by NEDECO (Netherlands' Engineering Consultants Foundation), another by the Hydraulics Research Station of Wallingford (Oxfordshire), and one by the Institute of Geological Sciences. Cost estimates exceed, according to the Dutch consultants, $6 billion and $8 billion for a single- and double-basin scheme, respectively. Although the running costs of the plant would be very low and its lifespan very long, the cost of energy produced would fall between 10 and 20 U.S. ¢, more than double the current price of electricity in Great Britain. Fig. 11.7.

CURRENT STATUS

Wilson (1977) calculated that a Severn River plant could produce 8,000–14,000 GW-hr at a competitive price; this represents 4–7% of

British annual consumption. An additional, but extraneous, argument in favor of constructing the plant is that it will be necessary, no later than 1985, to build a new road across the estuary, and that this road could pass atop of the electrical center. Shipping time savings would amount, at current traffic and costing rates, for 658,000 net registered ton ship-days, to about $222,000 (1977 U.S. $) or £110,000. The three 1977 reports on current feasibility of the Severn River barrage scheme were published concurrently with the United Kingdom government "Energy Paper Number 23."

Heaps, in 1968, had already focused attention on the impact of a barrage. The Dutch group, while estimating its construction feasibility, predicted a 1-m decrease of the tidal range. The Hydraulics Research Station estimated the change instead to be an increase of 1.4 m whose effects would affect the estuary, the Bristol Channel, the Irish Sea, and even the English Channel on the other side of the British Isles. This amounts to a difference of 2.4 m in predictions.

Cost estimates range from between $4,800 and $6,000 million (in 1977 U.S. $) for a single-basin scheme, and climb to between $6,000 and $8,000 million for a double-basin scheme.

The Dutch firm foresees 20 years of construction or nearly thrice the time needed to build the Rance River plant.

Pointing to the great difference of opinions, "Energy Paper Number 23" ventures that "a very great deal of effort [may] be required to reduce the uncertainty" as regards tidal range changes and thus "a definitive feasibility study of the barrage would need to take into account its possible effects over a very wide area." However, the British are pursuing further studies, and personal communications to the author indicate implementation plans are still very much alive (1978). (Fig. 11.6)

One change from prior plans is that instead of an electricity plant stream-upward from Bristol, the currently favored site lies downstream of Weston-Super-Mare, roughly 30 km closer to the sea. (Fig. 11.7)

The most frequently reported sites for the United Kingdom are the Severn River estuary (range 9 m; annual energy output, 20,000 GW-hr; capacity, 4,000 MW), the Solway Firth (5 m; 13,000 GW-hr; 5,000 MW), Norecambe Bay (6 m; 10,000 GW-hr; 4,000 MW), Camarthan (5.5 m; 7,000 GW-hr; 2,000 MW), and in Ulster Strangeford (3 m; 2,000 GW-hr; 200 MW) and Carlingford (3.5 m; 1,300 GW-hr; 120 MW). There are, of course, several other locations, but they have not

THE SEVERN PROJECT 285

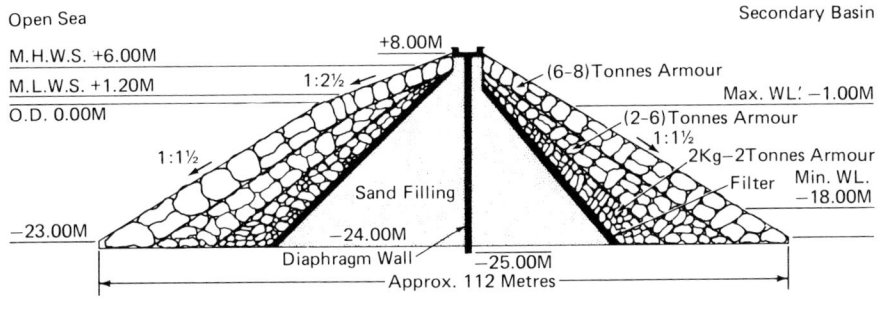

Section Through Proposed Severn Barrage

Fig. 11.6. Section through proposed barrage. (Shaw, *An environmental assessment of a Severn barrage*)

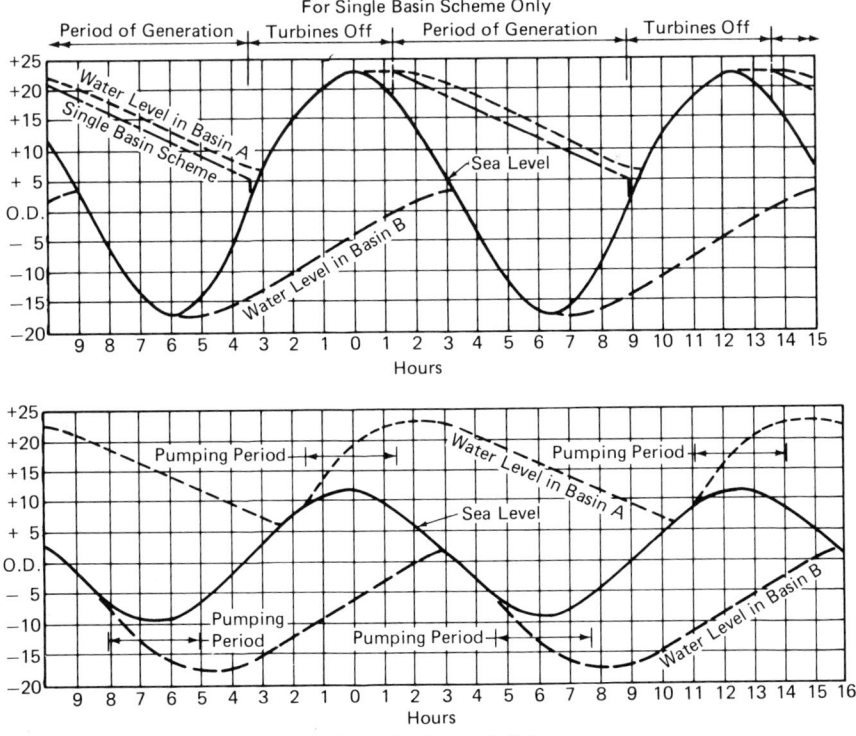

Fig. 11.7. A comparison of tidal power plant operation with schemes involving, respectively, one and two basins.

been reported, at least recently, as sites under consideration. Furthermore, A. J. Milne (1978) has recently discussed the possibility of the use of tidal power in sea lochs in Scotland.

CONCLUSIONS

Electrically compatible with thermal sources, a Severn tidal power station would be able to extend Britain's options as oil runs out, and provide storage and higher efficiencies from thermal and other benign sources. The Severn barrage could provide up to 10% of British electrical needs, and about one-third of their Department of Energy estimates of nuclear output in the year 2000. If built, it could be constructed largely from the wastes of colliery and china workings, and be in operation at the energy-critical time of 1990. Direct underground transmission links with an established electrical network would assure its integration in the National Grid. The Severn Barrage would ameliorate the tide range to amounts elsewhere normal in the United Kingdom, maintaining daily cleansing of beaches, reducing currents hazardous to navigation and wave action, and hence reducing the need for coastal protection. It would improve water clarity by lessened erosion of sediments, maintain salinities, and create opportunities for mariculture in conjunction with thermal power stations. Though there would be no likely restraint on salmon migration, one could expect some reduction in wading bird reserves. Riparian agricultural opportunities would be greatly improved. No industrial pollution risk, given the already necessary tightening of effluent discharge consent conditions, is foreseen. Aesthetically, the almost fully and permanently submerged structure, visually unobtrusive to the water line, would provide a unique opportunity to improve aesthetics and release land sterilized by colliery and china clay tips; however, while enhancing the Parrett bore at Bridgewater, it would prevent formation of the Severn bore. For tourists, the barrage would transform the Severn estuary to "Solent-like" appeal for water recreation and leisure center development. Although these observations are based on established facts, much more information is urgently needed on both the existing environment and the impact that a Severn barrage might permissibly make before firm judgments are reached on the preferred siting for, and mode of operation of, optional schemes.

REFERENCES

Collyns, G. S. 1951. The production of tidal power by the two-basin system, *Engineering* **166**:393–394, 427–429.
Gibson, A. H., 1933. Construction and operation of a tidal model on the Severn (London), *Appendix to the Report of the Severn Barrage Committee.*
Headland, H., 1949. Tidal power and the Severn barrage, *Proc. Inst. Elec. Eng.* **96(II)**, *No. 51*:427–451.
Ibid., 1950. **97(II)** (June).
Ibid., 1951. **98(I)** (March).
Heaps, N. S., 1972. Tidal effects due to water power generation in the Bristol Channel, in Gray, T. and Gashus, O. K. (Eds.), *Tidal Power.* New York: Plenum, pp. 435–455.
Hooker, A. V., 1970. Severnside of the future, *Proc. Inst. Civ. Eng.* **47**:337–348, **49**:467–486.
Jones, C., 1980. Quebec turns water into gold, *Christian Science Monitor* (July 30).
Kennedy, G. E. and Headland, H., 1957. Etudes de l'usine marémotrice de la Severn, *La Houille Blanche (IVes Journées de l'Hydraulique)* **II**:456–464.
Richards, B. D., 1948. Tidal power, its development and utilization, *J. Inst. Civ. Eng.* (Great Britain, April):104–144.
Shaw, T. L., 1978. The role of tidal power stations in future planning for electricity storage in U.K., *Int. Symp. Wave and Tidal Energy* **H2**:11–22.
Shaw, T. L., 1978. The status of tidal power, *Water Power* **24**, *No. 6*:29–34.
Shaw, T. L., 1977. Tides, currents, and waves, in Shaw, T. L. (Ed.), *An Environmental Appraisal of the Severn Barrage.* Bristol: The University of Bristol.
Vernon, K. R., 1974. Hydro (including tidal) energy, *Philo. Trans. Royal Soc. London, Series A,* **276**:485–493.
Wilson, E. M., 1964. A new approach to power from the tides, *New Scientist* **24**, *No. 415* (Oct. 29):290–291.
Wilson, E. M., 1965. Energy from the tides, *Science J.* **I**, *No, 5*:50–56.
Wilson, E. M., 1965. The Solway Firth tidal power project, *Water Power* **17**, *No. 11*:431–440.
Wilson, E. M., 1966. Feasibility study of tidal power from Loughs Stangford and Carlingford, with pumped storage at Rostrevor, *Inst. of Civ. Eng.* **34** (May):83–100.
Wilson, E. M., 1973. Energy from the sea—tidal power, *Underwater J.* **5** (Aug.):175–186.
Wilson, E. M., Severn, B., Swales, M. C., and Henery, D., 1968. The Bristol Channel barrage project, *Proc. Int. Conf. Coast. Engin. (London)* **XI**:1304–1325.
1933. *Report of the Seven Barrage Committee.* London: H. M. Stationery Office.
1945. *Report on the Severn Barrage Scheme.* London: Ministry of Fuel and Power.
1964. The estuarine barrage, *The Engineer* **218**, *No. 5677* (Nov. 13):787–788.
1966. Tidal power from the Bristol Channel, *The Engineer* **221**, *No. 5739* (Jan. 21):109–110.
1967. Tidal power from the Severn, *The Engineer* **223**, *No. 5802* (April):509–515.
1977. *Severn Barrage Seminar: Proc. Energy Paper 27.* London: H. M. Stationery Office.
1981. Canadians utilize hydroelectric power, *Journal of Commerce* (Feb. 25).
Department of Energy, 1977. *Tidal Power Barrages in the Severn Estuary; Recent Evidence on their Feasibility.* Energy Paper No. 23. London: H. M. Stationery Office.

12
Environmental Impact

Several impact studies have been carried out on tidal power projects, but actual observations exist only for the Rance River plant. No report has been issued dealing with the Kislaya plant, and the only comments available are those of Bernshtein (1974), who felt the plant would enhance oyster production and suggested to use the powerhouse as a herring trap. Concern had been voiced at one time about passage of ascending adult fish and descending salmon smolts in connection with the Severn River project, but ultimately, biologists concluded that such plants would hardly have any effects on the fisheries.

Physical changes brought about by the construction of a plant include the physical obstruction it constitutes, changes in current patterns and velocities, alteration in pools of tidal levels and ranges, reduced flushing, higher summer and lower winter temperatures, winter icing, a lesser vertical mixing with ensuing reduced dissolved oxygen content in deeper water layers and a lower rate of oxygen uptake, reduced surface layer salinity, and greater water stratification, erosion, and siltation. These modifications unavoidably would have repercussions in the local ecology. The matter is not whether but *how* much modification will be caused.

THE RANCE RIVER PLANT

Overall no unfavorable impact has followed the building of the Rance River plant. No environmental study had been made prior to its construction, nor was any in-depth study conducted since. Some reports exist, according to the EDF (Electricité de France) dealing with the influence such a plant could have on the physical milieus, where it would be implanted. The main concern had been to limit prejudices to the site and to ascertain that the high-velocity currents created by the plant would not, while nevertheless maintaining upstream of said plant a sufficient volume of water, perturbing the entrance to the harbor of St. Malo.

In the views of I. B. Thompson and his collaborators, the road atop the barrage has had no significant effect on achieving a greater cohesion between urban structures on either side of the Rance; this view is however not shared by Electricité de France personnel. The town of La Richardais, formerly entirely rural, had already been affected by the vicinity of the Dinard urban center. This influence has increased since construction of the barrage: about 20% of the territory of the town was lost to new roads and locales for cables, while socially it has been affected by the, albeit temporary, influx of barrage workers. A loss of tourism revenue has resulted from the disappearance of the beach but a new yacht marina may provide new income; the barrage gates are open more often than those of the harbor of St. Malo and the mooring fees are lower. Semidetached villas have been built in an area some 500 m from the barrage; they are occupied by top personnel of the electricity company. Some geographers held up the prospect that La Richardais may become a dormitory settlement because of its proximity to the new road.

The beach of St. Suliac was badly hurt, although with government help yachting, camping, and scenic promenades may provide touristic incentives.

Dinard and St. Malo were detached from one another by the river. Urban development, however, has become more integrated, because the barrage constitutes an artificial physical continuity that did not exist before its construction.

Basically, the impact of the Rance River tidal power plant has been very limited. The location of the plant was selected so that the high-speed currents generated by the plant would not perturb the entrance to St. Malo harbor and yet that a sufficient water volume be retained upstream from the plant. No silting problem has occurred due to two factors: first, the Rance hardly carries any solid materials, and second, its natural discharge is quite small compared to that of the tide.

Changes have affected some tidal characteristics. In the retaining basin, the tidal range has decreased while the high tide level is longer; the tidal maxima are less and the tidal minima are higher. No unfavorable consequences resulted for the riparians, and these new conditions have benefited tourists' activities.

Some navigation patterns have been modified: high-speed currents situated at the right of the powerhouse led to the enactment of a prohibited zone for navigation; on the other hand, the tide currents which permanently affected the area have been suppressed. One prohibited area is on the seaward side and the other on the estuary side. The control room is below water level and close to the Machine Hall. Therefore, a scanning television camera, with a screen on the control panel, is used to view the prohibited areas, the road, and the outdoor buildings.

Shipping to and from St. Malo harbor is dependent upon the lock, but conditions are improved, due to the reduction in the current speeds at spring tides, so that the number of boats passing through the lock has doubled from 1968 (5,287 boats) to 1973 (10,380 boats). In spring, EDF publishes the periods during which the pool level will be more than 4 m and 8.50 m. These are the minimum levels for shipping wishing to use the Rance and Le Chatelier Locks, respectively. Moreover, exact pool levels are published two days in advance.

Changes have occurred in the levels of the pool: the maximum equinoctial levels have been reduced from 13.50 to 12.80 m, and the minimum levels are higher. Furthermore, the high levels are more frequent and remain steady for 3–4 hr. This new pattern of levels has improved conditions for people living on the shores of the pool as regards flooding, yachting, etc., and it has effectively turned the estuary into a lake.

An unfavorable effect is the hazard of unpredicted surges on either side of the powerhouse as a result of a rapid shutdown of the units. Under certain conditions, this is prevented by the gate apparatus being immediately reopened by the computer to reduce the surge, with the blades at a right angle dependent upon the head, to keep the turbine speed below a predetermined value. Moreover, there is a system of red warning lights to signal the surge conditions at critical points in the estuary.

The plant having been constructed "in the dry," the estuary was closed during building; a new ecological equilibrium apparently has set in since the plant went into production, but environmental specialists admit that its "quality" is not less than its predecessor's. Some sand banks have been swept away by current effect, and others have been displaced or remain permanently between water levels in the pool, depriving some fishes, called *lanzons* by the French, of their habitat. The *lanzons* have disappeared, but other animal species have taken over, and the oyster culture has been substantially boosted in the pool. The river is not actually dammed, since sluices, open many hours a day, constitute a 900-m^2 passage, and fishes pass without being harmed, through the plant's turbines as well. As for life on the seaward side, the barrage has hardly had any influence on the river's mouth waters. It may be interesting to note that the plant does not pollute, and that pollution conditions in the pool, due to river-carried effluents, have not worsened since the cutoff was completed.

The closing of the estuary during the construction phase appears to have led to a new ecosystem equilibrium since the plant was put into operation. The open sea has not been influenced biologically by the barrage.

Since the Rance River has no solid transport, and because its discharge current is weak compared to tidal movements, no silting problem exists in the pool.

Economically, the impact has been most favorable as the plant spurred industrialization and permitted power export.

Theoretically, the building of tidal power plants does result in slowing down the rotation speed of the earth and hence the lengthening of the day; no such consequence has been observed here, but obviously, the multiplication of such plants could possibly have consequences that are difficult to estimate. The rotational speed seems to slow down a few seconds per century, which nevertheless represents about 2 billion hp. It is often said that tidal energy is an inexhaustible source of power; this is not absolutely correct because, in fact, any energy extracted from the tides comes from the rotational energy of the earth, so taking some of it away will slow down the movement. However, realistically, any practical rate of extraction will have a small effect only compared with the existing natural dissipation due to scouring, and wastage as heat, air movement, and noise.

In opposition to the care that would-be builders of tidal power plants in Canada, the United States, and Great Britain take in regard to environmental impact, no such study had been conducted for the Rance project. However, some research had been carried out concerning the potential influence that such a plant could have on the physical milieu where it would be built. The Electricité de France maintains that experience confirmed the forecasts made by those studies.

THE PASSAMAQUODDY PROJECTS

More thorough assessments accompanied the later United States-Canadian projects. When, in November 1948, the Passamaquoddy-Cobscook Bays plans were revived by the governments of the United States and

Canada, a succession of studies were made which culminated in a comprehensive report (Passamaquoddy Engineering Board Investigation, 1959). No great disadvantages for the fishing industry were foreseen (Passamaquoddy Fisheries Board Investigation, 1959), though some adverse effect on the fisheries would be probable. Herring, some claimed, would be wiped out inside the dam, but that catch represents only 2½% of the commercial catch of herring from St. Mary Bay (Nova Scotia) to Cape Elizabeth (Maine).

The proposed double-basin scheme would cause the water level in the upper pool to be always higher than the mean ocean level and, in the lower pool, always lower. Tide range would be 1.5 m in the high pool and 2 m in the low pool, instead of 5.5 m. A single-basin plant would reduce the range in Cobscook Bay from 5.5 m to 1.2–1.5 m. The basin level would not normally drop below one-third of the tide range. The tidal resonance patterns in the Bay of Fundy would hardly be influenced, and tidal amplitude would increase, at most by 31 cm at the Bay's head. However, currents in both pools and in a constricted area outside of them would undergo substantial changes. Fresh water would not gain as much salinity because less ocean water would enter the pools, and vertical mixing would be reduced since water velocity would drop during periods when gates are closed. This, in turn, would lead to more pronounced water stratification, resulting in sharper seasonal variations for surface waters inside the pools. These surface waters would be warmer in summer and, in winter, ice would form and salinity drop; such changes would be minimal in deeper waters, but due to a lesser degree of vertical mixing, oxygen concentration would decrease here from a near 100% to no less than 50%.

Erosion was not expected to increase, but some silting would not be excluded, which would affect fisheries in the pools. In his part of the 1959 Fisheries Report, Dow stated: "Anticipated reductions in erosion, scouring, and turbulence together with higher water temperatures resulting from impoundment and the less rapid transportation of soluble compounds and minerals elements carried into the area and by fresh water runoff should enhance the supply of fish- and shell fish-supporting nutrients, except for those species which would be denied access by engineer structures."

All fish species would not be similarly affected. Winter flounder would remain unaffected. Herring, according to some, would also be unaffected, but others fear a reduction. Cod, haddock, and pollock, which breed outside the pools but migrate in and out, would have major problems leaving them. Anadromous fish (e.g., salmon, alewives, smelt, and trout) would benefit from higher temperatures and lesser tidal ranges, provided fishways would facilitate their passage from pool to pool. Scallops would suffer from silting, but their numbers would increase in high-velocity deep channels. Clams would be restricted to a small area of the upper pool and more or less maintain themselves in the lower pool, but there would be a ten-year productivity interruption. Lobster larvae would benefit from the higher temperatures, as would shipworms, which would also be a plague for wooden ships, structures, and implements. Lobster ponds in the upper basin would have to be relocated, however, and fishing in the winter would be seriously hampered by ice formation.

The Stone and Webster study of March 1977 has not substantially modified the prior environmental assessment. The turbidity resulting from the placing underwater of the dam materials could temporarily reduce the phytoplankton population, and disturbance of the bottom habitat, including possible increased sedimentation, may have the same effect on the benthos. While fishways and apertures could facilitate the movement to and from the pools of resident and anadromous fish, the problem would be more acute for marine mammals. The mammals are whales (finback, minke, and right), seals (harbor and gray), and harbor porpoises.

Permanent covering of existing mud flats would doom clam and sea worm harvesting and cut down the living and breeding area of waterfowl. On the positive side, aquaculture of mussels, oysters, snails, sea trout, and Coho salmon would benefit.

THE HALF-MOON COVE PROJECT

A preliminary estimate made by Norman Laberge (1978) for the Pleasant Point Indian Project showed that exposure time of some low flats would be reduced due to a decrease of the tidal range affecting life in the intertidal zone. Some changes could be used to benefit aquaculture and thus compensate for the losses. A pilot project with 40,000 oysters was started in 1978 to assess the ostreicultural potential.

THE COOK INLET SCHEME

The Cook Inlet (Alaska) would undergo the same ecological changes as have been noted for Passamaquoddy. Pacific salmon spawning would be unhampered in the Matanuska and Knik River systems but severely hampered in the Sutsina system. There are no commercial fish or shellfish operations at present. Public access to the existing wilderness may adversely affect area land dwellers.

A single-basin plant in Knik Arm would reduce amplitudes inside the basin from 7.9 m to 1.5 m; a similar project in Turnagain Arm would reduce ranges there from 7.44 m at the entrance to 1.8 m in the pool. The two-basin scheme would modify ranges in Knik Arm (high pool) and Turnagain arm (low pool). They would be 2.75 m and 3.65 m, respectively. Sediments in suspension would accumulate at the rate of 3.3 cm annually in Knik Arm upon their arrival via debouching rivers. This would not happen in Turnagain Arm. Reduced mixing and increased stratification would extend the duration of the ice cover. Ice break-up problems might be reduced.

THE BAY OF FUNDY PROJECT

The Canadian study of the effects of tidal power development in the Bay of Fundy undertook to identify the impacts likely to result from construction and operation, and to provide a rough ranking of their relative importance.

The recent reassessment (1977) of the Fundy Bay project came to the conclusion that the scheme would lead to lower environmental pollution loading as compared to generation expansion programs without tidal power. Federal and provincial environmental agencies and numerous other scientific sources foresee no environmental problems that would prohibit development of any of the sites considered, though the impact would still be lesser for site $A8$ than for site $B9$.

A single-basin scheme in the Minas Basin would limit tidal variations from between 30 cm above mean sea level to 7.60 m, against the natural levels of from 7.90 m below mean sea level to 8.53 m above. With a double-basin arrangement, the level variations would go from 8.83 m below mean sea level to 9.15 m above. One could estimate at 5% the decrease in tidal amplitude outside a Minas Basin pool but within the Bay of Fundy. It was determined that tidal range would be reduced, possibly by as much as 15% below the chosen site, with an ensuing decrease in velocity of tide current up to 20% and unassessed modifications of sedimentation processes. Turbulence would be reduced and sediments would slowly accumulate upstream of the dam (more so if a single-effect scheme were selected). Currents and water levels would be affected, and erosion, including that along the coastline, would be accentuated.

No adverse effects on fisheries outside the retention basin are foreseen. To the contrary, clam production could improve as siltation would be reduced, but inside the pool it would be adversely affected. Fisheries are not important in the impoundment area; anadromous fish would hardly be affected, at most by some changes in food supply, but salmon smolts might meet large numbers of predatory fish (e.g., striped bass) as they exit from the powerhouse turbines.

Navigation is of secondary importance and no locks are needed for most Canadian sites. Navigation channels may have to be dredged to provide access to Moncton and on the seaward side of a plant built on the Minas Basin. Sewage from the city of Moncton would have to be treated if a Petitcodiac scheme were decided upon because of reduced flow and velocity and inadequate flushing.

All structures in contact with the waters at present will be affected by water level changes and these effects were considered in selecting operating levels. These would have mild unfavorable consequences for marshland drainage, while the protection to dykes afforded by the tidal power plant structures would greatly reduce the frequent storm damage.

In one of his studies, Waller (1970) concluded that construction of a plant "will depend to a very large extent on the value judgments that are made by politicians. They must decide when the point has been reached when society is prepared to accept . . . tidal power . . . to avoid pollution due to alternative power sources."

THE SEVERN BARRAGE

In 1977, the Severn Barrage Group, a consortium of several private enterprises, concluded that a barrage on the river would, in fact, be beneficial because adjustment of the tidal range would provide daily cleansing of the beaches. Currents now hazardous to navigation would be reduced and so would wave action, lessening coastal protection costs. Salinities would be maintained and the reduction of erosion would enhance water clarity. Wading bird reserves would somewhat shrink, but opportunities for mariculture would be good, while salmon migration would not be impeded. Pollution from industry would not be worsened since regulations on effluent discharges already have been tightened. The world famous Severn bore would disappear, but this would adequately be compensated by an enhanced Parrett River bore at Bridgewater. The report emphasizes then aesthetic improvements and use of the estuary for water recreation.

Although there are differences depending on the mode of operation, the precise site, and the scheme selected, the environmental study of the Severn barrage shows a decrease in maximum levels seaward for spring tides, in tidal ranges, and, with a two-basin system, in the creation of a diurnal tide. The sudden change in flow through the barrage causes seiching in the basin. With a single-basin scheme, this occurs at the end of the refilling period at a favorable time for increasing the power available from the ebb generation scheme. The estuary has good mixing in terms of salinity distribution; its high energy strongly influences movement and distribution of unconsolidated sediments, which are mostly a thin sand layer. Construction of a barrage would bring about sand redistribution and displacement of muds. Reduced current velocity might increase deposition rates, and long-range alterations such as tidal flats and salt marshes may form. Fine sediment present in large proportions could be deposited in channels and marginal areas. The alluvium carried seawards by the major rivers may build deltaic deposits, which would have to be dredged regularly so that flood discharges would not be impeded and no flooding would occur upstream. (Figs. 12.1–12.3)

Suggested are the maintaining of the low water level at its natural stage but a tidal range of 3–6 m instead of the present spring range of 12 m (e.g., at Avonmouth). This would protect the Severn River's several mud flats. These are of importance to fishes and birds. Their modification, relocation, or disappearance would have a serious impact. They are the habitat of the polychaete worms, crustaceans, and molluscs upon which flounder, bass, and whiting feed (as well as several species of birds). (Figs. 12.4, 12.5)

Deoxygenation and reduction of tidal mixing are linked; they would be harmful to invertebrate and fish life and possibly prevent migratory fish from making their full journey. Increased water stratification and a decrease in salinity may cause retreat toward the sea of numerous marine animals; continuous covering of existing flats may sound the death knell of plants and of some habitants of these flats who like a period of uncovering.

ENVIRONMENTAL IMPACT 297

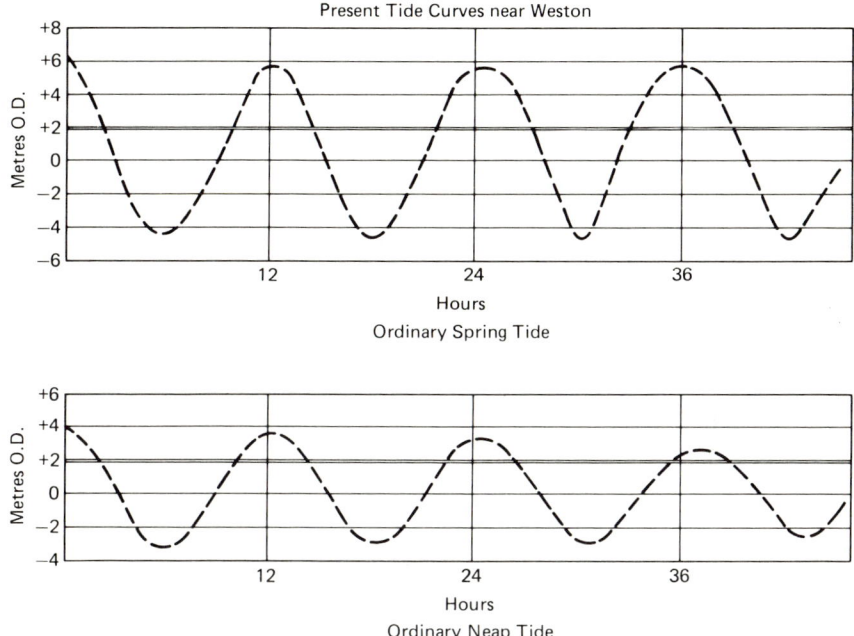

Fig. 12.1. Present tide curves near Weston-Super-Mare. (Shaw, *An environmental assessment of a Severn barrage*)

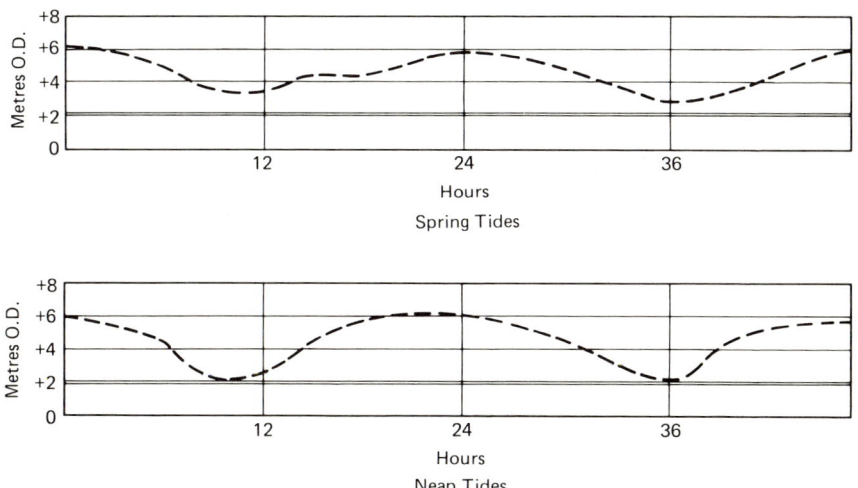

Fig. 12.2. Post-barrage tide curves, according to Shaw. (Shaw, *An environmental assessment of a Severn barrage*)

298 TIDAL ENERGY

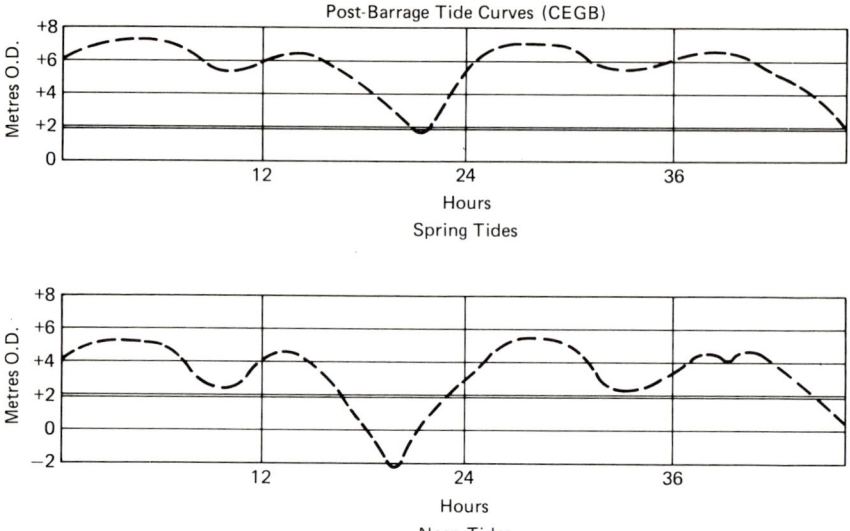

Fig. 12.3. Post-barrage tide curves, according to Central Electricity Generating Board. (Shaw, *An environmental assessment of a Severn barrage*)

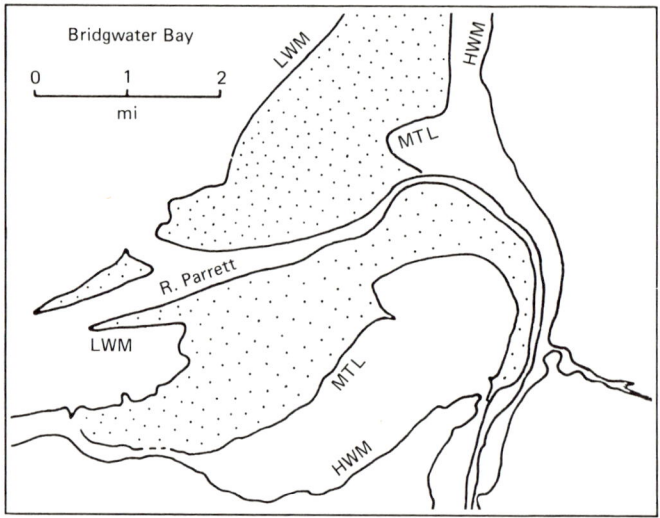

Fig. 12.4. Bridgwater Bay. Area permanently under water after eventual barrage construction. (Shaw, *An environmental assessment of a Severn barrage*)

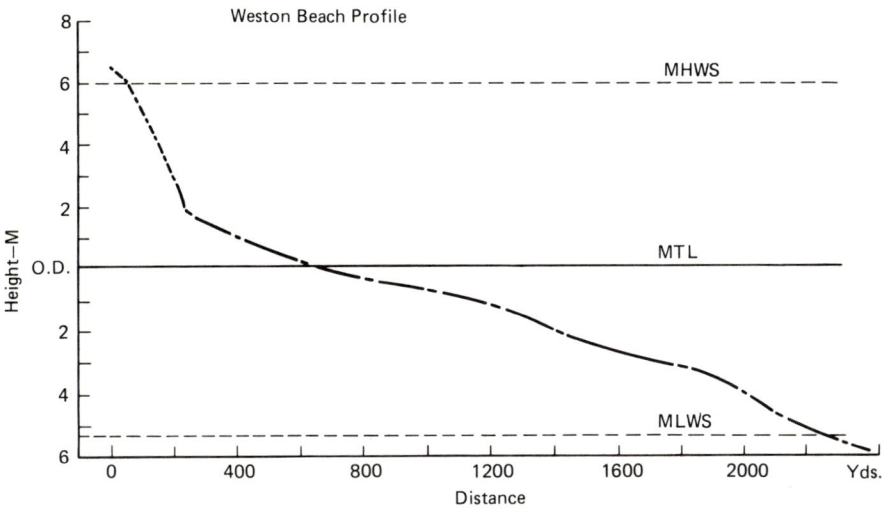

Fig. 12.5. Profile of Weston Beach. Submergence of part of the present intertidal area.

There is, at this time, no evidence that the present paucity of animal life is due to pollution and not to natural factors; nor is there reason to assume that, were a barrage built, biological degradation would ensue, provided pollution is not permitted to worsen.

The estuary is well oxygenated; however, turbulent flow over obstacles would reduce overall oxygen levels, although perhaps compensation may come by the aeration of water on passing upstream through the barrage. It is the dissolved oxygen that helps cope with the organic effluents. Currently, much of the man-made effluents is untreated. Thermal pollution is to be expected from the planned three nuclear and two thermal power stations within the barrage impoundment area, and this could lead to still further oxygen content decrease.

The bioenvironmental study of J. H. Andrews (1977) concluded to the unlikelihood of a significant effect upon wading bird populations if tidal ranges were not altered; it suggests a solution of small impact. As for wind fowl, it might benefit from a tidal power scheme, but may have to be protected from industrial growth and increased recreational activities.

While, most probably, smolts might manage to pass through the barrage, it is not so evident that the adult salmon would swim through it again on its way to the sea, and physiological or mechanical damage is not excluded. Special fish passages could be provided. In Scotland, however, experience has demonstrated that fish pass through much faster turbines.

A model study was made by Longhurst, Radford, and Uncles (1977). The barrage proponents would, in effect, halve both the number of tides per day and the amount of water throughput; this could reduce the tidal mixing by 25% but leave the estuary still "well mixed." Reduction of tidal energy would result in deposition of the presently suspended load; if deposited evenly of in bulk in channels, no problems are expected, but mud carried across the barrage would result in silting problems in time.

CONCLUSION

It has always been evident that construction of a tidal power plant, like any other man-made structure, would bring about environmental modifications. The changes are both physical and biological, and if one wants to go a step further, they have also a sociological impact. As Waller (1970) wrote: "A society that was unaware until recently of adverse environmental effects of technological change, or considered those effects unimportant . . . has begun to recognize that neither the physical resources contained in our environment, nor the capability of that environment to resist stress without failure, are unlimited." We can today make choices whether to go through with a technologically feasible project or dispense with it.

In the case of tidal power, the public attitude is today far different from what it was at the time the Rance River plant was built. This is why careful environmental impact studies are being conducted. From the preliminary assessments, however, it seems that no major environmental damage is to be expected and that, based on the actual experience of the Rance and the projections for the Quoddy, Fundy, and Severn projects, tidal power plants, preferably with some adjustments for the protection of fisheries and wildlife, can be safely built at present. The 1969 "Feasibility" study pointed to the costs of controlling environmental pollution in alternative sources of power. Water thermal pollution by nuclear and fossil fuel plants, and air pollution in addition by the latter, should be added when comparing costs of electricity: a tidal power kW-hr cost, then, 5.6 mills, and a fossil fuel kW-hr plus pollution control did amount to (3 mills + 1.4 mills =) 4.4 mills, for a nuclear power kW-hr, it was (1 mill + 1.1 mill =) 2.1 mills; the difference in costs dips to 1.2 and 3.4 mills, respectively, against the usually quoted higher cost differential of 2.6 and 4.6 mills.

Should we not stress that tidal power is at least free from causing water-borne or atmospheric pollution and be reminded that, on the other hand, fossil fuel power plants emit gaseous effluents, send into the atmosphere particulate matter, and raise the temperature of any water body into which cooling water may be discharged—(as do nuclear power plants)? In addition, coal-fired plants have to dispose of slag and fly-ash, and nuclear plants of radioactive waste.

In one of his several speeches on the millieu, Cole (1966) emphasized, appropriately, the greater environmental impact of nuclear powerplants: "Probably a billion times [the] amount of radioactivity [heretofore under human control; i.e., ten curies] has already been disseminated in the environment ... A plant of modest size ... near Oswego, New York ... will ... release to the atmosphere 130 curies per day."

Tidal power affects the environment even less than hydroelectric plants do, because there is no need to flood vast areas of land for a retaining basin to create a sufficient head for power production.

From the preceding thoughts, would it not appear that the time has come for society to call upon the use of tidal energy as an alternative source of power that will not sacrifice or abuse the fragile environment?

REFERENCES

Andrews, J. H., 1977. Impact on wading birds, in: Shaw, T. L. (Ed.), *op. cit. supra*, pp. 100–106.
Clark, R. H., 1972. Energy from Fundy Tides, *Canadian Geographic J.* **LXXXV**, No. 5:150–163.
Cole, L., 1966. Man's ecosystem, *Bioscience* **16**, 4, 243–248.
Gibrat, R., 1973. L'énergie marémotrice dans le monde: L'usine marémotrice de la Rance et l'environnement, *La Houille Blanche* **211**, No. 2–3:145–151.
Longhurst, A. R. *et al.*, 1977. Ecosystem models and the prediction of ecological effects, in: Shaw, T. L. (Ed.), *op. cit. supra*, pp. 83–92.
Maurin, G., 1977. Personal communication with author, dated Jan. 24 *(Electricité de France)*.
Shaw, T. L. (Ed.), 1977. *An Environmental Appraisal of the Severn Barrage.* Bristol (Great Britain): The University of Bristol, Department of Civil Engineering.
The International Passamaquoddy Engineering Board, 1959. *Investigation of the International Passamaquoddy Tidal Power Project-Report (Oct.).* Washington-Ottawa: The International Joint Commission.
The International Passamaquoddy Fisheries Board, 1959. *Report (Oct.).* Washington-Ottawa: The International Joint Commission.

Waller, D. H., 1970. Environmental effects of tidal power development, *Proc. Int. Conf. Utilization of Tidal Power.*

1969. *Feasibility of Tidal Power Development in the Bay of Fundy (Oct.), Board Report and Committee Report.* Ottawa: Atlantic Tidal Power Programming Board.

1977. *Reassessment of Fundy Tidal Power (Nov.).* Ottawa: Bay of Fundy Tidal Power Review Board.

1973. *An Energy Policy for Canada. Summary of Analysis.* Ottawa: Department of Energy, Mines and Resources.

13
The Future of Tidal Power

President John F. Kennedy, who attempted to revive the Quoddy-Fundy project two decades ago, stated: "Each day a million kilowatts of power surge in and out of Passamaquoddy Bay." The Bay of Fundy could provide 15–50 billion kW-hr/yr, a maximal saving of 28–87 million barrels of oil annually (or $10–33 million). Even though tidal power cannot solve the energy crisis, it is apparently an important supplementary source of power. (Tables 13-1, 13-2)

Why not more tidal energy plants?

THE ECONOMICS FACTOR

The main deterrent to harnessing tidal power is economics. Even at a time when coal was in short supply and economic conditions deteriorating, there were more opponents—or at least postponers—of tidal power plants' construction. Many tidal power projects apparently are uneconomic because of the large and expensive structures required to impound the water against a poor power return from the low maximum head or pressure available from the tides at most sites of the world. Water turbines can be built that will work on a very few feet head, but their cost is in inverse proportion to the head available because of the larger amounts of water required to produce power at low heads. Great progress, though, has been made toward the development of low-head machines that are more economical. Furthermore, since the design of reversible blade turbines that can also act as pumps, tidal basins can become at times storage houses for power by pumping water from the sea.

Yet, even with pumping, tidal energy is not always available at the best period of the day (i.e., peak hours). It is not possible to generate electricity at high tide, and therefore a tidal plant cannot be classified as a peaking station even if the plant factor and the annual cost/kW-hr is equivalent.

Table 13-1. World Oil Consumption, Based for the First Nine Months in Thousands of Barrels/Day. *Source: U.S. Department of Energy and Central Intelligence Agency (Adapted).*

NATION	1978	1979	PERCENT OF CHANGE
United States	18,738	18,417	−1.7
Japan	,998	5,122	2.5
Federal Republic of Germany	2,574	2,653	3.1
France	1,961	2,033	3.7
Canada	1,658	1,749	5.5
United Kingdom	1,651	1,687	2.2
Italy	1,480	1,532	3.5
Total	33,059	33,193	0.4

Table 13-2. World Oil Output in Billions of Barrels/Day in 1977.

AREA	1977 OUTPUT (Billion barrels/ day × 10^9)	METRIC TONS (Billion barrel = 0.1364 ton)	INCREASE OR DECREASE VERSUS 1976 (%)
Middle East	21,927,000	2,990,840	+8
Communist areas	12,962,000	1,768,000	+5.6
United States	7,966,000	1,086,560	−2.3
Africa	6,257,000	853,450	+12.1
South America	4,542,000	619,530	−6.7
Asia and Pacific	2,766,000	377,280	+10.9
Europe	569,000	77,610	+78.6
Total world	59,104,000	7,773,270	+7.3

The Rance plant has the same economic characteristics as those of peaking stations, without their flexibility.

The rentability of a tidal power energy plant is a function of the size of its basin, and the larger the basin, the lower the production costs, further reduced by greater tide level differences. But the larger the basin, usually the longer the dam to be built, and dams are very costly, even though when the length of the dam is doubled, the size of the basin quadruples.

Output of power per length unit of dam must, however, be checked against the width of the basin entrance, since too narrow an entry may cause constriction unless sufficient depth exists.

The location of the plant has often militated against construction: only half a dozen sites would be meaningful if tidal ranges of at least

7 m and distances to consumer areas of less than 200 km are considered. However, consumption areas shift, and high-voltage transmission and national grid systems have already considerably reduced the importance of an electric plant's geographical location. Furthermore, as Kay (1975) points out, tidal power is capable of being developed in very large units in favorable locations. It has the additional advantage of producing no noxious wastes, of consuming no exhaustible energy resources, and of producing a minimum disturbance to the ecologic and scenic environment. There are, accordingly, many social advantages and few disadvantages to the utilization of tidal power, wherever tidal and topographic factors combine to make this practicable.

Some segments of the economic community stress efficient allocation of resources and worry about choosing between competing demands for limited resources, placing major emphasis on discounted cash flow return. Kay feels that this attitude might obscure the real issue behind a barrage of possibly faulty numbers for interest rates, secondary benefits, transfer functions, exhaustion or increase in cost of alternate power sources, and pollution abatement requirements.

Suitable cost-benefit analyses have then been countered by questioning the public interest aspect of government participation in power development, a matter naturally of no concern in the Soviet Union, China, or western European countries. This query is usually brought to the fore by the investor-owned utilities.

It has been pointed out that perhaps the "difficulties with tidal power could be blamed on the think-big approach." Tidal flats could be suitable sites for modest plants. Taking the Cohasset (Massachusetts) Inlet as an example, Hickok calculated that 4,300 kW-hr/day, or enough to supply 172 households, can be generated; this represents 11% of the community needs at only 8 mills/kW-hr. The 5.67-hectare flat has a 21.3-m-wide inlet through which four times a day 227,000 hectoliters flow. The tidal flat could operate a 1,200-hp reversible water wheel; at 40% efficiency of tidal power extraction, 4,300 kW-hr/day are generated. These considerations might be useful for developing countries needing power, possessing suitable sites, and not in need of "gigantic" schemes.

Although, as pointed out, information on the Chinese achievements is scarce, we do know that they took into consideration the value of small schemes and constructed several. Yet small schemes may produce more expensive power, and larger-size ones have greater labor efficiency and a somewhat lower building price tag per kW of installed capacity. A 12½-MW reversible bulb unit costs only half the price of a 2-MW turbine (per kW installed) at a rated head of 4 m. However, one may reasonably expect progress in turbogenerating equipment—e.g., fixed-blade units with variable speeds could operate at very low heads—so that low-head, small-size plants may be designed that would bring about substantial cost savings.

DISADVANTAGES OF TIDAL POWER PLANTS

There are still other disadvantages to tidal power plants. Tides are intermittent and power production varies. The smallness of the heads of water under which turbine wheels must operate requires, at present, large (thus expensive) wheels, or gearing between turbine and generator. But electricity produced by a tidal power plant can be more economical than nuclear power and thermal power, unless coal and oil plants are ultramodern.

Construction and Operating Conditions

The French solved the power variation and intermittent production problems by using two-way pump turbines. Because electricity, as such, cannot be stored (but water can), and because there is a large difference in production costs at peak and at off-peak periods of conventionally generated electricity, the reversible turbines, with their facilities for storing power, have brought about a reopening of the already dust-covered dossiers of many an old tidal power plant proposal. Wilson (1973) wrote that a single basin with simple (i.e., not with reversible blades) turbines and pumped storage is less expensive and that costs could be shrunk further by eliminating cofferdam structures through mounting the machines in prefabricated cells to be towed to and sunk on a prepared dredged foundation. Use of caissons shortens construction time and cuts down on expensive interest, another extra cost of tidal power plant schemes. (Figs. 13.1-13.3)

This represents a newer approach than that followed in the Rance River plant construction. Taking advantage of head smallness, the engineer can ignore barrage impermeability and channel filling may be achieved by tipping rock rubble, avoiding cofferdamming, and using large concrete foundations. Then, following the Soviet approach, the dam and plant are completed by cell construction.

THE FUTURE OF TIDAL POWER 307

Fig. 13.1. Rance River plant. Construction inside a cofferdam. *(Source: Electricité de France)*

Fig. 13.2. Rance River plant. The cofferdam in process of being removed. *(Source: EDF)*

British manufacturers have mounted generators on the periphery of the actual turbine blades themselves, thus dispensing with the need for enlargements of water passage around the bulb (rim turbines).

Further construction cost reduction can be made possible by compacting units through a combination of power unit and spillway, and transportation problems can presently be eased by using the lower and lighter bulb-type generators. In a recent assessment of the future of large-scale tidal power projects, Wilson listed the following prerequisites; the term *large* is important because modest schemes (e.g., in developing countries) could be built even though all prerequisites would not be satisfied:

1. Large interconnected electrical transmission networks must make possible output absorption. (Apparently, this is one of the problems that led to the shelving of the Australian plans. But in Britain, high-voltage lines exist close to Solway Firth and the Bristol Channel.)

Fig. 13.3. In the dry construction behind cofferdam. *(Source: Electricité de France)*

THE FUTURE OF TIDAL POWER   309

Fig. 13.4. Kislaya plant. Powerhouse construction, on land, in the building dock.

2. Large floating caissons of concrete must be sunk satisfactorily in pre-prepared underwater foundations. (The Soviets have mastered this technique on Mezen Bay, and in the North Sea, the Dutch used it as well.) (Figs. 13.4, 13.5)
3. Robust and low-maintenance-need, low-cost electromechanical equipment of high hydraulic efficiency must be developed. (While no "solution" has been found yet to this problem, various types of equipment have been proposed and some have been developed. Most proposals involve a generator linked to a single propeller turbine. An American system uses a slant-axis machine driving a

Fig. 13.5. Kislaya plant. The powerhouse is being floated out of the building on its way to the site.

compressor through a speed-increasing gear, and a British system links a single propeller turbine to a hydrostatic pump on a common shaft.)

While production increased and thus lowered the cost of electricity in the Rance and Kislaya schemes by using reversible blade turbines, the extra cost has put in doubt their inclusion in future plants. The Canadian plan does not propose to use reversible blade turbines.

Though the bulb turbine has been praised, it is expensive. A turbine with a so-called rim generator (in this type, the rotor surrounds the turbine runner as a rim carried by the runner blades) would be less costly.

A less important disadvantage is the sedimentation process at a given site. This is not restricted to tidal plants, but exists as well for run-of-river schemes. It might be quite serious, though, in some geographical locations. Indeed, the flood tide carries into the basin a large quantity of materials which will not be returned to the sea by the ebb tide whose velocity is not sufficient to accomplish such a task. For instance, the rate of sedimentation at the Severn location is rapid and could possibly endanger neighboring harbors such as Bristol, Cardiff, and Newport.

Cost and Energy Output

The principal variables, besides operating conditions, affecting cost and energy output at a given site are the type and number of turbogenerators, the diameter of turbine runners, the operating head or range of heads of the turbine, the rate capacity of the generators, and the number of gates (sluices) installed to augment reservoir filling. (Figs. 13.1, 13.2, 13.4)

The builders of the Rance River scheme had access only to slow computers. Swales and Wilson (1966), using high-speed digital computers of the latest design, claim to have developed the most economic combination of turbogenerators and sluices, making it possible to reduce the price of tidal power produced energy to a level directly competitive with conventionally produced energy. Their study made for a site in Knik Arm, Alaska, and for ebb generation, could be extended for two-way generation if pumping is not involved.

Electronic computers have greatly assisted the determination of the power output from potential tidal projects. An equation simulating the tidal cycle, the number, and the hydraulic characteristics is fed into the computer. Because of the regularity of tides, power can be estimated for months ahead.

Since it is possible to predict months, even years, ahead the production of a tidal plant, it is possible to integrate a tidal power plant's production "into any electrical system so as to obtain the best use of the capacity and energy available from the tidal plant."

Wilson and Swales (1972) gave evidence that the overall system cost can be lowered to compete with any other means of power production, and offers, in addition, many other benefits.

REASSESSMENT

Factors in favor of the tidal plant are its steadiness; the simultaneous use of the dam as a road (as in France) or for a railroad; the improvement of navigation to estuary ports, which applies to British harbors; the production of cheap energy; its virtually inexhaustible supply; and the spending of capital in underdeveloped or moribund regions. An estimated one million dollars a year would be saved by mileage reduction through a Bay of Fundy tidal station roadway.

The tidal power plant has the advantage of longer life over a thermal plant and also over the river hydroelectric plant. Tides can be predicted with accuracy for the distant future, and the tidal plant cannot be affected by drought, flood, or ice jam.

Among cost saving features, the Canadian group is currently considering generation at 13.8 kV (instead of 5 kV), thereby eliminating intermediate step-up transformers, placing the switchyard inside the powerhouse (using SF-6 gas-insulated equipment) and thus reducing the number of circuits needed for power transmission to shore, and grouping 285-mVA transformers in pairs, which reduced their number and makes a better use of the capacity of the 345-kV breakers.

It is evident that an economic analysis must consider the tidal power plant's competitiveness with alternative energy production schemes, but comparisons are difficult. As soon as an estimate is completed, construction materials and labor, due to unrelenting inflation, have become more expensive. Nonrenewable energy sources either become exhaustive or their price increases. Technology steadily improves. No two sites offer identical conditions.

Lebarbier (1975) stated that the cost of a tidal power plant can only be offset against that of a conventional thermal power plant, in a system "where there is an energy, rather than a power shortage, a situation that will prevail in France through the mid-eighties because not enough hydro-storage plants are capable of operating at full load during the sixteen weekday hours which correspond to the intermediate load zone on the daily load duration cycle."

But since the Rance River plant was constructed, there have been some second thoughts about nuclear power, and nuclear plant construction costs have also risen. There are new ever-higher prices of oil, oil supplies are subject to sudden curtailments, and experience has shown that average annual operating costs of the other types of hydroplants are higher than those of the Rance plant, even though overall costs are lower due, in part, to a lesser amortization of construction expenses. These expenses may run as high as 90%, and the investment is actually a protection against inflation, particularly in view of the tidal power's plant lifetime of 75–100 years.

Another suggestion made in recent years is to use tidal power plants in pumped storage mode. No extra energy would be produced, but economic gain would be attained through load smoothing by use of main power at off-peak times to "overfill" the basin. The "heat pump" effect could lead to a net energy gain as basin level height would be artificially increased in periods when high tide occurred at strategic times during off-peak hours.

A study commissioned by the government of the United States and carried out by the Boston firm of Stone and Webster through 1977 concluded that tidal power was already cost-effective in the Passamaquoddy Bay region, using then-existing technology, considered on a lifetime basis. The projects recommended for implementation would be exclusively located in the United States. This is not a rejection of a cooperative undertaking with Canada, but a time-saving device because no international agreements would have to be worked out and agreed upon, and, on the other hand, Canada has outstanding options of its own, entirely within its own territory.

FATE OF SOME PROJECTS

Construction of a United States facility remains much further from realization than such an undertaking in Canada, Korea, or elsewhere. An Alaska plant is even more unlikely because of hydropower available and untapped. Yet, as the energy crisis unfolds, tidal power, a proven possibility, should not be written off as a power source of the future. Considering that every bit helps in a pinch, one can stop a minute to remember that a "mini-tidal power plant" of 4 MWe at Half-Moon Cove would produce annually 18 million kW-hr and save 43,000 barrels of oil!

Although realization of the Chausey Islands project remains highly hypothetical, its high potential deserves some comments. Originally designed in the sixties, with the Rance plant as a prototype, it was to have an installed capacity of 12,000 MW and an annual output of 25–27 billion kW-hr. The gigantic undertaking requires gates, dykes, a power plant, a retaining basin of 580 km^2, some 35 km of sea construction, a 100-km dam, etc. Three hundred turbines, each of 40-MW unit capacity, were suggested. A second proposal, fathered by Albert Caquot (Constans, 1978) envisioned two basins of 1,100 km^2 each, with the power plant installed between them, and sluice gates over a length of 50 km. Its installed capacity is 6,000 MW, with an annual production of 34 billion kW-hr. Caquot linked to his project construction of a port in the upper pool and a 12,150-hectare industrial zone on a platform between the two retaining basins. Continuous generation would be achieved. (Figs. 13.6, 13.7)

Costs are, however, so staggering that the project appears irrealizable without substantial government participation. On the basis of FF 5.80 to U.S. $1, they are currently estimated to exceed $3.5 billion ($7.62

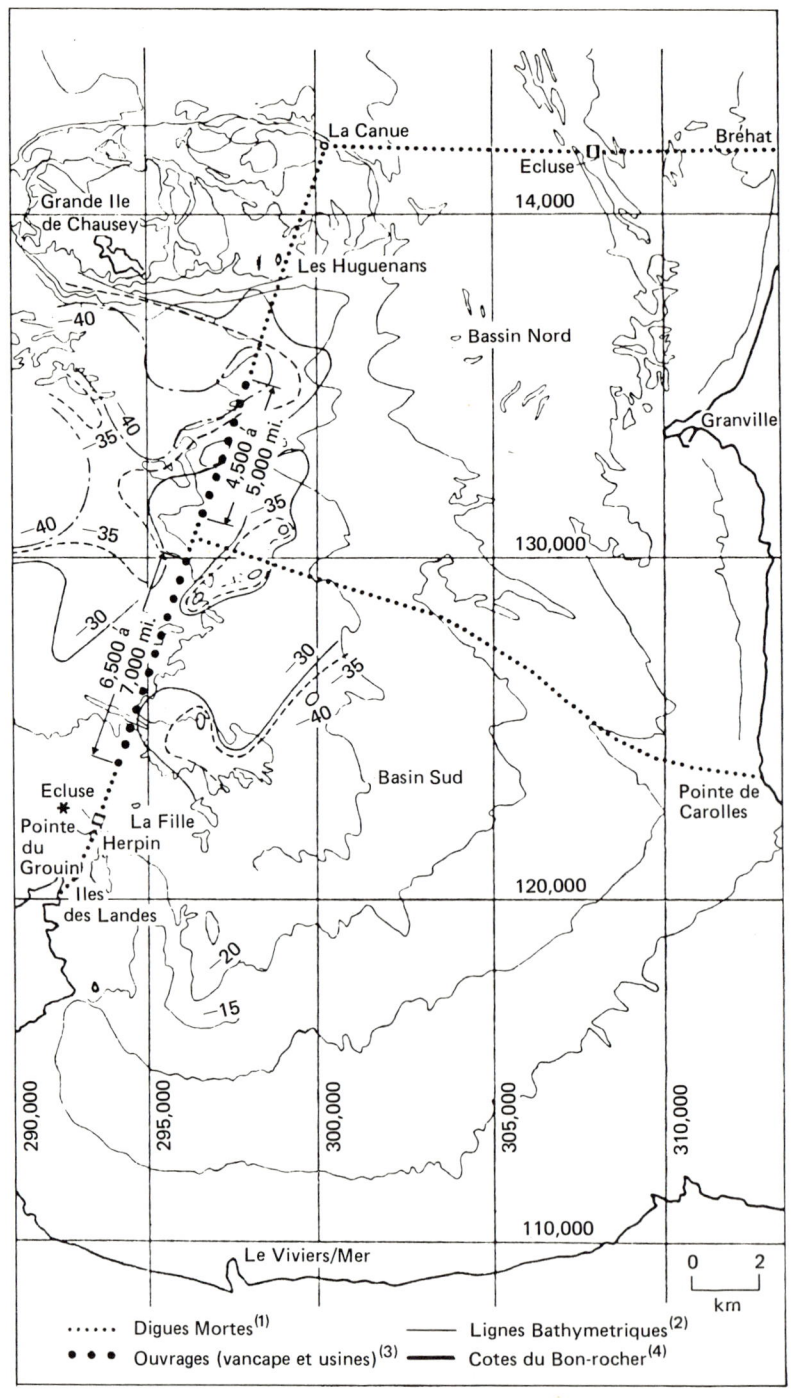

Fig. 13.6. The Chausey Islands tidal project: location. ("L'usine marémotrice de Chausey," *La Production d'Electricité d'Origine Hydraulique*. Paris: La Documentation Française, 1976)

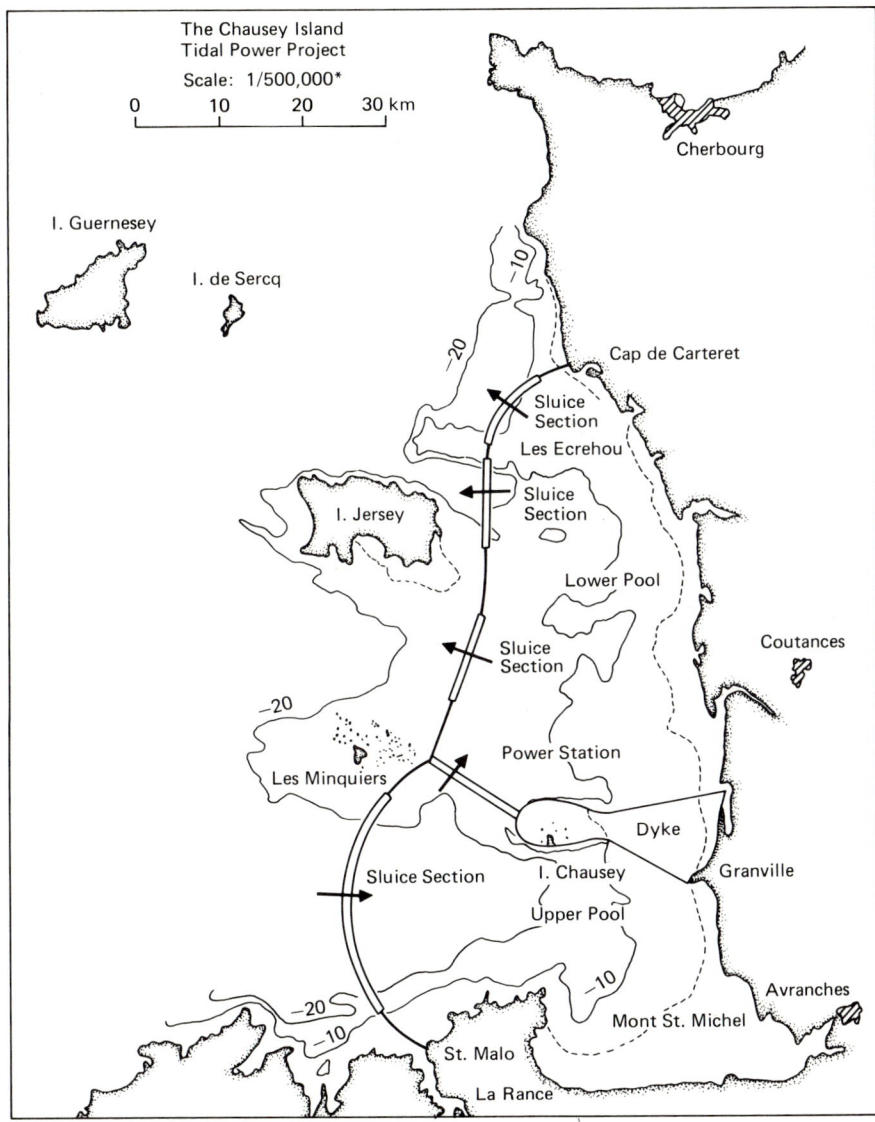

Fig. 13.7. General layout of the Chausey Islands tidal power project. (Lebarbier, *Nav. Eng. J.*, April 1975)

according to private sources) or perhaps more than double that of equivalent nuclear facilities. One redeeming factor is that when a negative decision was reached, oil was plentiful and its price low by today's standards; furthermore, nuclear power was then looked upon favorably.

The Cobscook Bay tidal power project was the subject of an economic analysis in March and July 1979, conducted by the U.S. Corps of Engineers. An initial analysis of a representative tidal project using a life-cycle methodology, based upon total cost escalation, after project construction including inflation, led to positive net benefits. The alternative to linked or multiple-pool tidal power schemes is nuclear power: the analysis showed that schemes designed for dependable capacity are too expensive, and thus tidal power projects should instead be designed to maximize energy. The report also seems to show that single-pool schemes with a large impounded bay and a relatively small installed capacity yield the greatest economic (i.e., energy) advantage.

The Federal Energy Regulatory Commission has a cut-off between combined cycle (31 mils/kW-hr) and a nuclear power (7 mils/kW-hr) at 40% capacity factor; this makes tidal power plants uneconomical at a greater than 40% capacity factor. The net benefits mentioned above seem to be reduced: assuming price increases for oil of the current magnitude, tidal power, while providing net benefits during several years in the high escalation rate case, does not provide net benefits over the life of the project. Why? The report mentions high initial cost and the lack of dependable capacity and the recent funding of alterative (and, in many cases, less expensive) forms of energy. This makes tidal power, though more competitive, still not attractive.

However, when ancillary benefits are added to the power benefits, there is a considerable improvement of the benefit to cost ratio. The ancillary benefits include mariculture, area development, and recreation. One can also allow "credits" for relative price shift analysis, and use "real" fuel escalation rates. Finally, if the Commission would allow a 50% capacity factor instead of the 40% one, the benefit to cost ratio would also improve. Based upon these possibilities, the report concluded that the study should not be discontinued, but instead be reconsidered.

Had the Passamaquoddy project been completed in pre-World War II days, its estimated cost, over a 100-year lifespan, would have been $2.4 million, but it would be generating power at less than 10/kW-hr. Current construction of a plant would run from $22 million to $916 million for annual outputs of some 16–790 million kW-hr.

THE ENERGY PINCH

Comparing tidal power to fossil fuel-generating stations, Lawton (1972) calculated that, with furnace efficiency of 50%, heat content of 140,000 BTU per U.S. gallon (9.590 kcal/l), 24.6 kW-hr will be the equivalent and, with oil at 3.52¢, is exclusive of maintenance and equipment. A modern fossil fuel-fired thermal unit of 500 MW provides 9,200 BTU/kW-hr (2.320 kcal/kW-hr), with 6 million BTU/barrel (9,540 kcal/l). The yearly output of a single-basin scheme in the Bay of Fundy would save 10.1 million barrels of oil per year (about 1.6 million kl).

A 1977 study conducted by the Eurocean group based in Monaco concluded that currently a tidal power plant requires an investment of $800–1,000/kW, and expected energy costs of between 25 and 30 mills/kW-hr.

An extremely important consideration when deciding whether a conventional or a tidal power plant should be built is the life-cycle cost analysis covering at least the economic life of both projects. One can foresee that very probably the initial higher cost of the tidal power scheme will be wiped out as fossil fuel costs keep rising.

The United States is widely cited as the largest per capita user of energy. Perhaps, then, the best way to grasp the extent of energy use and its sources is to take the United States as the example.

In 1973, the United States total use was 75.6 quadrillion (Q) of BTU (75.6×10^{15}). This amount came from 31% (23.6 Q) from natural gas, 30% (22.7 Q) from domestic oil, 18% (13.5 Q) from coal, 16% (12 Q) from imported oil, 4% (2.9 Q) from hydropower, and only 1% (0.90 Q) from nuclear sources. By 1976, natural gas represented only 28%, while hydropower had climbed to 5% and nuclear power to 3%. However, the share of imported oil had soared to about 42% of the total United States oil use, or close to 19.5% of the total energy consumption. (Table 13-1)

How was this huge amount of energy used? To industry goes 41%; 24% is for transportation, and 15% is for commercial users. The remaining 20% goes to residential customers: 10.6% for home heating; 3% for water heating; 2.5% for lighting and minor aims; 1.4% for refrigeration and 1.2% for air conditioning; 0.9 for cooking; and 0.4 for clothes driers.

Numerous seminars, workshops, and studies have been devoted to finding a solution to closing the gap between the need and the demand for energy, and the future supply. Consumption, they show, leaps forward by about 2⅓% annually. Total barrels/day produced was 59,-

104,000 in 1977, an increase of 7.3% from 1976, though both the United States and South America had a smaller output. (Table 13-2)

According to estimates, by 1986, the United States would consume 104.2 Q, of which 34% (35.4 Q) would be oil, 24% (25.3 Q) natural gas, 20% (20.7 Q) coal, 4.5% (4.6 Q) hydropower, 14% (14.7 Q) nuclear power, and 3.5% (3.5 Q) synthetic oil and gas, oil shale, and geothermal and solar energy. Self-sufficiency could only be attained if all oil used would be of domestic origin.

Apparently, there is an urgent need to find alternative sources of energy before the year 2000. While some predict a decreased production rate of oil between 1983 and 1987, even the most optimistic predictions (e.g., Emery, 1974, and Wood, 1979) underscore that supplies may amply suffice while we develop alternate sources of power. "In such a context," writes J. A. Constans (1978) of Eurocean, "there is no doubt that not only is energy from the sea a promising and realistic concept, but it is also able to valuably complement the existing energy supply at local, national and regional levels." Tidal energy is one of these sources and may help satisfy the need for power. Whether man will extend his quest for sources and build, for instance, more tidal power plants will

Table 13-3. Schedule of Korea Tidal Power Project. *(Source: Y. Chun, Korea Tidal Power Project, Phases II and III, Seoul, South Korea.)*

ITEM \ YEAR	1978	1979	1980	1981	1982	1983	1984	1985	1986
Phase I									
Identify tidal power potential and sites	8								
Report review and site selection	4								
Contract of engineering services		3							
Phase II Basic design		12							
Phase III Detailed design				24					
Phase IV Construction works						60			Operation

ultimately depend less on the abilities of engineers than on capital necessary, the competition from other sources of energy and how fast how bad the "energy pinch" becomes.

REFERENCES

André, H., 1976. Operating experience with bulb units at the Rance tidal power plant and other French hydro-power sites, *IEEE* Chap. 1102-3-PWR.

Bonnefille, R. and Chabert-Dhières, G., 1957. Etude d'un modèle tournant de mer littorale. Application au modèle de l'usine marémotrice des îles Chausey, *La Houille Blanche* **XXII**, *No. 6*:651–658.

Charlier, R. H., 1969. Harnessing the energies of the ocean: A review and bibliography, *Marine Tech. Soc. J.* **3**, *No. 3*:13–32; **3**, *No. 4*:59–81; *Ibid.* (1970); "A postscript," **4**, *No. 2*:63–65.

Cotillon, J., 1974. La Rance: Seven years of operating a tidal power plant in France, *Water Power* **26**, *No. 10*:314–322.

Duclos, M. I., 1958. Projets d'usines marémotrices en France, *Bull. Soc. Franc. Electric.* **8**, *No. 85*:24–39.

Duhoux, L., 1964. Fermeture de la Rance. Déroulement des travaux et analyse des observations, *La Houille Blanche* **XIX**, *No. 4*:491–508.

Emery, K. O., 1974. Provinces of Promise, *Oceanus* **XVII** (Summer):14–19.

Gibrat, R., 1976. The current revival of tidal power studies, *A.S.T.E.O. Proc.* (March 5).

Haswell, C. K. et al., 1972. Pumped storage and tidal power in energy systems, *J. Power Div. Amer. Soc. Civil Eng.* **98**, *No. 10*:201–202.

Hermès, P., 1950. Les projets d'aménagement de la baie du Mont Saint-Michel pour l'utilisation de l'énergie des marées, *Le Génie Civil* **127**, *No. 16*:310–313.

Lebarbier, C. H., 1975. Power from the tides. The Rance tidal power station, *Naval Eng. J.* **87**, *No. 2*:57–68.

Ministère de l'Industrie et de la Recherche (France), 1976. L'usine marémotrice de Chausey, in *La Production d'Électricité d'Origine Hydraulique. Rapport de la Commission de la Production Hydraulique et Marémotrice* (Paris, La Documentation Française). Dossiers de l'Energie 9, Chap. IV:41–49.

Shaw, T. L., 1976. Tidal power closing the gap, *Water Power and Dam Construction* (May):24–27.

Shaw, T. L., 1977. A policy for tidal energy, *Marine Policy* **1**, *No. 1*:61–69.

Simonet and Jacquinet, 1954. Les usines marémotrices. Projet de la Baie du Mont Saint-Michel, *Information Géographique* **4**:147–148.

Vantroys, L., 1957. Le régime des marées dans la Manche, *La Houille Blanche* (4es Journées de l'Hydraulique) **1**:176–181.

Wood, P. W. J. 1979. New slant on potential world petroleum resources, *Ocean Industry* **14**, *No. 4*:59–72.

1973. Six ans d'exploitation de l'usine marémotrice de la Rance [Session des 16 & 17 Nov. 1972, Soc. Hydrotechn. de France], *La Houille Blanche* **2/3**:125–270.

Glossary

A

abutment caisson: see *caisson*

amphidromic point: a no-tide or nodal point on a chart of cotidal lines from which the cotidal lines radiate

amplitude (tidal): the semi-range of a constituent tide; sometimes used as synonym of range

anadromous fish: fish whose life cycle includes reaching maturity in the ocean and spawning in fresh water

annual capacity factor: quotient of average output and installed capacity (in MW)

aquaculture: raising of living species in water

apogean tide: tides of decreased range occurring monthly near the time of the moon's apogee

apogee: that point on the orbit of the moon (or any other earth satellite) farthest from the earth (opposed to perigee)

B

backwater effect: water held back from the main flow forming an inlet approximately parallel to the main body and connected thereto by a narrow inlet

base load plant: see *load plant*

bay: 1) a recess in the shore or an inlet of a sea between two capes or headlands; not as large as a gulf but larger than a cove; 2) an inward bend in the ice edge formed either by wind or current

bed: smallest division of a layered rock series separated from overlaying and underlying material by a clear change in character

biomass: amount of living matter per unit of water surface or volume expressed as weight

bottom: any ground covered by water

bore: a high breaking wave of water, advancing rapidly up an estuary (bores can occur at the mouths of shallow rivers if the tide range at the mouth is large; they can also be generated in a river when tsunamis enter shallow coastal water and propagate up the river)

bulb turbine: ogive-shaped axial flow turbine that can function as turbine and pump

buttress: support for wall under superstructure

C

caisson: watertight "box" used as a gate; also such a "box" wherein men can do construction work under water

caisson (abutment): a caisson designed to be placed at each end of the main structure to retain the rock-fill dam

caisson (tidal station): a large reinforced prestressed concrete structure that can be constructed on dry land at a suitable site of the coast, then launched, floated, and towed into position

change of tide: a reversal of the direction of motion (rising or falling) of a tide (sometimes applied to a reversal in the set of a tidal current)

cofferdam: watertight temporary structure built on a river or lake bottom keeping out water during the laying of the dam foundation

conjunction: when moon and sun are directly in line with the earth and the moon is between the earth and the sun

corrosion: gradual deterioration of materials by chemical processes due to contact with seawater

cotidal line: a line connecting (on a map or chart) points where the average interval between the moon's transit over the meridian of Greenwich and the time of the following high water, is the same, expressed in either lunar or solar time

countercurrent: a secondary current, usually setting in a direction opposite to that of the main current

crest: 1) the highest part of a wave; 2) sometimes: part of wave above sea water level

cubature: total volume of a water retention area (m^3, ft^3, hl, etc.)

current: a horizontal movement of the water, which may be either tidal or non-tidal.

D

declination: angular distance north or south of the celestial equator

delta: depositional area near a river's mouth

dependable peaking capacity: amount of power (MW) that can be produced continuously for a minimum of 4 hours to meet daily peak load requirements upon demand

deposition rate: amount of sediment deposition per chosen unit of time

discharge: rate of flow of a river at a given instant, expressed as volume per unit of time (m^3/sec)

diurnal: daily, especially pertaining to actions completed within 24 hours and which recur every 24 hours; thus, most reference is made to diurnal cycles, variations, ranges, maxima, etc.

diurnal constituent: any tide constituent whose period approximates that of a lunar day (24.84 solar hours)

diurnal inequality: (also called *daily inequality*) the difference in heights and durations of the two successive high waters or of the two successive low waters of each day; also, the difference in speed and direction of the two flood currents or the two ebb currents of each day

diurnal range: the amount of variation between the maximum and minimum of any element during 24 hours

diurnal tide: a tide in which there is only one high water and one low water each lunar day *double ebb:* an ebb current having two maxima of speed separated by a smaller ebb speed

double effect: see *reversible operation*

double tide: (or *agger, double high water, gulder*) a high water consisting of two maxima of nearly the same height separated by a relatively small depression, or a low water consisting of two minima separated by a relatively small elevation

dyke: embankment constructed to close off a river's natural course

dead dyke: such a dyke containing no powerhouse or machinery

E

ebb axis: the average direction of the tidal current at strength of ebb

ebb current: the tidal current associated with the decrease in the height of a tide (ebb currents generally set seaward, or in an opposite direction to the tide progression; erroneously called ebb tide)

ebb interval: the interval between the transit of the moon over the meridian of a place and the time of the strength of the following ebb tidal current

ebb strength: the ebb tidal current at the time of maximum speed, usually associated with the lunar tide phases at springs near perigee and/or maximum river discharge

ebb tide: see *ebb current*

effluents: materials being discharged into a waterway, bay, etc. from sewer, sewage tanks, etc.

electric grid: general electric transmission system of a region

embayment: shoreline indentation forming a bay

energy: electrical power output expressed in kW-hr

equilibrium tide: the hypothetical tide due to the tide-producing forces under the equilibrium theory; tide relating to the attractions of celestial bodies, particularly the sun and the moon

equinoctial tide: tide occurring when the sun is near equinox (during this period, spring tide ranges are greater than average)

erosion: the wearing away of material by such agent as ice, water, wind

estuary: 1) the segment of a stream or river influenced by the tide of the water body into which it flows; 2) the mouth of a river where tide and river current meet

F

fairway: main traveled part of a river

falling tide: ebb tide, tide cycles between high water and following low water

firm power output: electric power that can be generated under all conditions

fishway: arrangement allowing fish to pass a dam

flap gate: gate positioned so as to allow water into the holding basin and requiring no mechanical means of operation

flow: the combination of tidal and non-tidal current, which represents the actual water movement

G

geothermal: see *thermal plant*

geothermal gradient: the change in temperature of the earth with depth (in °C/m, °F/ft, m/°C, ft/°F) *gravitation:* mutual attraction between masses of matter

gravity: force that tends to pull bodies toward the earth; the resultant of gravitation and centrifugal force

greater ebb: the stronger of two ebb currents, occurring during a tidal day (usually associated with tidal currents of mixed type)

greater flood: see *greater ebb* and apply to flood current

H

half tide: the condition or time of the tide when at the level midway between any given high tide and the following or preceding low tide

harmonic: a sinusoidal quantity whose frequency is an integral multiple of the frequency of a periodic quantity to which it is related

head: in hydraulics, the vertical distance between the surface of a liquid and another point in the column; thus, a measure of the force exerted at the lower point by the weight of the column

higher high water: the higher of two high waters occurring during a tidal day where the tide exhibits mixed characteristics

harmonic analysis (or theory): 1) a statistical method for determining the amplitude and period of certain harmonic or wave components in a set of data with the aid of Fourier series; 2) the method by which the observed tide or tidal current at any place is separated into elementary harmonic constituents

hydro (electric) plant: plant producing electricity by forcing a river to flow through a powerhouse in a dam barring it

I

igneous rock: rock formed by solidification of molten material or magma

impounded area: see *retaining basin*

inlet: a short, narrow waterway connecting a bay or lagoon with the sea (when it is a natural inlet maintained by tidal currents, the name *tidal inlet* or *tidal outlet* is applied)

installed capacity: total nameplate rating of turbogenerators (in MW)

isobar: a line connecting (on a map) points having the same barometric pressure (reduced to a common datum)

isobath: a line connecting (on a map) points at equal depth

isocotidal: see *cotidal line*

isogonic: a line connecting (on a map) points having the same magnetic declination

isohyet: a line connecting (on a map) points having the same amount of precipitation for a specific period

isohypse: a line connecting (on a map) points of equal altitude

isotherm: a line connecting (on a map) points of equal temperature

J

jetty: a structure that influences a water current or protects a harbor or river entrance (US); a pier for berthing vessels (UK)

K

kinetic energy: when referring to a progressive oscillatory wave, the sum of the energy of motion of the particle within the wave

knot: unit of speed used in navigation, also in reference with current speed (= 1 nautical mile/hr)

L

lagging of the tide: the periodic retardation in the time of occurrence of high and low water due to changes in the relative positions of the moon and the sun (the opposite effect is called *priming of the tides*)

load plant: 1) *peak:* annual capacity factor between 0.05 and 0.30; 2) *intermediate:* same between 0.31 and 0.60; 3) *base:* same between 0.61 and 0.095

loch: an inlet or arm of the sea, often nearly landlocked

lunar day: time of rotation of the earth with respect to the moon, or interval between two successive upper transits of the moon over the meridian of a point. (= ±1.035 solar day)

M

mariculture: raising of living species of and in seawater

marigram: a graphic record of the rise and fall of the tide in the form of a curve that shows the time of any stage of the tide represented by abscissas and the height in ft (or m) by ordinates

maximum tractive force (of the tide): see *periodicity*

mean neap range: the average semi-diurnal range occurring at the time of quadrature (it is smaller than the mean range where the type of tide is either semi-diurnal or mixed and is of no practical significance where the type of tide is diurnal)

mean range: the difference in height between mean high water and mean low water, measured in ft (or m)

metamorphic rock: rock that has undergone structural and mineralogical changes due to modifications of temperature, pressure, chemical environment

mixed tide: the type of tide in which a diurnal wave produces large inequalities in heights and/or durations of successive high and/or low waters (this term applies to the tides intermediate to those predominantly semi-diurnal and those predominantly diurnal)

mooring: securing to the bottom

moulins à marée: tide mills

mud flat: a muddy or sandy coastal strip, usually submerged by high tide

N

neap tide: tide of decreased range, occurring about every two weeks when the moon is in quadrature

nodal point: see *amphidromic point*

non-tidal current: current not caused by the tide-producing forces; lacks the periodicity of tidal currents

O

opposition: relative position of two celestial bodies with their celestial longitudes differing by 180°, as when the moon and sun are in line with the earth and on opposite sides of the earth

oscillatory wave: a wave in which each individual particle oscillates about a point with little or no change in position (in a progressive oscillatory wave only the form advances, the particles move in closed orbits)

ostreiculture: raising of oysters in special *oyster parks*

P

peak load plant: see *load plant*

peaking power: power needed over and above constant generation to meet daily upsurges in electrical demand

perigean tide: tides of increased range occurring monthly near the time of the moon's perigee

perigee: the point in the orbit of the moon (or any other earth satellite) nearest to the earth; opposed to apogee

periodicity: the periodicity of tide-producing forces ranges from 3.1 hours to 1,600 years (the maximum tractive forces occur once every 1,600 years; the next occurrence is expected in the year 3300)

pumping process: in tidal schemes, using pumps to over-fill or over-empty the retaining basin

pumped storage (in tidal scheme): pumping water to a storage area during peak generation, so that it can flow back at times of need, thus allowing a full 24-hour cycle of generation

Q

quadrature: the position in the phase cycle when the two principal tide-producing bodies (moon and sun) are nearly at a right angle to the earth; the moon is then at quadrature in its first quarter or last quarter

R

resonance: in tides, the water movement resulting from the natural period of oscillation of a body of the tide-producing forces

retaining (or impounding) basin: the basin in which the water of the incoming (flood) tide is retained; it is closed by a dam or barrage

reversible operation: in a tidal power system, operation in which power is produced utilizing both ebb and flood current

rim turbine: turbine in which the rotor surrounds the turbine runner as a rim

rock-fill dam: dam or barrage constructed of large rocks placed on outer faces with a core of dredged sand material to reduce free movement of water through the structure

S

salt marsh: flat, poorly drained coastal swamps that are flooded by most high tides

sediment: particulate organic and inorganic matter which accumulates in a loose unconsolidated form (it may be chemically precipitated from solution, secreted by organisms, or transported by air, ice, wind, or water, and deposited)

sedimentary rock: rock formed by the accumulation of sediment

seiche: 1) a standing wave oscillation of an enclosed or semi-enclosed water body that continues pendulum fashion, after the cessation of the originating force, which may have been either seismic, atmospheric, or wave induced; 2) an oscillation of a fluid body in response to a disturbing force having the same frequency as the natural frequency of the fluid system (tides are now considered to be a seiche induced primarily by the periodic forces caused by the sun and the moon)

slipforming: method of concrete construction which allows form work to slide up the placed concrete, thus permitting continuous construction

sluiceway: the passage way or artificial channel to let the water through

sluiceway (gated): opening, in a structure, fitted with either flap gates, or mechanically operated gates, to control water levels

spillway: passageway to permit the excess water being carried off

stratification of waters: a situation in which two or more layers of water are arranged according to their density with the lightest on top

spring tide: tide of increased range which occurs about every two weeks when the moon is new or full.

stand of tide: the interval at high or low water when there is no appreciable change in the height of the tide; its duration will depend on the range of the tide, being longer when the tide range is small and shorter when the tide range is large (where a double tide occurs, the stand may last for several hours even with a large range of tide)

syzygy: the two points in the moon's orbit when the moon is in conjunction or opposition to the sun relative to the earth; time of new or full moon in the cycle of phases

T

thermal plant: plant producing electricity by conversion of heat energy whether from coal or petroleum (called geothermal when heat is provided by steam or heat from the earth's interior)

tidal basin: a basin affected by tides, particularly one in which water can be kept at a desired level by means of a gate

tidal current: current produced by the tide-generating forces

tidal flat: see *mud flat*

treme: method of accurately placing material below water level by means of a cylindrical tube (the tube's foot is so arranged that water cannot enter and material is placed undiluted by water)

tropic tide: occurs twice monthly when the effect of the moon's maximum declination north and south of the equator is the greatest

tidal movement: the movement which includes both the vertical rise and fall of the tide, and the horizontal flow of the tidal currents (this movement is associated with the astronomical tide-producing forces of the moon and the sun acting upon the rotating earth)
tidal range: the difference in height between consecutive high and low waters (where the type of tide is diurnal, the mean range is the same as the diurnal range)
turbidity: reduced water clarity resulting from the presence of suspended matter
turn of the tide: see *change of tide*

U

undercurrent: a current flowing beneath a surface current in a different direction or at a different speed (a *countercurrent* flows in the opposite direction of the surface current and at a different speed)
upswelling: the rising of water from a lower to a lesser depth (of importance to OTEC type plants)

V

venturi tube: a short tube with a constricted throatlike passage that increases the velocity and reduces the pressure of a fluid forced to pass through it

W

wave crest: see *crest*
weir: 1) dam, to back water, across a river; 2) an obstruction to divert water through an aperture

Y

yield strength: a measure of the resistance to plastic deformation of a material subjected to a specified type of loading

General Bibliography

Ailleret, P., 1966. The place of tides in the development of the concepts of power generation, *Rev. Franç. de l'Energie* **XVII**, 183, 642–659.

André, H., 1978. Ten years of experience at the "La Rance" tidal power plant, *Oc. Manag.* **4**, 2–4, 165–178.

Andrews, J. H., 1977. Impact on wading birds. In Shaw, T. L. (ed.), *An environmental appraisal of the Severn Barrage*. Bristol, The University, pp. 100–106.

Anthony, R. J., 1979. The changing times on tidal power, *Envir. Sci. & Techn.*, 530–532.

Arnaud, C., 1958. *Le monde a faim de kilowatts*. Paris: Del Duca.

Back, P. A. A., 1978. Hydroelectric power generation and pumped storage schemes utilizing the sea, *Oc. Manag.* **4**, 2–4, 179–206.

Banal, M. and Bichon, A., 1981. Tidal energy in France. The Rance tidal power station. Some results after 15 years of operation, *Int. Symp. Wave and Tidal En.* (Cambridge), 327–338.

Barnier, L., 1968. Power from the tides, *Geographical Magazine* **50**, 1118–1125.

Barr, D. I. H., 1977. Power from the tides and waves. In Lenihan, J. (ed.), *Marine Environment* (vol. 5 of *Environment and man*). New York: Academic Press.

Barrett, M., 1981. Integrating tidal and wave power into the U.K. electricity systems, *Int. Symp. Wave and Tidal En.* (Cambridge).

Barton, R. (ed.), 1968. Hydrospace, *Quarterly Review of Ocean Management*.

Bay of Fundy Tidal Power Review Board, Management Committee, 1976, *Preliminary report, Stage 1 of the Phase 1 Study Program*. Ottawa: Min. Mines.

Behrman, S., and Thurlow, G. G., 1977. Use of colliery shale for construction. In Shaw, T. L. (ed.), *An environmental appraisal of the Severn Barrage*. Bristol, The University, pp. 132–135.

Bernstein, L. B., 1961. *Central tidal-power stations in contemporary energy production*. Moscow: State Publishing House.

———. 1961. "Tidal Power—A Russian view," *Canadian Consulting Engineer* (May).

———. 1961. *Prilivniye elektrostantsu v sovremyennoy energetikye*. Moskva-Leningrad: Gosud energeticheskoye izdatyel'stvo.

———. 1964. The Rance River tidal-power plant (in Russian), *Gitsrotekhnicheskoye Stroityel'stvo* **VI**, 46.

———. 1965. *Tidal energy for electric power plants*. Jerusalem, Israel Program for Scientific Translation.

Bernshtein, L. B., 1965. *Tidal energy for electric powerplants*. Translated from the Russian by Jerusalem, Israel Program for Scientific Translation, Springfield, Va., NTIS (378 p.).

Bernshtein, L. B., 1974. Russian tidal power station is precast offsite, floated into place, *Civ. Eng.* **44**, 4, 46–49.
Bernshtein, L. B., 1974. Kislogubsk: A small station generating great expectations, *Water Power* **26**, 5, 172–177.
Berryman, M. S., 1979. Tidal energy and the energy crisis: an assessment of technology and the interrelationship. In *Marine Technology '79: Ocean Energy*. Washington: Marine Technology Society, pp. 107–116.
Bigourdan, G., 1920. Un moyen économique d'utiliser la force des marées, *Comptes-Rendus de l'Académie des Sciences (France)* **171**, 211–212.
Bird, E. C. F., 1978. Energy regimes and the Australian coast, *Ocean Management Conf.* **4**, 2–4.
Boardman, W. F., 1963. Quoddy and Rankin rapids as a multiple purpose project: the many surveys and reports have not yet produced a proper evaluation of Quoddy, or Quoddy-Rankin, as a multiple purpose project, *Public Utilities Forthnightly,* **71** (April 11), 19–25.
Boisnier, G., 1921. *Utilisation de l'énergie des marées.* Paris: Annales des Ponts et Chaussées.
Boisnier, G., 1921. *Utilisation de l'énergie des marées.* Rennes: Impr. Oberthur (96 p).
Bonnefille, R., 1963. Etude énergétique de la marée dans le golfe de St. Malo à partir des observations en nature, *Bulletin du Centre de Recherches des Etudes des Côtes,* **IV**, 153–165.
Bonnefille, R., 1976. Les réalisations de l'Electricité de France concernant l'énergie marémotrice: *La Houille Blanche* **31**, 2, 87–149.
Bonnefille, R., and Chabert-Dhières, G., 1967. Etude d'un modèle tournant de mer littorale. Application au modèle de l'usine marémotrice des îles Chausey, *La Houille Blanche* **XXII**, 6, 651–658.
Bonnefille, R., and Jeannel, M., 1964. Etude du modèle réduit de la coupure de la Rance, *La Houille Blanche* **XIX**, 4, 481–488.
Bourges, Y., 1966. The Rance tidal power scheme and the Saint Malo region, *Rev. Franç. de l'Energie,* **XVII**, 183, 861–863.
Bouteloup, J., 1950. *Vagues, marées, courants marins.* Paris: Presses Universitaires de France.
Braikevitch, M., 1972. Straight flow turbine. In Gray and Gashus (eds.) *Tidal Power.* New York: Plenum, pp. 415–434.
Brindze, R., 1968. *The rise and fall of the seas.* New York: Harcourt, Brace & World, pp. 72–82.
Butler, G., and Isen, H. C. K., 1966. *Corrosion and its prevention in waters.* New York: Van Nostrand Reinhold.
Cabanius, J., and Svilarich, E., 1966. The Rance project and its contribution to hydroelectric technology, *Rev. Franç. de l'Energie,* **XVII**, 183, 847–860.
Caillez, H., and Faral, M., 1966. General description of the tidal generating station and its electromechanical equipment, *Rev. Franç. de l'Energie,* **183**, 768–809.
Cattaneo, F., 1923. Rapport sur l'utilisation des marées, *Congrès Intern. de Navig.* (Landres) **XIII**, *Sect. 2, Comm. 3.*
Caquot, A., 1966. The definitive cut-off project, *Rev. Franç. de l'Energie,* XVII, 183, 712–721.
Caquot and Defour, 1937. *Utilisation perfectionnée de l'énergie des marées.* Paris: Presses Universitaires de France.
Carlisle, N., 1967. *Riches of the sea: the new science of oceanology.* New York: Sterling Press, pp. 95–102.

Casacci, S. X., 1961. Advances in low-head machines, *Water Power,* **13**, 2, 62–67; **13**, 3, 104–108; and **13**, 4, 152–157.
Casacci, S. X., Duport, J. P., and Pariset, E. F., 1961. Research developments and results concerning bulb units: applications to river and tidal power plants, *Trans. Engineering Inst. Canada,* **IV**, 2.
Casacci, S. X., and Chapus, E. E., 1969. The bulb turbine, *Proc. Winter Meet. Assn. Mech. Eng.*
Casseau, M., 1962. Classification des cycles d'une usine marémotrice, *Mémoires et Travaux de la Société Hydrotechnique de France* **I** (supplément), 155–162.
Cattaneo, F., n.d. *La transformazione della forza del mare in energia elettrica (Le forze motrici del mare).* Genoa Stabilimento grafico editoriale [Piazza Saull, 5–2].
Center for Compliance Information, 1977. *The energy source book.* Aspen, Colorado: Aspen Systems Corporation.
Charlier, R. H., 1968. Tidal power, *Oceanology Intl.* **III**, 6, 32–35.
Charlier, R. H., 1968. Marea si foamea de kilowatsi, *Progresele Stiintei* **IV**, 11, 481–485.
Charlier, R. H., 1977. Energy from the sea. In B. L. Gordon (ed.), *Marine Resource Readings.* Washington: University Press of America, pp. 115–161.
Charlier, R. H., 1969. Tidal energy, *Sea Frontiers.* **XVI**, 6, 339–348.
Charlier, R. H., 1969. La mer et la soif des kilowatts, *Revue de l'Université libre de Bruxelles* (Aug-Sept,), 17–33.
Charlier, R. H., 1969. Harnessing the energies of the ocean, part I, *Marine Technology Soc. J.* **3**, 3, 13–22.
Charlier, R. H., 1969. Harnessing the energies of the ocean, part II, *Marine Technology Soc. J.* **3**, 4, 59–81.
Charlier, R. H., 1970. Harnessing the energies of the ocean—a postscript, *Marine Technology Soc. J.* **4**, 2, 62–65.
Charlier, R. H., 1970. French power from the English Channel, *Habitat* XIII, 4, 32–33.
Charlier, R. H., 1975. Comments to 'Power from the tides' by C. Lebarbier, *Naval Eng. J.* **87**, 3, 58–59.
Charlier, R. H., 1978. Tidal power plants: sites, history and geographical locations, *Proc. Int. Symp. Wave & Tidal Energy (Canterbury)* **I**, 1, A1, 1–6.
Charlier, R. H., 1980. Tides and turbines, *Sea Frontiers* **26**, 6, 355–362.
Charlier, R. H., 1981. Tidal power, *Living Alternatives* **II**, 8, 31–34.
Charlier, R. H., 1981. Energy from the ocean: a look at tidal power, *Alternative Sources of Energy* **X**, 50, 23–27.
Chojniki, T., 1972. Calculs des marées terrestres théoriques et leur précision. *Polska Akad. Nauk. Zaklad. Geofizyki* **55**, 3–42.
Civiak, R., 1978. Tidal power, *US Environmental Data and Information Serv.* **9**, 5, 9–11 (Sept.).
Clare, R. and Oakley, A. J., 1981. The towing and positioning of caissons in a tidal barrage, *Int. Symp. Wave and Tidal En.* (Cambridge), 177–190.
Clark, R. H., 1971. Recent tidal power investigations in the Bay of Fundy, Paper 2.2-55, *Proc. 8th World Energy Conf.* Bucharest, Roumania (June. 27–July 2).
Clark, R. H., 1972. Energy from Fundy tides. *Canadian Geographical J.* **LXXXV**, 5, 150–163.
Clark, R. H., 1972. Fundy tidal power, *Energy International* **9**, 11, 21–26 (Nov).
Clark, R. H., 1972. La baie de Fundy, *2000-Grands Aménagements Mondiaux*, No. 83, 9–11.

Clark, R. H., 1973. La energia das marés na baia de Fundy, *Boletin Geografico*, No. 235 (Jul./Aug.).
Clark, R. H., 1976. Progress of reassessment of exploiting tidal energy, *E.I.C. Congress Proc.* **90**.
Clark, R. H., 1978. Reassessing the feasibility of Fundy tidal power, *Water Power and Dam Construction*.
Clark, R. H., 1978. Power from the tides: *Geos*. Fall, 12–14.
Clark, R. H., 1978. The economics of Fundy tidal power, *Proc. Int. Symp. Wave & Tidal Energy (Canterbury)* **I**, 1, E3, 41–54.
Clark, R. H., 1979. Tidal power, *Proc. Int. Conf. on Future Energy Concepts* [London 1/30–2/1, 1979].
Clark, R. H., 1980. Organization and management for tidal power studies, *Proc. Int. Symp. on Tidal Power*, Seoul, Korea (Nov. 14–15, 1978), Korean Inst. for Res. and Devel., pp. 105–110.
Clark, R. H., 1980. Tidal power development, an international perspective, paper presented to *Int. Conf. on Performance of Concrete in a Marine Environment*, American Concrete Institute at St. Andrews-by-the-Sea (August 22, 1980), unpublished.
Clark, R. H., 1981. Prospects for tidal power, *17th Sesion, Int. Conf. on Long-term Energy Resources, UNITAR*, Montreal (Nov. 26–Dec. 7, 1979).
Clark, R. H., 1981. Fundy tidal power—a retrospective view, paper submitted December 1980, *International Journal of Ambient Energy*.
Clark, R. H., and Karas, A. N., 1979. Studies of tidal power from Bay of Fundy, *Int. Conf. on Future Energy Concepts, Institution of Electrical Engineers*, London (Jan. 30–Feb. 1, 1979), pp. 143–151.
Clark, R. H., and Walker, R. L., 1976. Progress of feasibility reassessment of exploiting tidal energy, *Proc. 90th EIC Congress*, Halifax, Nova Scotia (October).
Cochrane, S. R., and Wilson, E. M., 1981. The Stangford Lough tidal energy project, *Int. Symp. Wave and Tidal En.* (Cambridge, 315–326).
Cohn, P. D., and Welch, J. R., 1969. Power sources. In *Handbook of Ocean and Underwater Engineering* (Myers, Holm, and McAllister, ed.). New York: McGraw-Hill, pp. 6-32 to 6-33.
Coleman, R. S., et al., 1976. Bibliography on pumped storage to 1975, *IEEE Trans., Power Apparatus and Systems* PAS-95, 3, 839–850.
Collins, J., 1977. Tidal energy, *LC Science Tracer Bulletin* TB 77-8. Washington Library of Congress, Science and Technology Division, (8 p.).
Comyns, R. A., 1977. The use of China clay sand in construction. In Shaw, T. L., (ed.), *An environmental appraisal of the Servern Barrage*. Bristol, The University, pp. 146–150.
Considine, D. M. (ed.), 1977. *Energy technology handbook.* New York: McGraw-Hill, 1, 884 p.
Constans, J., 1978. Present and future possibilities of energy production from marine sources, *Proc. Int. Symp. Wave & Tidal Energy (Canterbury)* **I**, 2, C1, 1–22.
Constans, J., 1978. *Energy.* Monaco: Eurocean (Association Européene Oceanique), pp. 203–236.
Constans, J., 1979. *Marine sources of energy.* New York: Pergamon.
Cotillon, J., 1974. La Rance: Six years of operating a tidal powerplant in France, *Water Power* **26**, 10, 314–322.
Creager, W. P., and Justin, J. D. (eds.), 1950. *Hydroelectric handbook,* 2d Ed. New York: Wiley, 1, 151 p.

Daborn, G. R. (ed.), 1977. *Fundy tidal power and the environment.* Wolfville, Nova Scotia: Acadia University Institute, 304 p.
Dalton, F. K., 1961. Tidal electric power generation, *J. Royal Astronomical Society of Canada,* **55**, 1, 22–33; and **55**, 2, 57–72.
Danel, P., 1960. L'évolution de l'équipement des basses chutes, *Revue de l'Enseignement Supérieur* **1**, 103–116.
Daric, G., 1957. Schéma de fonctionnement d'une centrale sousmarine équipression à fluide auxiliaire, *La Houille Blanche* (IVes Journées de l'Hydraulique) II, 694–701.
Davey, N., 1923. *Studies in tidal power.* London: Constable.
Davis, C. V., and Sorensen, E. E. (eds.), 1969. *Handbook of applied hydraulics,* 3d Ed. New York: McGraw-Hill.
de Bélidor, B. F., 1737. *Traité d'architecture hydraulique.* Paris: Ecole d'artillerie et du génie. Conservatoire National.
Debes, M., 1945. Utilisation de l'énergie des marées, *Revue de l'Institut Technique des Bâtiments et des Travaux Publics,* (no. d'avril).
Decelle, A., 1966. Twenty-five years of efforts, *Rev. Franç. de l'Energie,* **XVII**, 183, 636–641.
Decoeur, 1890. *Appareil Hydraulique avec nouveau modèle de turbine pour l'utilisation continue de la force des marées:* Brevets Nos. 205–339 (29 avril 1890).
Deniaux, B., 1974. Tidal power generation in the French bay of Mont Saint-Michel, *Marine Affairs J.,* **2**, 97–115.
DePalma, J. R., 1968. An annotated bibliography of marine fouling for marine scientists and engineers, *J. Ocean Techn.* II, 4, 33–44.
De Rouville, A., 1957. General report on the utilisation of the tidal mechanical energy, *La Houille Blanche* II, 435–455.
Derrington, J., 1978. The use of concrete caissons for river barrages, *Proc. Bristol Colston Symp.* (April).
Derrington, J. A., 1978. Principles of design and construction for marine structures for wave/tidal/ocean thermal energy, *Proc. Int. Symp. Wave & Tidal Energy (Canterbury)* I, 1, G2, 13–26.
De Vos, F. J., 1957. Raisons pour lesquelles aucune usine marémotrice ne sera insérée dans le nouveau projet d'endiguement dit "Deltaplan," en Hollande, *La Houille Blanche* (IVes Journées de l'Hydraulique) II, 465–471.
Dhaille, R., 1957. Technique et rentabilité des dièdres à houle: *La Houille Blanche* (IVes Journées de l'Hydraulique) II, 421–429.
Dietz, R. S., 1972. Mineral resources and power. In Idyll, C. P. (ed.), *Exploring the ocean world: A history of oceanography.* New York: Crowel, pp. 164–195.
Dubois, R., 1962. Les essais du groupe marémoteur expérimental de St. Malo, *La Houille Blanche* **XV**, 2, 131–140.
Dubrow, M. D., 1969. Tidal power. *Encyclopedia of the Earth Sciences.*
Duhoux, L., 1964. Fermeture de la Rance; Déroulement des travaux et analyse des observations, *La Houille Blanche* **XIX**, 4, 491–508.
Easton, W. R., 1921. *Report on North Kimberley District.* Perth, Australia: North West Department Government.
Escande, L., 1967. Recherches hydrauliques récentes, *Arch. Hydrotechn. Polska,* **14**, 1, 3–13.
Fallon, P., 1964. The Rance estuary tidal power project, *Public Utilities Forthnightly* 70 (December).
Faral, M., 1973. Les différents types de protection contre la corrosion mis en oeuvre à l'usine marémotrice de la Rance, *La Houille Blanche* **211**, 2–3, 67–72.

Fentzloff, H. E., 1972. The tidal power plant San Jose, Argentina. In: Gray and Gashus (eds.), *Tidal Power*. New York: Plenum.
Fichot, E., 1923. *Les marées et leur utilisation industrielle*. Paris: Gauthier-Villars.
Firth, J. N. M., 1977. Pollution. In Shaw, T. L. (ed.), *An environmental appraisal of the Severn Barrage*. Bristol, The University, 93–99.
Fixel, A. E., 1969. Tidewater power generation, *Official Gazette U.S. Patent Office*, (February 11), **859**, 2, 415.
Fjelstad, J. E., 1965. Internal waves of tidal origin, *Geophysica Norvegica* **25**, 5, 1–73.
Fletcher, B. N., 1977. The geology of the area of the proposed barrage. In: Shaw, T. L., (ed.), *An environmental appraisal of the Severn Barrage*. Bristol, The University, 122–127.
Fox, W., Brooks, B., and Tyrwhitt, J., 1976. *The Mill*. Toronto: McClelland & Stewart.
Frieberger, A., and Collogen, C. P., 1964. "A laboratory methodology for studying marine fouling," *Feder. Soc. for Paint Techn., Digest XXXVI*, **77**, 1198–1209.
Friedlander, G. D., 1964. The Quoddy question: time and tide, *IEEE Spectrum* **I**, 9, 96.
Furst, G. B., and Sud, S., 1977. Raw tidal energy absorption capability of a power system. *Proc. Summer Meet. IEEE (Mexico City)*, July.
Furst, G. B., and Swales, M. C., 1978. Review of optimization and economic evaluation of potential tidal power developments in the Bay of Fundy, *Proc. Int. Symp. Wave & Tidal Energy (Canterbury)* **I**, 1, E2, 22–40.
Garrett, C., 1977. Predicting changes in tidal regime: the open boundary problem, *J. Phys. Oceanography* **7**, 2, 171–181.
Gauthier, M., 1962. A new approach to tidal power plant calculations, *La Houille Blanche* **XV**, 2, 259–275.
Gibrat, R., 1953. L'énergie des marées, *Bulletin Société Française d'Electricité* **VII**, 3, 283–332.
———. 1955. *Les usines marémotrices*. Paris: Electricité de France.
———. 1956. L'usine marémotrice de la Rance, *Revue Franç. de l'Enérgie* (avril).
Gibrat, R., 1957. Cycles d'utilisation de l'énergie marémotrice, Soc. Hydrotech. Fr. (IVes J. de l'Hydraulique) *La Houille Blanche* **II** (Supplém.) 488–497.
Gibrat, R., 1962. The first tidal power station in the world under construction by French industry on the Rance River, *French Technical Bulletin*, **2**, 1–11.
———. 1962. Source de l'énergie des marées: énergie cinétique de la terre ou énergie thermique du soleil?, *La Houille Blanche* **XV**, 2, 255–266.
Gibrat, R., 1966. *L'energie des marées*. Paris: Presses Universitaires, 230 p. **XVII**, 183, 660–684.
Gibrat, R. V., 1966. Energiya prilivov. V knigye: *Myezhdunarodnaya Assotsiatsiya po giravlicheskim issledovaniyan, XI Congress*, 1965, tom **6**, Lennigrad, 223–242.
Gibrat, R., 1966. *L'énergie des marées*. Paris: Presses Universitaires, 230 p.
Gibrat, R., 1973. L'énergie marémotrice dans le monde: l'usine marémotrice de la Rance et l'environnement. *La Houille Blanche* **211**, 2–3, 145–151.
Gibrat, R., and Auroy, F., 1956. Problèmes posés par l'utilisation de l'énergie des marées, *Fifth World Power Conference* (Vienna) **12**, 111, H/22, 4299–4328.
Gibson, A. H., 1933. Construction and operation of a tidal model on the Severn, *Appendix to the Report of the Severn Barrage Committee*.
Gibson, H. C., 1966. The biological implications of the proposed barrages across Morecambe Bay and the Solway Firth, In Lowe-McConnell, R. H. (ed.), *Manmade lakes*. London: Academic Press.

Gibson, R. A., and Wilson, E. M., 1978. Tidal energy integration using pumped-storage, *ASCE, J. Power Div.*
Glasser, G., and Auroy, F., 1966. Research into the devlopment of the bulb unit, *Rev. Franç. de l'Energie,* **XVII**, 183, 722–767.
Godin, G., 1973. *The tidal power potential of Ungava Bay and its possible exploitation in conjunction with the local hydroelectric resources.* Canada, Marine Sci. Directorate, Manuscript Report Series 30.
Godin, G., 1972. *The analysis of tides.* Toronto: University of Toronto Press, 264 p.
Gordon, F. R., 1964. *Secure Bay—Walcott Inlet tidal power scheme.* Gov. of West. Austr. "Preliminary geological report," Record No. 1964/6 (May 4).
Gorlov, A. M., 1980. Small scale tidal energy consumption. In: Oktay, U, (ed.), *Energy resources and conservation related to the building environment.* London: Pergamon Press (pp. 492–498).
Gorlov, A. M., 1979. Some new conceptions in the approach to harnessing tidal energy, *Proc. Miami Int. Conf. Alternat. Energy Sources II,* 1171–1795.
Gougenheim, A., 1967. The Rance tidal energy installation, *J. Inst. Nav.* **XX**, 3, 229–236 (July).
———. 1976. L'utilisation de l'énergie des marées, *Cahiers océanographiques* **XIX**, 4, 277–293.
Gougenheim, A., and Romanovsky, V., 1957. Les remontées d'eau profonde, (IVe Journées de l'Hydraulique) *La Houille Blanche* **II**, 712–719.
Grant, A. D., 1981. Power generation from tidal flows for Navigation buoys, *Int. Symp. Wave and Tidal Navigation En.* (Cambridge), 117–128.
Gray, T. J., and Gashus O. K., (eds.), 1972. *Tidal Power.* New York: Plenum, 630 p.
Great Britain House of Commons. Select Committee on Science and Technology, 1977. *Power in the Severn Estuary.* London: Stationery Office, 70 p.
Greenberg, D. A., 1979. A numerical model investigation of tidal phenomena in the Bay of Fundy and Gulf of Maine, *Marine Geodesy* **2**, 2, 167–187.
Griffin, D. M., 1974. *Energy from the ocean: an appraisal.* Washington, D.C.: Naval Res. Lab. (Memo. Rep. #2803), 43 p.
Guillaumin, M., and Larquier, M. de, 1973. Exploitation de l'usine de la Rance: méthode et résultats, *La Houille Blanche* **211**, 2–3, 131–144.
Halacy, D. S., Jr., 1977. *Earth, water, wind and sun. Our energy alternatives.* New York: Harper, pp. 59–71.
Harrah, B. K., and Harrah, D., 1975. *Alternate sources of energy.* Metuchen, NJ: Scarecrow Press, 201 p.
Haswell, C. K., 1977. Civil engineering aspects. In Shaw, T. L. (ed.), *An environmental appraisal of the Severn Barrage.* Bristol, The University, pp. 128–131.
Haswell, C. K., Huntington, S. W., Shaw, T. L., Thorpe, G. R., and Westwood, I. J., 1972. Pumped storage and tidal power in energy systems, *Proc. A. Soc. Civ. Eng.* PO **2**, 201–220.
Heaps, N. S., 1968. Estimated effects of a barrage on tides in the Bristol Channel, *Proc. Inst. Civ. Eng.* **40**, 495–509.
Heaps, N. S., 1972. Tidal effects due to power generation in the Bristol Channel. In Gray and Gashus (eds.), *Tidal Power.* New York: Plenum, pp. 435–456.
Headland, H., 1949. Tidal power and the Severn Barrage, *Inst. Elec. Eng., Proc.* **96**, II, 51, 427–451.
———. 1950. Tidal power and the Severn Barrage, *Inst. Elec. Eng., Proc.* 97, **II**
———. 1951. Tidal power and the Severn Barrage, *Inst. Elec. Eng., Proc.* 98, **I**
Heronemus, H. E., Mangarella, P. A., McPherson, R. A., and Ewing, D. L., 1974. On the extraction of kinetic from oceanic and tidal river currents. In Stewart, J. B., Jr.

(ed.), *Proc. MacArthur workshop on the feasibility of extracting useable energy of the Florida current.* NOAA Atlantic Oceanographic and Meteorological Laboratories (Miami).

Hughes, E. M., and Glanville, R., 1974. Tidal power, *Proc. Internal Symp. on Altern. Energy Sources Centr. Elect. Bd.* 36–45.

Huntsman, A. G., 1928. *The Passamaquoddy Bay power project and its effect on the fisheries.* Saint John (New Brunswick), The Telegraph Journal.

International Passamaquoddy Engineering Board, 1959. *Investigation of the international Passamaquoddy tidal power project: report to the International Joint Commission.* Washington, D.C.: The Board.

International Passamaquoddy Fisheries Board, 1959. *Passamaquoddy fisheries investigations.* Report to the International Joint Commission. Washington/Ottawa: The Board, 53 p.

International Joint Commission, 1961. Reports on Passamaquoddy Tidal Power Project, *Dept. of State Bull.* **44** (May 22).

International Joint Commission, 1961. Rules against feasibility of Passamaquoddy project, *Electrical World,* **155**, (May 8), 44.

International Joint Commission, 1961. *Investigation of the International Passamaquoddy Tidal Power Project. Report of the International Joint Commission, Docket 72;* Washington: The Commission (April).

Issaacs, J. D., and Seymour, R. J., 1973. The ocean as a power resource, *Int. J. Environ. Stud.* **3**, 4, 201–205.

Jefferys, E. R., 1981. Dynamic models of tidal estuaries, *Int. Symp. Wave and Tidal En.* (Cambridge), 69–86.

Jeffreys, J., 1920. Tidal friction in shallow seas, *Philos. Trans.* **239**.

Jones, I. E., 1968. The Rance tidal power station, *Geography* **53**, 11, 412–415.

Kagan, B. A., 1974. Dissipation of tidal energy in the Arctic seas, *Akad. Nauk SSSR Bull. Atm. & Ocean. Phys. Ser.* **9**, 6, 375–376.

Kagan, B. A., 1977. Global dissipation and exchanges of energy between ocean and earth tides, *Akad. Nauk SSSR Bull. Atm. & Ocean. Phys. Ser.* **13**, 7, 485–490.

Kammerlocher, L., 1957. Groupes générateurs hydroélectriques immergés, type bulbe. Développement et évolution constructive, *Rev. Génér. de l'Electricité,* **66**, 7, 342–360.

———. 1958. Innovations technologiques dans la conception des groupes bulbes turbines-pompes immergés, *Soc. Hydrotechnique de France* (V^{es} Journées de l'Hydraulique) Aix-en-Provence, V, 2.

———. 1960. La station marémotrice expérimentale de Saint-Malo, *Rev. Génér. de l'Electricité,* **69**, 5, 237–261.

Karas, A. N., 1977. System planning for Bay of Fundy tidal power developments, *Proc. Summer Meet. IEEE (Mexico City),* July.

Kay, R., 1975. Comments to "Power from the tides by C. Lebarbier," *Nav. Eng. J.* **87**, 3, 57–58.

Keiller, D. C. and Thompson, G., 1981. One dimensional modelling of tidal power schemes, *Int. Symp. Wave and Tidal En.* (Cambridge), 19–32.

Kennedy, G. E., and Headland, H., 1957. Etudes de l'usine marémotrice de la Severn, *La Houille Blanche* (IVes Journées de l'Hydraulique) II, 456–464.

Keyerleber, K., 1973. Passamaquoddy: A good idea is hard to kill, *The Nation* **216**, 275–276.

Keyerleber, K., 1977. Tidal power: neglected energy resource, *Key Biscayne (Florida) Nat. Symp. on Energy and the Oceans; Conf. Coursebook* 7–58.

Kirby, R., and Parker, W. R., 1977. Sediment dynamics in the Severn Estuary: background for studies of the effects of a barrage. In Shaw, T. L. (ed.), *An environmental appraisal of the Severn Barrage*. Bristol, The University, pp. 41–52.
Kohl, J. (ed.), 1976. *Energy from the oceans, fact or fantasy?* Conf. Proc., Jan. 27–28, 1976, Raleigh, N.C., Center for Marine and Coastal Studies, North Carolina State University. Sea Grant publication UNC-SG-76-04, 110 p.
Kuznetsov, E., 1979. *Analysis of a flexible barrier for a tidal power plant. A summary report to Northeastern University.* Columbus, OH: Battel Institute.
Laba, J. T., 1964. Potentials of tidal power on the North Atlantic coast in Canada and the U.S., *Coastal Eng. (Proc. 9th Conf. Coastal Eng.)* 832–857.
Laberge, N., 1978. *Discussion papers of the Half-Moon Cove tidal power project.* Perry, Maine: Pleasant Point Reservation.
Lambert, M., and Legrand, M., 1973. Bilan de la protection cathodique à l'usine marémotrice de la Rance, *La Houille Blanche* **211**, 2–3, 257–262.
Lawton, F. L., 1972. Tidal power in the Bay of Fundy, In Gray and Gashus, *Tidal Power*. New York: Plenum, pp. 1–104.
Lawton, F. L., 1974. "Time and tide," *Oceanus* **XVII**, summer, 30–37.
Lebarbier, C. H., 1975. Power from tides: The Rance tidal power station, *Nav. Eng. J.* **83**, 2, 57–71.
Lebarbier, C. H., 1975. Power from the tides. Discussion, *Nav. Eng. J.* **87** 3, 52–56.
Leborgne, M., Comportement des métaux à l'usine marémotrice de la Rance, *La Houille Blanche* **211**, 2–3, 251–256.
Lee, S. T., and Deschamps, C., 1977. Mathematical model for economic evaluation of tidal power, *Proc. Summer Meet. IEEE (Mexico City)*, July.
Lefrançois, J., 1973. Fonctionnement de l'usine de la Rance: comportement du matériel électromécanique, *La Houille Blanche* **211**, 2–3, 162–170.
Legendre, R., 1949. Les ressources énergétiques de la mer, *Bulletin Institut Océanographique Monaco* **947**, 1–16.
Le Grand, R., and Lambert, M., 1962. Mesures électrochimiques appliquées à l'étude de la protection cathodique des ouvrages de la Rance, *La Houille Blanche* **XV**, 2, 177–186.
Le Grand, Y., 1957. Energie électromagnétique des océans, *La Houille Blanche* (Comptes-Rendu des IVes Journées de l'Hydraulique, 1956) **I**, 225–228.
Leicester, R. J., Newman, V. G., and Wright, J. K., 1978. Renewable energy sources and storage, *Nature* **272**, 518–521.
Le Mehaute, B., 1976. *A preliminary assessment of the tidal power potential at two sites in the vicinity of Cutler, Maine:* Arlington, Virginia: Tetra Tech, Inc., 43 p. (available NTIS AD-A023 824).
Lewis, J. G., 1963. The tidal power resources of the Kimberleys, *J. Inst. Eng., Australia*, **35**, 12, 333–345.
Licheron, S., 1962. La lutte contre la corrosion du matériel des usines marémotrices, *La Houille Blanche* **XV**, 2, 166–176.
Little, C., 1977. Possible biological effects. In Shaw, T. L. (ed.), *An environmental appraisal of the Severn Barrage:* Bristol, The University, pp. 61–71.
Longhurst, A. R., Radford, P. J., et al., 1977. Ecosystem models and the prediction of ecological effects. In Shaw, T. L. (ed.), *An environmental appraisal of the Severn Barrage*. Bristol, The University, pp. 83–92.
MacCellan, H. J., 1952. Energy consideration in the Bay of Fundy system, *J. Fisheries Board of Canada* **XV**, 2, 1935.
Macmillan, R. H., 1966. *Tides*. New York: Elsevier, pp. 172–179.

Renne, R. R., 1966. The future of water resources, *Oceanology Int.* **I**, 2, 67–71.

Richards, A. F., 1976. Extracting energy from the oceans. A review, *Marine Techn. Soc. J.* **10**, 2, 5–24.

Richards, B. D., 1948. Tidal power, its development and utilization, *J. Inst. Civ. Eng. (Great Britain),* 104–144 (April).

Rigaud, M., 1926. *Les réserves d'énergie.* Paris: Gauthier–Villars.

Robinson, I. S., 1981. Surges in tidal basins—can it increase output? *Int. Symp. Wave and Tidal En.* (Cambridge), 53–68.

Romanovsky, V., 1950. *La mer, source d'énergie.* Paris: Presses Universitaires de France.

Rouch, J., 1961. *Les marées.* Paris: Payot, Bibliothèque Scientifique, 230 p.

Rouzé, M., 1959. *Energie des marées.* Monte-Carlo: Editions du Cap.

Sanhes, J., 1962. Protection contre la corrosion marine de la station marémotrice expérimentale de St. Malo, *La Houille Blanche* **XV**, 2, 195–204.

Saunders, D. W., 1975. Kimberley tidal power revisited, *J. Inst. Eng. Austral.* **47**, 11, 47–55.

Savage, J. A., 1975. *Potential of tidal power and Gulf Stream power sources.* Austin: Governor's Energy Advisory Council, 49 p.

Savery, S. P. A., 1977. Natural aggregates supply considerations. In Shaw, T. L. (ed.), *An environmental appraisal of the Severn Barrage:* Bristol, The University, pp. 136–145.

Saylor, J. P., 1965, The Passamaquoddy boondoggle: economic feasibility of utilizing high tides near the Maine-New Brunswick border to generate electric power: *Public Utilities Fortnightly,* 71 (January 17), 15–22.

Schureman, P., 1975, *Tide and current glossary:* (Revised edition by Steacy Hicks) Washington, D.C.: National Ocean Survey (25 p.)

Scott, W. E., 1976. Australia takes new look at tidal energy, *Energy Intl.* **13**, 9, 41–43.

Sebo, S. A., 1975. Ocean powers, *Maritime Studies & Management* **2**, 4, 202–214.

Secretary of the Interior, 1964. *The international Passamaquoddy tidal power project and Upper Saint John River hydro-electric power development.* Washington, D.C.: Department of the Interior.

Seifert, A., 1948. Gezeitenkraftwerk in Wilhelmshafen, *Arch. Energiewirtschaft* **IV**, 209.

Select Committee on Science and Technology, Energy Resources Subcommittee, 1975. *Tidal power for electricity generation. A memorandum from the Central Electricity Board:* London: HMSO (Part IV).

Select Committee on Science and Technology, Energy Resources Subcommitte, 1977. *The exploitation of tidal power in the Severn estuary.* London: HMSO (4th rep.).

Seoni, R. M., 1977. Major electrical equipment proposed for tidal power plants and the Bay of Fundy, *Proc. Summer Meet. IEEE (Mexico City),* July.

Severn, B., and Campbell, R. O., 1978. Prefabricated caissons for tidal power development, *Proc. Int. Symp. Wave & Tidal Energy (Canterbury)* **I**, 1, G1, 1–12.

Severn Barrage Group, 1976. *Proposal for the project. Definitive study of the Severn Barrage.* London: David Mappin (Offshore) Ltd.

Shaw, T. L., 1974. Tidal energy from the Severn Estuary, *Nature* **249**, 5459, 730–733.

Shaw, T. L., 1975. Tidal power and the environment, *New Scientist* **68**, 972, 202–4/206.

Shaw, T. L., 1977. A policy for tidal energy, *Marine Policy* **1**, 61–69.

Shaw, T. L., 1977. Tides, currents and waves, In Shaw, T. L. (ed.), *The exploitation of tidal power in the Severn Estuary.* Bristol, The University, pp. 1–34.
Shaw, T. L., 1978. Tenth world energy conference, *Water Power & Dam Construction,* January, 58–62.
Shaw, T. L., 1978. The status of tidal power, *Water Power & Dam Construction,* June, 29–34.
Shaw, T. L., 1978. The role of tidal power stations in future scenarios for electricity storage in the U.K., *Proc. Int. Symp. Wave & Tidal Energy (Canterbury)* I, 1, H2, 11–22.
Shaw, T. L., and Thorpe, R. G., 1971. Integration of pumped storage with tidal power, *Proc. Am. Soc. Civ. Eng.* PO 1, 159–180.
Shaw, T. L., and Westwood, I. J., 1974. Optimising pumped storage with tidal power in an estuary, *Amer. Soc. Mech. Eng.* Paper in 74-WA/pwr-7, 7 p.
Shaw, T. L., and Wheater, H. S., 1978. Some observations on the virtues of integrating tidal power in the U.K. electrical network. *Proc. Bristol Univ. Colston Res. Symp.*
Shepard, F., 1949. Evidence of world-wide submergence, *J. Mar. Res.* VII, 661–676.
Sibley, A. K., and McNiece, W. H., 1960. Harnessing the tides, *Military Engineer,* 52, January/February, 1–6.
Skillman, J. M., and Wheater, H. S. 1977. The prospects of tidal power, *Inst. Civ. Eng., Proc.* (Part 1—Design and construction) 62, 701–705.
Smith, L., 1959. The Quoddy project stirs again, *Public Utilities Fortnightly,* 64 (November 5), 753–765.
Smith, L., 1961. The status of power supply in Maine, *Public Utilities Fortnightly,* 68, (December 7) 873–882.
Smith, P. C., 1978. Circulation, variability and dynamics of the Scotian shelf and slope, *Fisheries Research Bd. J. (Canada)* 35, 8, 1067–1083.
Sogreah, 1959. *Usina maremotriz del golfo de San José Anteproyecto.* Grenoble, France: Société d'Etudes et des Applications Hydrauliques.
Sogreah, 1959. *Construcción del Canal San José.* Grenoble, France: Société d'Etudes et des Applications Hydrauliques.
———. 1959. *Anexo, Calculo aproximado de la influencia de la rugosidad de la inercia del agua sobre la caida turbinable.* Grenoble, France: Société d'Etudes et des Applications Hydrauliques.
Sogreah, 1963. *Tidal power plant for Collier Bay.* Grenoble, France: Société d'Etudes et des Applications Hydrauliques (Report No. R8527 for the Ministry of Industrial Development of the Government of Western Australia).
Sogreah, 1965. *Collier Bay tidal power development: Secure Bay.* (Report No. R9011 to the Public Works Department of the Government of Western Australia). Grenoble, France: Société d'Etudes et des Applications Hydrauliques.
Stokes, C. J. and Street, R. D. J., 1981. Turbine caissons for the Severn barrage, *Int. Symp. Wave and Tidal En.* (Cambridge), 167–176.
Swales, M. C., and Wilson, E. M., 1966. Optimisation of tidal power generation, *Water Power* 20, 3, 109–114.
Tanner, R. G., 1979. Tidal energy in the Bay of Fundy. In *Marine technology '79: ocean energy.* Washington, D.C.: Marine Technology Society, pp. 91–99.
Terry, R. D., 1966. *Ocean Engineering.* Energy sources and conversion, undersea construction, vol. 3. No. Hollywood, California: Western periodicals.
Thompson, I. B., et al., 1967. *The St. Malo Region.* Nottingham (England) Geography Field Group. Regional Studies No. 12, pp. 8, 16, 19, 63, 69–72, 75–76, 95.

Tinkler, J. A., 1977. Drainage and land quality. In Shaw, T. L. (ed.), *An environmental appraisal of the Severn Barrage*. Bristol, The University, pp. 53–60.

Trites, R. W., 1959. *Probable effects of proposed Passamaquoddy power project on oceanographic conditions*. Intl. Passamaquoddy Fisheries Bd. Rep. to Int. Joint Commission, Ch. 7, Appendix 1.

Tuthill, A. H., and Schillmoller, C. M., 1965. *Guidelines for selection of marine materials*. New York: International Nickel Company.

Tuttel, J., 1978. Watermolens eevwenoud en eevwig boeiend, *Aard & Kosmos* **8**, 9 en 10.

U.N. Department of Economic and Social Affairs, 1957. *New sources of energy and economic development*. New York: United Nations, 150 p.

U.S. Army Corps of Engineers, 1964. *Supplementary engineering report to the Report on the International Passamaquoddy Tidal Power Project and Upper Saint John River Hydroelectric Power Development:* Waltham, Mass.: U.S. Corps of Engineers, New England Division.

U.S., Congress, House Committee on Foreign Affairs, 1953. *Survey of Passamaquoddy tidal power project*. Hearings before the subcommittee of the Committee on Foreign Affairs, House of Representatives. H. J. Res. 112, H.J. Res. 113, and H. J. Res. 114 to authorize and direct the International Joint Commission on United States–Canadian Boundary Waters to make a survey of the proposed Passamaquoddy tidal power project and for other purposes. July 14 and July 22, 1953. Washington, D.C.: Government Printing Office (83rd Cong., 1st sess.).

U.S., Congress, 1955. *Passamaquoddy international tidal power project*. Report on S. J. Res. 12, a resolution requesting the Secretary of State to arrange for the International Joint Commission, United States and Canada, to conduct a survey of the proposed Passamaquoddy tidal power project. Washington, D.C.: Government Printing Office (84th Cong., 1st sess., H. Rept. 1152).

U.S., Congress House, 1965. Communication from the President of the United States, Lyndon B. Johnson. Conservation of the Natural Resources of New England. Report on the Passamaquoddy-St. John River Basin Power Development, together with a recommendation for the immediate authorization of the Dickey-Lincoln School Project on the St. John River. Washington, D.C.: Government Printing Office (89th Cong., 1st sess., H. Rept. 236).

U.S., Department of the Interior, 1961. *Report to Passamaquoddy-Saint John River Study Committee*. The International Passamaquoddy Tidal Power Project and Saint John River, United States and Canada, Load and Resources Study.

U.S., Department of the Interior, 1964. Office of the Secretary. Fact Sheet—August. (Release: Wood 343-3171.) *The Passamaquoddy international tidal power project and Upper Saint John River Hydroelectric Development*.

U.S., Energy Research and Development Administration, Technical Information Center, 1976. *Solar energy: A bibliography*. Oak Ridge, Tennessee, 2 vols.

U.S., Federal Power Commission, 1941. *Passamaquoddy tidal power project*. Letter from the chairman of the Federal Power Commission, transmitting in response to S. R. No. 62 (76th Cong.) a report on the Passamaquoddy tidal project, Maine. Washington, D.C.: Government Printing Office (76th Cong.).

U.S., Library of Congress. Congressional Research Service, Science Policy Research Division, 1978. *Energy from the ocean: Report prepared for the Subcommittee on Advanced Energy Technologies and Energy Conservation Research, Development and Demonstration of the Committee on Science and Technology*. Washington, D.C.: Government Printing Office (Chapter 4, Energy from ocean tides, pp. 175–222).

U.S., Passamaquoddy-Saint John River Study Committee, 1963. *The International Passamaquoddy tidal power project and Upper Saint John River hydroelectric power development*. Report to President John F. Kennedy in response to letter of May 20, 1961, submitted by Stewart L. Udall, Secretary of the Interior. Washington, D.C.: Printing Office, 95 p.

U.S., Senate, Committee on Public Works, Subcommittee on Flood Control, Rivers and Harbors, 1964. Passamaquoddy-St. John appendix material compiled on conjunction with hearing on S. 2573, a bill to authorize the international Passamaquoddy tidal power project. Committee Printing, 88th Cong., 2nd sess.

U.S., Senate, 1964. *Passamaquoddy-St. John Hearing,* August 12, 1964, on S. 2573. 88th Cong., 2nd sess.

Vaidyaraman, P. P., and Brahme, S. B., 1977. "Tidal power generation in India," *Inst. Eng. J. (India)* part EL 57:200-206.

Valembois, J., 1957. Possibilité de captage de l'énergie de la houle au moyen de résonateurs: *La Houille Blanche* (IVes Journées de l'Hydraulique) **II** 418-420.

Valembois, J., and C. Birard, 1954. Les ouvrages résonants et leur application à la protection des ports: *Proc. 5th Conf. Coast. Eng. (Berkeley).*

Van London, A. M., 1954. *The mode of action of anti-fouling paints: interaction between anti-fouling paints and sea water.* Amsterdam, Netherlands: Paint Research Institute (Report 62C).

Vantroys, L., 1957. Nature de l'énergie des marées: *La Houille Blanche* (IVes Journées de l'Hydraulique) **I**, 133-141.

———. 1957. "Le régime des marées dans la Manche," *La Houille Blanche* (IVes Journées de l'Hydraulique) **I**, 176-181.

———. 1957. Perturbation apportée au régime des marées par le fonctionnement d'une usine marémotrice, *La Houille Blanche* (IVes Journées de l'Hydraulique) **I**, 188-199.

———. 1958. Le remous d'un ouvrage dans une mer à marée: *Bull. Inf. Comité Central d'Océanogr.* **X**, 8, 9, and 10.

Vernon, K. R., 1974. Hydro (including tidal) energy: *Philos. Trans. Roy. Soc. London Ser. A,* 276, 485-493.

Vincent, M., 1924. *Réflections sur l'utilisation future des énergies naturelles: vagues, chutes hydrauliques et barométriques, chaleur solaire.* Paris: Fischbacher.

Voyer, M., and Penel, M., 1957. Les calculs de la production d'une usine marémotrice, *La Houille Blanche* (IVes Journées de l'Hydrauliques) **II**, 472-487.

Wailes, R., 1941. Tide mills in England and Wales: *Junior Inst. Eng., J. and Record of Trans.,* **51**, 91-114.

Walker, H., 1965. France meets the sea in Brittany, *National Geographic* **CIII**, 4 (April).

Waller, D. H., 1970. Environmental effects of tidal power development. In Gray and Gashus (eds.), *Tidal Power.* New York: Plenum, pp. 611-625.

Witherell, R. G. and Debelius, C. A., 1981. Preliminary assessment of Cook Inlet tidal power, *Int. Symp. Wave and Tidal En.* (Cambridge), 421-435.

Warnock, J. C., and Tanner, R. G., 1978. Selection of optimum sites for tidal power development in the Bay of Fundy, *Proc. Int. Symp. Wave & Tidal Energy (Canterbury)* **I**, 1, E1, 1-22.

Wayne, W. W., Jr., 1977. *Tidal power study for the U.S. Energy Research and Development Administration,* final report. Boston: Stone and Webster Engineering Corp., 2 vols.

Wayne, W. W., Jr., 1977. The current status of tidal power: *Key Biscayne (Florida), Nat. Symp. on Energy and the Oceans,* Conf. Coursebook 7-58.

Wayne, W. W., Jr., 1978. Tidal power possibilities in the United States, *A.S.C.E., Pittsburgh* Preprint 3199.

Wertheim, J. K., 1961. Studies on the electrical potential between Key West, Florida and Havana, Cuba, *Trans. Amer. Geophys. Union* **35**, 872–875.

Wheathley, J. D., 1977. Impact on social and recreational habits. In Shaw, T. L. (ed.), *An environmental appraisal of the Severn Barrage*. Bristol, The University, pp. 151–154.

Wheeler, S. J., 1981. Optimization of tidal power schemes, *Int. Symp. Wave and Tidal En.* (Cambridge), 237–248.

Wick, G. L., 1977. Prospects for renewable energy from the sea, *Mar. Techn. Soc. J.* **11**, 5–6, 16–21.

Widdern, H., Cardinal von, 1952/53. La turbine tubulaire, *EscherWyss Bulletin* **XXV/XXVI**, 22–30.

Wilson, E. M., 1964. A new approach to power from the tides, *New Scientist*, **24**, 415, 290–291.

Wilson, E. M., 1965. Energy from the tides: *Science Journal I*, 5, 50–56.

———. 1965. The Solway Firth tidal-power project, *Water Power*, **17**, 11, 431–440.

———. 1966. Feasibility study of tidal power from Loughs Stangford and Carlingford, with pumped storage at Rostrevor, *Inst. Civ. Eng.*, **34**, 83–100.

Wilson, E. M., 1968. The Bristol Channel barrage project: *Proc. Conf. Coastal Eng.* **XI**, 83, 1304–1325.

Wilson, E. M., 1972. Tidal energy and its development. In *Conf. on Engineering in the Ocean Environment*, 3d ed. *Ocean '72*. New York: Institute of Electrical and Electronic Engineers, **0**, 47–56.

Wilson, E. M., 1973. Energy from the sea—tidal power, *Underwater J.* **5**, 4, 175–186.

Wilson, E. M., 1977. Tidal energy and system planning, *Consulting Engineer (London)* **41**, 4, 25.

Wilson, E. M., and Gibson, R. A., 1978. Studies in retiming tidal energy, *Proc. Int. Symp. Wave & Tidal Energy (Canterbury)* **I**, 1, H1, 1–10.

Wilson, E. M., and Swales, M. C., 1972. Tidal power from Cook Inlet, Alaska. In Gray and Gashus (eds.), *Tidal Power*. New York: Plenum, pp. 239–248.

Winters, A., 1972. A desk study of the Severn Estuary, *Proc. Math. & Hydraul. Model. of Estuar. Poll., Water Poll.* Technical Paper No. 13, 105–113.

Wishart, S. J., 1981. A preliminary survey of tidal energy from five U.K. estuaries, *Int. Symp. Wave and Tidal En.* (Cambridge), 299–314.

Won, T. S., 1975. The tidal power resources and their power generation projects of the western coast of Korea, *Proc. Pacif. Sci. Assn., Pac. Sci. Cong.* **XIII**, I, 162.

Yen, J. T., and Isaacs, J., 1978. Dynamic dam for harnessing ocean and river currents and tidal powers, *Proc. Ann. Mar. Techn. Soc./IEEE Comp. Conf., Ocean '78: The ocean challenge*. **IV**, 582–584.

Yong, W. J., 1977. Tidal power projects on the west coast of the Republic of Korea, *World Energy Conf. (Istanbul)* **X**, 1–11.

1899. Les moulins à marée de New York, *Revue Scientifique,* **IV**, 11, 1, 30.

1902. Le moteur à marée en Californie, *Revue Scientifique,* **IV**, 17, 8, 253.

1933. *Report of the Severn Barrage Committee*. London: H. M. Stationery Office.

1933. *Appendix to the Brabazon Severn Barrage committee report: Report of the expert coordinating subcommittee*. London: H. M. Stationery Office.

1941. *Passamaquoddy Tidal Power Project, Maine*. Washington, D.C.: Federal Power Commission.

1945. *Report on tidal power, Petitcodiac and Memramcook estuaries*. Frederictown: Government of the Province of New Brunswick.

1946. *Rapport sur les études par la Société d'Etudes pour l'utilisation des marées.* Paris: Electricité de France.
1946. *Report on tidal power, Petitcodiak, and Memramcook estuaries, Province of New Brunswick.* Ottawa: King's Printer.
1959. Passamaquoddy feasible for United States, engineers say, *Electrical World,* **152**, 54.
1962. The Rance tidal power plant, *La Houille Blanche* **XV**, 2, 117–129.
1963. The closure of the Rance estuary, *La Houille Blanche* **XVIII**, 7, 789–798.
1964. The estuarine barrage, *Engineer,* **218**, 5677, 787–788.
1965. White House offers Marine Powers Plan, *New York Times* July 11, 1 and 51.
1965. Power and water, *Congressional Record* (August 17), A4580 (daily edition). [Editorial from *Patriot Ledger,* Quincy, Mass.]
1965. Border power project announced for Maine, *Washington Post* July 11, A6.
1966. Tidal power from the Bristol Channel, *Engineer* **222**, 5739, 109–110.
1966. Tidal power comes to France, *Engineer* **202**, 5228, 17–24.
1966. The Rance tidal-power station, *Engineer* **222**, 5784, 856–860.
1966. The Rance tidal-power station, *Engineer* **222**, 5785, 891–895.
1966. France harnesses sea tides for electrical power, *France Actuelle,* **XV**, 16.
1967. Canada sparks Fundy tidal power study, *Engineering News Record* 179, 22, 25–27.
1967. Power from the Rance and the Rhine, *Engineer* **223**, 5790, 74–75.
1967. Tidal power from the Severn, *Engineer* **223**, 5802, 509 and 513.
1967. Bay of Fundy tidal power study, *Engineer* **223**, 5809, 786.
1967. Rance tidal power plant, *Materials Protection* **VI**, 1, 46–47.
1967. France harnesses the tides, *Ocean Industry* **II**, 2, 5–8.
1969. *Handbook of ocean and underwater engineering.* New York: McGraw-Hill.
1967. The Rance tidal power scheme, *Water Power* **19**, 1.
1969. *Feasibility of tidal power development in the Bay of Fundy. Board and Committee Report.* Ottawa: Atlantic Tidal Power Programming Board.
1973. Six ans d'exploitation de l'usine marémotrice de la Rance, *La Houille Blanche* **211**, 2–3, 1–66; 125–270.
1973. *An energy policy for Canada. Summary of analysis.* Ottawa: Dept. of Energy, Mines & Resources.
1974. Tidal power may now make sense, *Business Week* **2356**, 115–118.
1975. The Severn Barrage: its possible implications, *Estuar. & Brackish Water Sci. Assn.* **11**, 7, 8–14.
1976. *Severn Barrage study.* Hydraulics Research Station.
1976/7. *Water resources development project, St. John River Basin, Dickey-Lincoln School Lakes design.* Memo. Nos. 3–4A. Walton, Mass.: Army Corps of Engineers, New England Division.
1977. *Reassessment of Fundy tidal power.* Bay of Fundy Tidal Power Review Board, Ottawa.
1977. *The energy source book:* Aspen, Colorado, The Center for Compliance Information, Aspen Systems Corp. (Part 3).
1977. *Tidal power barrages in the Severn estuary. Recent evidence of their feasibility.* London: H.M.S.O. (Energy Paper No. 23).
1978. *Severn Barrage seminar.* September 7, 1977. London: H.M.S.O.
1964. "The Quoddy question: time and tide," *IEEE* **I**, 9, 96–118.
1964. Constructors harness the tides in France, *Engineering News Record* **173**, 11, 32–34.
1967. Rance tidal power plant, *Materials Protection* **6**, 1, 46–47.

1968. English electric and tidal power, *Engineer* 226, 5877, 382.
1976. L'usine marémotrice de Chausey. In *La production d'électricité d'origine hydraulique. Rapport de la Commission de la production d'origine hydraulique et marine*. Paris: La Documentation Française **IV**, 41–49.
1977. British scheme to harness the tides may yet go ahead, *Ocean Industry* **12**, 8, 107–109.

Index

Aber Wrac'h, 55, 65
Abidjan, 16
Africa (tidal power sites), 150, 175
Agostoli, 45, 55
Alaska (tidal power sites), 219, 313 (*See also* Angoon, Knik Arm, Cook Inlet)
American Gas Association, 11, 14
amphidromic points, 101, 321
amplitude (tidal), 76, 79–81, 105, 280, 281, 321
Anderson, J. H., 18
Andrews, J. H., 299
Angoon Project, 141, 223
Annapolis River, 118, 257
Apel, J., 35, 51
Aqua Power barge, 36, 37
Asia (tidal power sites), 150, 168–175
Australia (tidal power sites), 150, 159–168, 181, 308

Baird, W. F., 33, 50
Barjot, 17, 23
basin (retaining), 119–126
basin (multiple . . . schemes), 123–126, 165
basin (single . . . schemes), 120–123
Bathen, K. H., 22, 49
Beau, C., 16
Bélidor, B. F. (de), 64, 74
benefit-cost ratio (Fundy Bay scheme), 268–272
Berns(h)tein, L., 66, 69, 118, 178, 206, 210, 217, 288
Bigourdan, G., 65, 73
Birdham Mill, 59, 60
Boggia, M., 16

Boisnier, G., 65, 73
Bolsena (Lake), 16
bore, 101, 261, 286, 296, 321
Bouteloup, J., 35, 51
Bracciano (Lake), 16
Brazil (tidal power sites), 156, 182
Bristol Channel, 150, 156, 177, 275, 277, 279, 308
British Columbia (tidal power sites), 158, 252–255
Bromley-by-Bow, 55, 59
bulb turbine, 114–115, 117, 118, 160, 165, 171, 194, 201, 208, 209, 239, 241, 246, 256, 264, 306, 308, 321
Busum, 65, 66

caisson (and caisson construction), 114, 140, 142, 169, 210, 214, 259, 306, 309, 310, 322
Campbell, 16
Caquot, A., 64, 74, 313
Carew, 58, 62
Caspian Sea, 9
cathodic protection, 135
Cattaneo, F., 65, 74
Chalcis, 45, 53, 55, 74
Chalibert Island, 187, 188, 190
Chausey Islands project, 180, 181, 185, 190, 313–316
Chun, Y., 318
Chignecto Bay, 105, 245, 256, 257, 264
China (tidal power sites), 52, 150, 171–172, 182, 305
Chlorella, 11
Chocon River plant, 156, 157, 158, 181

347

Clark, R. H., 138, 147, 148, 253
Claude, G., 8, 10, 16, 23, 65
Cobequid Bay, 253, 254, 257, 260, 264, 272
Cobscook Bay, 67, 155, 224, 226, 229, 232, 236, 237, 238, 239, 244, 245, 247, 292, 316
Cockerell, C., 31
cofferdams (and construction in . . .), 112, 139, 208, 306, 307, 308, 322
Cole, L., 301
Columbia River, 11, 115, 200
conjunction, 77, 322
Constans, J. A., 118, 138, 313, 318
contouring rafts, 31, 32
convergence, 97
Cook (Captain James), 219
Cook Inlet, 101, 219, 223, 293–294
Cooper, D. P., 67, 224, 225, 229, 245
Coriolis force (and effect), 94, 97, 98, 101, 282
Corona del Mar, 14
corrosion (protection against), 134–135, 210
Coyne, 33
Cumberland Basin, 129, 139, 140, 142, 143, 145, 147, 164, 245, 253, 257, 260, 262, 263, 272
current inversion, 104

Danube River, 55
d'Arsonval, A., 16
Davey, N., 65, 275
Dead Sea, 8, 11
Delta Plan, 63, 175, 177, 178–180, 183, 210, 214, 259
Developing countries, 14
Dhaille, R., 33, 50
Dickey-Lincoln scheme, 129, 240, 242, 243
Domesday Book, 55
Dorning, M., 16
double effect (in tidal power generation), 106, 200–203
Dover, 45, 53, 55
Dow, 292

East Greenwich, 57
East Medina, 57
Easton, W. R., 160, 181
ebb (current), 102, 323
economic feasibility, 142–146
economics (of tidal power), 137–148, 194, 200, 229, 230, 234, 248–249, 278–280, 282–284, 305

electricity generation, 111–136
electromagnetic energy, 5–6, 48
Emery, K. O., 318–319
environmental impact, 112, 288–302
equivalent cost, 140–141
ERDA and ocean energy, 45–48
Euripus Channel, 45
Eurocean, 317, 318
Europe (tidal power sites), 150, 175–181, 182, 183

Fagnoni, M., 159
Fentzloff, H. E., 152, 153, 155, 181
Fichot, E., 65, 74
Flemings, 62
flood (current), 102
Florida current, 35, 36
Flowers, A., 13, 49
forces of attraction, 84–89
France (tidal power sites exclusive of Rance River station), 52, 177, 180–181, 185, 190, 313–316
Frobisher Bay, 252
Froenkel, P. L., 36, 38
Fundy (Bay of), 66, 80, 97, 98, 99, 101, 105, 110, 133, 134, 137, 141, 142, 147, 158, 225, 226, 227, 228, 229, 236, 253, 256–274, 281, 294, 300, 303, 306, 307, 308, 311
Furst, G. B., 255, 257, 265, 274

geothermal energy, 39–40, 51, 323
Gérard, D. R., 19, 49
Geysers (The), 39
Gibrat, R., 66, 69, 70, 107, 185
Gibson, R. A., 129, 135
Gomella, C., 18, 49
Gorlov, A., 126, 127, 128, 136
Gougenheim, A., 17, 49
Great Britain (tidal power sites, exclusive of Severn River), 177, 178, 183, 284, 286
Great Salt Lake, 9
Gregory, 58, 74
Griffin, O. M., 21, 49
Gulf Stream, 22, 35

Half-Moon Cove project, 72, 127, 128, 245–249, 293
Hamburg, 53
Headland, H., 275, 276, 278
Heaps, N. S., 284

Hebrides, 32
Heronemus, H. E., 18, 41, 51
Horstman, 14
hydrogen economy, 2, 12

Ijsselmeer (or Zuiderzee), 279
India (tidal power schemes), 150, 172–175, 176, 182
Isaacs, J., 8, 33, 34
Isle of Man, 62

Jeffreys, J., 5, 48
Jordan River, 11

Kaplan (turbines), 36, 114
Katchalsky, A., 8, 49
Kay, R., 305
Kelvin (Lord), 93
Kennedy, G. E., 278
Kennedy, J. F., 241, 242, 303
Keohole Point, 22
Kimberleys, 111, 144, 145
Kislaya (bay and plant), 111, 113, 114, 140, 165, 181, 202, 206–218, 219, 259, 264, 279, 309, 310
Knik Arm, 156, 164, 221–223, 294
Korea (tidal power sites), 168–170, 181, 313, 318
Kuroshio (current), 35

Laberge, N., 248, 249, 293
Lacombe, H., 4, 33, 48, 50
Laplace, 92
Lavi, A., 18, 19, 22
Lawton, F., 106, 138, 141, 148
Lebarbier, C., 116, 117, 135, 136, 198, 199, 200, 201, 202, 203, 204, 205, 312, 315
Le Grand, Y., 4, 48
Leishman, J. M., 27
Leitz, F., 10, 49
Lewis, J. G., 165, 181
locks and gates, 113–114, 189
London Bridge, 56
Lyaknitskii, V. E., 66

Macrocystis pyrifera, 11, 13, 14, 16
Maitland, W., 56, 74
Mariano, 64, 74
Marine Farm Project, 11
mascaret, 101

Masson, 18
Maunsell (Report), 161–164, 165, 166, 181
Maynard, F., 65, 74
Metz, W. D., 22, 49
Meuse River, 16
Milne, A. J., 286
Minas Basin, 105, 129, 156, 232, 245, 254, 256, 257, 264, 294, 295
Minquiers (Les), 65, 185
Mitsui, A., 11, 49
Mont St Michel, 70, 80
Morello, A., 34, 50
Mosonyi, E., 69, 74
moulins à marée, 52–64, 251, 325
Musgrove, P. J., 35, 38
Muskie, E. S., 241

National Grid (U.K.), 276, 286
NEDECO, 283, 284
NEPOOL, 139, 146, 265, 268
Netherlands (tidal power sites) (see also: Delta Plan), 178–180, 183
New Zealand (tidal power sites), 165, 183
Nizeri, N., 16, 17
nuclear penetration (Fundy area), 266, 267

ocean bioenergy, 11–16, 46, 49
ocean currents energy, 34–37, 47, 51
Offshore Wind Power System, 41
offshore winds, 41, 47, 51
optimization, 194, 200–202
OTEC, 19–24, 46, 47, 48, 49
Othmer, D., 19, 50
Ougrée, 16

Pacem in Maribus (Convocation), 6
Pacifica, 24
Parenty, H., 50, 65, 74
Passamaquoddy, 55, 63, 67, 113, 115, 145, 146, 156, 158, 164, 197, 219, 224–245, 291–293, 303, 313, 316
Pattle, R. E., 8, 49
Pembroke, 59, 62
plant operation (Rance River), 191–193
plastic barrier, 126–128
Plougastel, 53
powerhouse, 114–118, 206, 210, 211, 213, 221, 233, 259, 261, 273

QAM, 13
Quay Mill, 60

Rance River tidal power plant, 21, 53, 55, 57, 70, 112, 113, 114, 130, 131, 134, 135, 139, 156, 160, 164, 181, 184–205, 212, 233, 234, 239, 245, 248, 258, 259, 264, 279, 284, 288–291, 300, 304, 310–313
range (tidal), 70, 76, 77, 79–81, 105, 106, 149, 150, 236, 280, 281
Raynor, C. J., 160, 181
Remenieras, G., 35, 51
resonance, 69, 97, 98, 112, 326
Reynolds, A., 24, 33
Rigaud, M., 65, 74
rim turbine, 115, 119, 223, 273, 308, 326
Roels, O. A., 19, 49, 50
Roosevelt, F. D. (President), 67, 68, 224, 230, 245
Russell Rectifier, 30

Salem (MA), 63
Saint-Guily, B., 4, 48
salinity power, 6–11, 46, 49
Salter Duck, 30, 31
salt water batteries, 6
San Clemente Island, 14
San José Gulf Project, 152–157, 180
Saunders, D. W., 167, 181
seiche, 97, 98, 327
Severn River, 70, 73, 99, 101, 111, 112, 113, 177, 197, 248, 295–300, 310
Shaw, T. L., 279, 280, 281, 297, 298
Shepody Bay, 253, 257, 260, 264, 272
site selection, 187–190, 258–264, 265
site value, 70
Slade's Spice Mill, 63
Slipper Mill, 60
Smith, L., 28
Société de l'Energie des Mers, 16
socioeconomic impact, 196–198, 234, 242, 243, 248, 249, 266, 282–283
SOGREAH, 156, 159, 160, 165, 169
Sorensen, K. E., 264
Sourirajan, S., 8
Spring Creek, 63
St. Malo, 45, 55
Steelman, G., 35, 51
Sternmann, B., 17
Stewart, H. B., 35, 51
Stone & Webster Study, 293, 313
storage, 123–126, 128–134, 165, 184, 191, 223, 233, 239, 264, 306, 326

St. Osyth, 56, 58, 61
Straflo turbine, 115, 118, 119, 223, 273
Subrahmanyam, K. S., 173, 182
Swales, M. C., 219, 220, 250, 255, 257, 265, 274, 311

Tanner, R. G., 142, 256, 274
thalassothermal energy, 8, 16–24, 46, 49
tidal current, 36–38, 75, 99–100, 104–105, 187, 289, 295, 327
tidal function, 72
tidal output integration (Fundy), 259, 265, 268
tidal species, 90
tidal "strength", 80
tide, equilibrium, 83, 323
tide-generating forces, 81–83, 88–90
tide mills, 52–64, 251
tides, 75–110
  astronomical, 78
  basins and estuaries, 94
  in geometrically shaped basins and estuaries, 94
  diurnal, 93, 105, 322
  lunar, 78
  semi-diurnal, 72, 93, 94, 98, 105, 106, 186
  theory of, 92
  types, 90–92, 93–94
tractive force, 81
transmission costs (of electricity), 167, 276, 282
Trieste (bathyscaph), 7
Tsukada, S., 11
two-pool scheme, 69, 72

Udall, S. L., 242
Ungava Bay, 251–252, 281
USSR (tidal power sites, exclusive of Kislaya Guba), 52, 150, 181, 206, 212–214, 215, 219
utilization factor, 69

value of power, 141–142
Van Wyck Mill, 64
Vigo, 53
Vitrivius, 54
von Arx, W., 24, 26, 35

Wailes, R., 55, 58, 59, 60, 61, 74
Wallace, R. H., 39, 51
Waller, D. H., 295, 300

Warnock, J. C., 142, 256, 274
Water Low Velocity Energy Converter, 35
wave energy tapping, 24–34, 47, 50, 282
Weinstein, J., 10, 49
Wick, G., 2, 6, 7, 8, 9, 24, 50, 105
Wickert, G., 64, 74
Wilcox, H., 12, 15

Wilhelmshafen, 65
Wilson, E. M., 129, 132, 133, 183, 219, 220, 250, 283, 306, 311
Wood, P. W. J., 318, 319
Woodbridge, 55
Wootton Bridge, 60
World Oil [Energy] Consumption (tables), 3, 304